CELL STRUCTURE

For Ian and Helen

CELL STRUCTURE

AN INTRODUCTION TO
BIOMEDICAL ELECTRON MICROSCOPY

K. E. Carr BSc PhD MIBiol
Department of Anatomy, University of Glasgow

P. G. Toner MB ChB DSc MRCPath MRCP(Glasg)
University Department of Pathology,
Glasgow Royal Infirmary

THIRD EDITION

CHURCHILL LIVINGSTONE
EDINBURGH LONDON MELBOURNE AND NEW YORK 1982

CHURCHILL LIVINGSTONE
Medical Division of Longman Group Limited

Distributed in the United States of America by
Churchill Livingstone Inc., 1560 Broadway, New York,
N.Y. 10036, and by associated companies, branches and
representatives throughout the world.

First Edition 1968
Second Edition 1971
Third Edition 1982

ISBN 0 443 02324 7

British Library Cataloguing in Publication Data
Carr, Katharine E.
    Cell structure – 3rd ed.
    1. Cells   2. Electron microscopy
    I. Title   II. Toner, Peter G.
    574.87′028   QH585

Library of Congress Catalog Card No. 81–67939

Printed in Great Britain by
William Clowes (Beccles) Limited, Beccles and London.

Preface to the Third Edition

During the last 10 years, the electron microscope has become increasingly employed in almost every corner of the biomedical field, while the range and availability of ultrastructural technology has greatly expanded. *Cell Structure* has evolved into its third edition under these continuing environmental pressures.

The most significant expansion in the use of electron microscopy has been its increasing integration into human medicine, through the pathology laboratory. The insights into cell structure and function provided by the basic scientist can now help to solve clinical problems. Conversely, the wealth of clinical material now readily available from wards and operating theatres can help to broaden the horizons of research and give new relevance to the work of the ultrastructurally orientated basic scientist. A new chapter has been added to explore this area in greater detail, explaining in simple terms the practical relevance of ultrastructural knowledge to the medical laboratory specialist. Human material has been used for over half of the illustrations of this third edition.

The image of electron microscopy has also been changed, both literally and figuratively, by new technology. The biggest single change has been the integration of scanning systems, particularly surface scanning, into biological and clinical research. This has particularly influenced the teaching of structural concepts, helping students to break free from the limitations of two dimensions. Other recent technical advances include the growing importance of X-ray microanalysis in association with electron columns, and the continuing development of high voltage electron microscopy. Specimen preparation has also advanced,

with freeze etching and cytochemical methods playing an increasing part in research. Techniques such as these have encouraged a more functional approach to the ultrastructural study of tissues and cells. An attempt has been made in this book to give recognition to these many developments, without losing sight of the continuing central role of conventional transmission electron microscopy.

We recognise that this book is now likely to be more used by undergraduates and postgraduates in biological sciences, and by medical specialists in training, than by the medical undergraduates who provided us with the stimulus to write the first edition. As a result, this edition has become larger than its predecessors, while continuing to follow the general pattern of the previous editions. The text has been almost entirely re-written, with the addition of new topics where appropriate to the overall aim. The space devoted to electron micrographs has almost doubled, allowing the illustration of a wider range of topics and techniques. Most of the micrographs are new. The line drawings, favourably received in previous editions, have been largely redrawn and have been added to. The reading list has been expanded and updated. Despite its increased size we hope that this edition will prove a concise and useful source of basic ultrastructural information for anyone exploring the ultrastructural world for the first time, at whatever level of specialisation.

We are grateful to many of our colleagues for their generosity in allowing us to use their micrographs and for other assistance. Particular thanks are due to H. S. Johnston, G. Bullock, C. Skerrow, I. A. R. More, A. A. M. Gibson, A. L. C. McLay and A. Jack. Dr R. V. Kristć has kindly given us permission to use four of his superb line drawings. We are indebted to Mrs M. Johnston for her creative skill with the other drawings.

We would like to thank our technical colleagues for their unstinting support over many years, particularly J. D. Anderson, J. Ito, C. Watt and D. McSeveney, along with the staff of the EM Units of the Royal Infirmary and Western Infirmary Departments of Pathology and of the Department of Anatomy. Thanks are also due to the past and present members of the clerical staff for their efforts with the various drafts of the manuscript. The help of Mrs Peedle, Mrs Main and Mrs Thomson is particularly appreciated. Finally we are grateful to Professor R. J. Scothorne and Professor R. B. Goudie for the use of the facilities of their respective Departments.

Glasgow, 1982

K.E.C.
P.G.T.

Preface to the First Edition

This book is intended as a simple introduction to biological electron microscopy. In it we have set out to do three things: firstly, to present the fine structure of the cell and a number of the more interesting specialisations of cell structure; secondly to provide enough technical information to satisfy the first questions of the more interested student and to indicate to him the potential uses and limitations of electron microscopy; finally to help the beginner to approach the examination and interpretation of an unknown micrograph in a systematic way.

We have not attempted to compile a comprehensive reference work on fine structure since texts of this type are already available. Nor have we tried to present a manual of technique but instead have limited this section to give background information upon which an interested student might subsequently build. We have assumed that the study of fine structure will form part of a more general biological training and our limited treatment of functional aspects is not intended to take the place of the detailed study of biochemistry.

We have become convinced of the need for a book of this kind from our contact with students at the early stages of their medical studies and also from our experience of the needs and interests of students attending extramural classes on biological electron microscopy at Glasgow University. We believe that a working knowledge of fine structure may soon be as important to the biologist as a knowledge of histology and that a systematic introduction to the subject is best provided in the present form, rather than as supplements in a larger text of anatomy or histology. We hope that the book will prove of interest not only to the medical and biology students at whom it is primarily aimed, but also to those now past their student days who have not been exposed to any formal teaching of the elements of fine structure. We would like to feel that this book might help any, who for this reason regard fine structure with misgivings, to feel more at ease when confronted with the increasing numbers of electron micrographs appearing in the pages of the scientific press.

We are indebted to Professor G. M. Wyburn for the use of the facilities of the Department of Anatomy and for his advice and helpful criticism, not only during the preparation of this book, but on many occasions in the past. The electron micrographs with which the book is illustrated were taken by us using the Philips E.M. 200 electron microscope of the Department of Anatomy at Glasgow University. Miss Jean Hastie and Miss Pauline Semple assisted in collecting and processing the tissues and in preliminary screening. Miss Margaret Hughes gave invaluable photographic support and prepared all of the final prints. Miss Jane Young of the Department of Anatomy and Mr D. Lang of E. & S. Livingstone produced the line drawings with skill and care. We are most grateful for the assistance provided in these different ways, without which our own work would have been immeasureably increased. We would also like to thank the staff of E. & S. Livingstone for their co-operation and assistance at all stages in the production of the book and we are most grateful to Mr F. Dubrey of Scottish Studios for the care he has taken with the reproductions of our electron micrographs.

A number of our friends and colleagues have given us their help and criticism. We are particularly grateful to Dr J. P. Arbuthnott and to Dr I. A.

Carr, and we would like also to thank Drs R. B. Goudie, W. A. Harland, E. Arbuthnott, D. Graham, K. C. Calman, J. S. Dunn, A. R. Henderson, A. M. MacKay, R. F. Macadam, Mr A. Martin, Miss J. Rentoul, and Mr R. Young for their comments at different stages. Professor J. R. Anderson, Western Infirmary Department of Pathology, Glasgow University, has kindly given his encouragement and interest. We accept all responsibility for the remaining shortcomings in the text and for inadequacies in the micrographs, but we hope that they will not prevent the book from being of use to those with an interest in cell structure.

P.G.T.
K.E.C.

Glasgow, June 1968

Contents

THE CELL AND ITS COMPONENTS

The purpose of this section is to introduce the reader to the place of the electron microscope in the study of cellular structure and to present, in a compact form, the fundamentals of the structural organisation of the subcellular components. Where possible, simple functional considerations are introduced, without an attempt being made to trespass on the territory of the cellular biochemist or the molecular biologist.

1

The study of biological structure

1.1 MORPHOLOGICAL SCIENCE

The study of structure is an essential basis of biology. Structural studies, however, can be pursued by different techniques, each suited to a particular level of detail. In the past, the texture of different tissues could be revealed by simple visual examination and dissection, but the introduction of the light microscope suddenly extended the horizons of the early anatomists beyond the range of the unaided eye. This forced a radical revision of previous concepts of scale and dimension: the obstetrician's fingerbreadth and the inch gave way to the millimetre and the micron.

Since then, the light microscope has been a mainstay of biological research and medical practice. Magnifications of up to 2000 times show details far beyond the reach of unaided vision, but with one important limitation. No matter how good the microscope and the specimen may be, the detail it can display is limited to half the wavelength of the illuminating beam. This limitation lies in the physical nature of light itself. For this reason two particles less than 0.2 μm apart in the specimen will not be distinguished, or resolved, as separate images but will appear to fuse into a single blob. Such details are said to be beyond the *limit of resolution* of the light microscope.

Despite this limitation, the foundations of cellular biology have been laid by light microscopy. From it came the theory of the cell as the unit of life, the description of the nucleus and cytoplasm and the identification of the elementary subcellular components, such as the mitochondria and the Golgi apparatus.

At the same time, scientists in other disciplines had been pursuing the study of structure with other techniques. X-ray diffraction, first introduced by physical chemists, provided new insights into the three dimensional structure of complex macromolecules, such as DNA, myoglobin and lysozyme. Aided by modern computer expertise, it is now a routine tool in structural biochemistry. Structure and function at the molecular and atomic level have become the province of the biochemist, the biophysicist, the molecular biologist and the physical chemist.

1.2 THE TECHNIQUE OF ELECTRON MICROSCOPY

The *electron microscope* provided the missing link between the details of tissue organisation as seen by the light microscope and the details of molecular architecture as revealed by the new biochemical and biophysical techniques. By using electrons instead of light, restrictions on resolution could be removed. The wave-length of the electron beam in the operating conditions of the electron microscope is many times smaller than the wave-length of visible light. The difference in scale is so great that a new range of units of measurement has become common currency in biology. In the now standard international system of units of measurement (SI units) the electron microscopist thinks in terms of nanometres (nm), each one representing one thousandth of a micron. It is this unit which is used in the following chapters to describe the dimensions of many subcellular components. An old unit still popular with the electron microscopist is the Ångstrom Unit (Å) originally used as the unit of wave-length in optical spectroscopy. The Ångstrom unit represents one ten-thousandth part of a micron. Hence one nanometre (nm) is equal to 10 Å. The wavelength of visible light limits the resolution of the optical microscope to about 200 nm: the practical resolution limit of the modern electron microscope is better than 0.2 nm.

The first electron microscope, constructed in the early 1930s in Germany, was soon followed by production models which could resolve details of structure beyond the range of the light microscope. Its development was delayed by the war of 1939–45, but even after that it was some time before this new technique made a significant impact on biological science. There were two main reasons for this delay. The first was the technical complexity of the instrument, which demanded of its operator a greater than average skill in engineering and electronics. The operator of an early electron microscope, more often a physicist than a biologist, spent many hours in repairing and maintaining the machine. The second reason was that the techniques used to prepare tissues for light microscopy were not suitable for the new demands of the electron microscope. These techniques had changed little in a century. The cells or tissues to be examined were preserved using chemical fixatives such as formalin and supported by embedding in blocks of paraffin wax. Histological sections, thin slices of the supported tissue, were then cut on a microtome with a metal knife, mounted on a glass slide and stained with coloured dyes.

These techniques, successful as they were and still are today for histology, were not compatible with electron microscopy, for reasons discussed in Chapter 14. However, between 1948 and 1954, several important technical advances in specimen preparation were introduced, including the use of plastic embedding media, glass knives, new sophisticated microtomes and specially controlled buffered fixatives containing osmium tetroxide. At the same time new refinements in microscope design led to greater reliability, simpler operation and less need for maintenance. These various factors combined to produce a sudden breakthrough in the study of biological fine structure.

1.3   THE IMPACT OF ELECTRON MICROSCOPY

In a few years the ultrastructural foundations of the new science of *cell biology* had been laid and a further revolution had taken place in our concepts of the organisation of living things. The first result was a sudden end to some of the controversies which for years had enlivened histology and cytology. The cytoplasm was shown to have a complex and variable fine structure and the existence of the Golgi apparatus was confirmed (Plates 19, 20). The intestinal striated border was shown to consist of a mat of well-defined microvilli (Plates 14, 113). A continuous epithelial lining was demonstrated in the pulmonary alveoli (Plates 57, 58). The intercalated disc of cardiac muscle was revealed as a zone of cell contact, disproving the syncytial theory of cardiac muscle (Plates 89, 90). The cell membrane was shown to have a true structural identity (Plate 4) and the myelin sheath was found to be a membrane specialisation (Plates 100, 101).

After the first phase of 'instant' problem-solving came the long and detailed task of cataloguing the newly available details of subcellular organisation. Electron microscopic examination has demonstrated various basic features common to all cells. This has led to a new understanding of cellular structure and function. Specific anatomical appearances associated with particular cell functions can now be rationally explained by combining the results of fine structural and biochemical investigations (Chapters 5–11). It is now possible to determine the composition and function of the different components of the cell and to assess the contribution of each to cellular metabolism (Chapters 2–4): examples include the link between the granular endoplasmic reticulum and the function of protein secretion (Plates 3, 15) and between mitochondrial structure and cellular energy requirements (Plates 21, 89).

Since the middle of the 19th century, when cellular pathology began to develop as a separate discipline, the light microscope has played an increasing part in medical research and subsequently in the diagnosis of disease. Naturally enough, the scope of present day diagnostic pathology is determined largely by insights and experience gained through such past and present light microscopic research. Now, however, the place of the electron microscope in *diagnosis* is becoming more clearly defined, as a result of an increasing interest in the ultrastructural features of disease. Current trends suggest that the electron microscope will play a growing part in practical service pathology as well as in pure research, as discussed in Chapter 13.

Over the last 200 years, the horizons of biology, medicine and pathology have been widened beyond recognition by light microscopy: the electron microscope, less than half a century after its invention, has already proved equally revolutionary. But despite this, it is important to recognise that the electron microscope and the light microscope will continue to have equally essential and complementary roles, which are in no way in competition. Just as the light microscope extends the range of the unaided eye, so the electron microscope extends the range of optical techniques, each new method gathering new information. With the help of the unaided eye, the tissues can be dissected; light microscopy can reveal their architecture; electron microscopy can expose the structure of the individual cell; biochemistry and physical chemistry can define the molecular patterns of life. There are many techniques for the investigation of biological structure; each one must be used with discretion to solve an appropriate problem.

Unfortunately, the more elaborate and specialised the technique, the less accessible it becomes. The student, who takes the light microscope for granted, only rarely has the opportunity to become personally involved in the operation of the electron microscope. In practical terms, the differences between light and electron microscopy are reflected in the different amounts of operator time which they consume. It would take weeks of work with the electron microscope to record on photographic film the area of tissue examined in minutes by the light microscopist. Moreover, the modern electron microscope is very expensive and requires highly skilled technical maintenance.

In the practical sense, it is obvious that such a machine can be available to only a few, but this is less serious than it would appear at first glance. The light microscope gives its best performance to the eye of the operator: the electron microscope on the other hand can only produce its best results on photographic emulsion. The EM operator uses visual control mainly for the selection of suitable fields to be taken as a permanent record of the specimen. It is, therefore, quite possible for the student to achieve a working knowledge of fine structure from text books without actually operating the instrument.

In the following chapters, an attempt will be made to familiarise the beginner with the main landmarks of the ultrastructural world, with an emphasis on the correlation of fine structure and cellular function.

2

Biological membranes and the cell surface

2.1 THE FINE STRUCTURE OF MEMBRANES

2.1.1 The membrane concept

A membrane can be defined as a tenuous partition or interface between two phases of the substance of the cell, or between the cell and its environment.

The membrane previously studied in the greatest detail by cytologists is the *cell membrane*, or *plasma membrane*, which forms the outer boundary of the cell, separating the cytoplasm from the extracellular environment. For many years it was realised that some form of partition, too thin to be observed directly using the light microscope, must exist at the cell boundary. The swelling of red blood cells when placed in hypotonic solutions indicates the presence of a semi-permeable cell membrane which allows the passage of water into the cell, while resisting the outflow of large molecules from the cytoplasm. The resilience and elasticity of the cell membrane can be demonstrated by micromanipulation. If the membrane is torn or punctured the escape of cell contents can be observed. Thus there is convincing evidence in favour of a structural partition between the interior of the cell and its surrounding environment.

Early workers had concluded that membranes contained an important lipid component, since fat solvents were found to destroy the membrane and fat-soluble substances were shown to pass across membranes much more readily than hydrated molecules. Chemical analysis of bulk samples of cell membranes obtained by differential centrifugation has shown that their main constituents are protein, phospholipid and cholesterol, with traces of polysaccharide material.

2.1.2 Electron microscopy of membranes

The demonstration by the electron microscope of the extent and specialisation of biological membrane systems has made it clear that membranes are essential in the organisation of cell structure and function. Electron microscopy of fixed tissue confirms the presence of a limiting membrane, with a basic pattern common to all cells.

After conventional fixation and staining, membranes appear at low magnification as thin dense lines owing to their affinity for the heavy metal atoms used in these procedures. Their thickness varies between 7.5 and 10.5 nm, far beyond the limits of resolution of light microscopy. On closer examination at higher resolution, most membranes have a three layered or *trilaminar* structure (Plates 4, 48b). In a typical case the two dense outer layers, each 2.5 nm thick, are separated by a pale unstained interspace of similar dimensions.

Two three-layered images associated with membranes can at times be confused. The first, at low magnification, is the electron microscopic appearance of two parallel cell membranes of adjacent cells with an intervening pale intercellular space (Plate 14). The second, at high magnification, is the trilaminar appearance of a single *unit membrane* in which the two dense components and their pale interspace can be seen (Plate 4). Confusion between these two images will not arise if the magnification of the micrograph is carefully noted.

Although this trilaminar pattern can be demonstrated in most biological membranes under ideal conditions, the exact dimensions of the different components may vary considerably and modifications of the common pattern can be seen in special situations. The cell membrane may, for example, be of different thickness at different points on the same cell. At the sides and base of the intestinal epithelial cell the membrane has a total thickness of about 8.5 nm while at the absorptive apical surface of the cell it is increased in thickness to about 10.5 nm and is very easily resolved into its three component laminae. The pale interspace is about 4 nm wide and the inner dense lamina often seems thicker and more dense than the outer, giving the membrane an asymmetrical appearance. It is likely that some of the differences in thickness and symmetry of the trilaminar structure of membranes in different sites are due to the presence of molecular components with different functional properties.

Variations are also seen in cytoplasmic membranes. The membranes of the Golgi apparatus and the limiting membrane of the lysosome are thicker than those of the mitochondria and the endoplasmic reticulum. In some micrographs the central pale lamina of the membrane appears broken by septa which connect the two dense external laminae, forming what at times appear as globular subunits.

It is important to remember that the dimensions and detailed fine structure of a membrane as seen by the electron microscope depend on specimen preparation techniques. For example, the detailed structure of a membrane fixed with potassium permanganate differs markedly from the patterns seen after osmium tetroxide and glutaraldehyde fixation. The electron image merely records the position of heavy metals introduced into the specimen by fixation and staining procedures. While this may well reflect some underlying pattern of the once living biological structure, the inherent limitations of the technique must be realised. The recent upsurge of interest in *freeze-etched* specimens of membranes (14.6.2) is due to the freedom of this technique from fixation and staining artefacts. The details of the frozen membrane are likely to approximate closer to the living state than in any fixed specimen (Plate 1).

2.1.3 Membrane models

The arrangement of the lipid and protein subunits of biological membranes has been a matter of dispute for many years. In the earliest popular theoretical model of membrane structure, the Davson–Danielli *lamellar* model, two protein monolayers were proposed as forming the outer and inner membrane surfaces. The lipid molecules, in a continuous double layer, were seen as forming the filling of the sandwich, as shown in Figure 1. Each lipid molecule was arranged with its hydrophilic polar end in association with the protein component and its hydrophobic non-polar end extending inwards at right angles to the plane of the membrane. The lipid molecules would thus lie parallel to each other, with a narrow central gap between their opposed non-polar ends.

Early work on membrane ultrastructure seemed to offer support for some such model, with the well-defined trilaminar pattern apparently reflecting the hypothetical lamellar array of molecules. The two dense outer layers visualised by electron microscopy might correspond to the polar ends of the lipid molecules with their associated protein monolayers, while the pale interspace might represent the non-polar lipid component. The lamellar lipoprotein 'unit' became popular as a possible common molecular basis for all membranes, as implied in the '*unit membrane*' hypothesis.

An alternative molecular layout was suggested

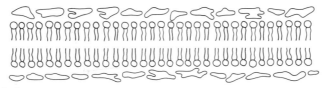

Fig. 1 *Simple lamellar hypothesis*
This highly schematic diagram shows the general pattern of molecular layout proposed in a simple version of the lamellar theory of membrane structure. Two rows of phospholipid molecules have their non-polar ends extending into the centre of the membrane. The circles, which represent the hydrophilic polar groups of the phospholipid molecules, are shown in association with an external surface coat of extended protein molecules, represented by the irregular shapes on the outside of the membrane model. This simplified model is no longer considered valid.

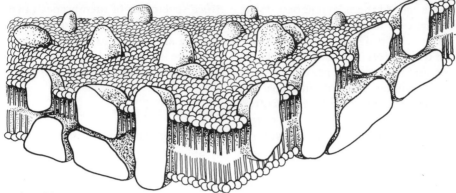

Fig. 2 *Fluid mosaic model*
This perspective diagram shows the essential features of the fluid mosaic model of membrane structure, the hypothesis most widely accepted at the present time. Globular protein molecules are embedded in a bilamellar phospholipid layer. Some proteins extend throughout the thickness of the membrane, while others are associated with either the outer or the inner surface of the membrane. This model explains many puzzling properties of membranes which could not be fully accounted for by earlier hypotheses.

in the *micellar* theory. Lipid micelles tend towards spontaneous aggregation, an attractive property for hypothetical membrane building-blocks. Associated protein molecules might coat or replace units of the micellar array, allowing for the carrier and transport mechanisms known to exist within membranes.

These earlier models, however, had various shortcomings. The known wide variations in the proportions and nature of membrane lipids and proteins were difficult to reconcile with any unit hypothesis. The membrane proteins, moreover, are more globular than extended in their configuration, and do not easily fit into a simple lamellar model.

The most popular current hypothesis is the *fluid mosaic* of Singer and Nicholson. They propose a basic bilamellar lipid matrix, in which integral globular protein molecules float freely as if in a sea (Fig. 2). Some of these molecules are exposed to the exterior and others to the interior of the cell, while others again are envisaged as traversing the entire membrane. This model allows for most of the known properties of membranes, including the molecular asymmetry of the external and internal surfaces and the free mobility of *receptor molecules* on the cell surface. In particular, this model best explains the appearances of freeze-fractured membrane surfaces (Plate 1), which show the presence of transmembrane macromolecules, particularly at areas of specialisation such as gap and tight

junctions. These appear as distinctive arrays of particles each measuring 2 to 4 nm in diameter, corresponding to the protein molecules which confer distinctive functional properties on these areas of structural specialisation of membranes. Finally, this hypothesis is flexible enough to cope with the enormous range of variation in the chemical composition of different membranes.

The appearances of membranes in high resolution transmission electron micrographs of thin sections are in themselves insufficient to resolve the conflicting claims of different molecular models. The difficulties in the interpretation of such micrographs in molecular terms are still too great to allow us any confidence in the results. Freeze-etched preparations, on the other hand, as mentioned above, represent the unfixed and virtually unprocessed state of the cell membrane, and can therefore be seen as the closest available approach that morphologists can make to meaningful molecular visualisation. For this reason, freeze-etching techniques are now of great importance in studies of membrane biology.

2.1.4   The significance of membranes

Membranes can be thought of as having both a structural and a metabolic significance. On the one hand, membranes partition the cell from its neighbours and from the extracellular environment and are used in the construction of the principal cell organelles. This structural role may be thought of as a *barrier* function. Membranes also have an *organising* function in relation to cell metabolism. A membrane can be made to serve different integrated biochemical functions simply by the incorporation of the relevant enzyme systems into its molecular structure. Enzyme molecules incorporated in the membrane allow the barrier and organising roles to be combined. Many of the biochemical reactions which take place at membrane surfaces are limited primarily by the total *surface area* available rather than by the availability of substrates. Hence the presence of unusually elaborate membrane systems is a common accompaniment of specialised cell function.

There are many examples of membranes which form both structural barriers and organised functional partitions in the cell. The cell membrane itself is the most obvious example, since it is across this interface that all exchange occurs between the cell and its environment. The overall thermodynamic equilibrium of the cell is critically dependent on the surface area of the cell membrane, which in turn sets the limits to the potential volume of any cell.

Within the cytoplasm the membranes of the granular endoplasmic reticulum form storage spaces and channels for the isolation and transport of newly synthesised protein secretions (Plate 16). The membrane which surrounds each secretion granule released from the Golgi apparatus (Plate 37) not only isolates the granule prior to its discharge but may also modify its contents by continued enzyme action. Membranes surrounding the lysosome resist the action of its powerful hydrolytic enzymes (Plate 63). The structural role of this membrane is of particular importance, since experimental rupture of lysosomes can lead to cell damage and even cell death.

Efficiency of function in specialised cells is often gained by accommodating the maximum active membrane surface within the minimum cytoplasmic volume. Thus the granular endoplasmic reticulum is well developed in protein-secreting cells (Plate 3), while the mitochondria, themselves made out of membranes, are large and numerous, with a particularly elaborate internal structure, in cells which consume large amounts of energy in their metabolic processes (Plate 21).

2.2 SPECIALISATIONS OF THE CELL SURFACE

2.2.1. Contour and function

Specialisations of surface contours can be related to specific functions. An increase in the basal surface area of epithelial cells produced by deep *infolding* of the cell membrane is found where there is particularly active transport of ions across the cell. Examples are seen in the cells of kidney tubules (Plate 51) and in the acid-secreting parietal cells of the stomach (Plate 21). Other examples of elaborate cell surfaces are found at the interface between bone matrix and the osteoclast (Plate 83) and at the sub-synaptic surface of skeletal muscle at the motor end plate (Plate 103).

The numerous finger-like *microvilli* at the luminal surface of the intestinal epithelial cell can increase its surface area by a factor of 20 or more (Plate 14). This enormous surface not only provides attachments for a mosaic of enzymes related to *terminal digestion* but also provides the area required for the efficient *absorption* of the ultimate products of digestion. The functional specialisation of this membrane may account for its unusual thickness. In transitional epithelium and in keratinised squamous epithelium, unusually thickened surface membranes are apparently associated with marked impermeability, linked to a protective function (Plate 76).

The cell membrane, which maintains the integrity of the cytoplasm, also controls or influences the passage of all materials into and out of the cell whether by *diffusion*, by *active transport* across the membrane, or by the process of membrane invagination known as *endocytosis*, as described below. Among the enzymes associated with the metabolic activity of the cell surface is adenosine triphosphatase (ATP-ase) which releases energy from ATP for use in cellular activity. This enzyme provides the energy required to drive the sodium pump, one of the fundamental ion transport systems of the cell, essential for the maintenance of the normal *membrane potential*.

Adenylate cyclase is another membrane-associated enzyme now thought to be of great importance. It is thought to be a transmembrane enzyme which is able to interact with a number of other cell surface membrane proteins, such as glucagon receptors. Its function is the conversion of ATP to cyclic AMP (cAMP). The intracellular concentration of the product rises when the activity of this enzyme is stimulated. Its physiological antagonist is phosphodiesterase (PDE), which lowers the level of cAMP by converting it to the noncyclic form. The rising and falling levels of cyclic nucleotides are now thought to represent a fundamental trigger mechanism. Thus the stimulation of adenylate cyclase sets in motion, through the rising level of cAMP, the particular tissue-specific response for which the cell is programmed. Inhibition of adenylate cyclase or stimulation of phosphodiesterase has the opposite effect, diminishing the physiological responsiveness of the effector cell. The early work on this system demonstrated its importance as the trigger to hormone-provoked endocrine activity, adenylate cyclase being activated by the attachment of hormones to receptor sites on the surface of the target endocrine cell. Similar systems are now known to be involved in the control of carbohydrate metabolism in the liver, through the opposing functions of glucagon and insulin. No doubt the cyclic nucleotides will prove to have even wider implications for cell biology than are at present documented.

Thus the metabolic and the structural roles of the cell membrane are interdependent and of equal importance. The innumerable variations of structure and function in different membrane systems are one facet of the infinite adaptability which has ensured a key role for membranes in the living world.

2.2.2 Endocytosis and membrane interchange

This is the process which enables cells to engulf substances from the environment, through active membrane invagination. This can be observed as a dynamic process in living cells in tissue culture, particularly in macrophages and amoebae, by the use of phase contrast light microscopy. The electron microscopist, however, must rely on fixed tissue to provide images of endocytosis, such as indentations of the cell surface, leading to the formation of isolated endocytic vacuoles within the cell. Two major forms of endocytosis are recog-

nised, the uptake of fluid materials by *pinocytosis* (cell drinking) and the uptake of particulate materials by *phagocytosis* (cell eating). The opposite phenomenon, *exocytosis*, involves the discharge of substances from cells by a process which is essentially the reverse of the endocytosis mechanism.

In pinocytosis, the cell membrane becomes deeply invaginated at some point on the cell surface, forming a narrow membrane-lined pinocytosis channel which remains in continuity with the exterior of the cell and contains extracellular fluid material. The walls at the neck of the channel then fuse, leaving its blind end as a pinocytic or *pinocytotic vacuole* isolated in the cytoplasm. The contents of this vacuole are still surrounded by the membrane which originated as part of the cell surface. In a sense, therefore, the vacuole contents lie outside the cell, as long as the limiting membrane remains intact. The fluid contents of the vacuole are then progressively absorbed into the cytoplasm and the membrane of the vacuole is dismantled. When certain substances known as *inducers* are dissolved in the extracellular fluid, they are capable of causing a marked increase in the rate and volume of pinocytotic uptake.

Additional information gained from tracer experiments can indicate both the direction and the quantitative significance of an observed pinocytotic phenomenon. Studies on macrophages in vitro have shown that each cell forms at least 125 pinocytotic vesicles every minute, equivalent to an intake of 0.43 per cent of the total cell volume, using 3.1 per cent of the total cell surface area in the process. This raises the question of the origin and fate of the cell membrane. The macrophage can engulf the equivalent of its entire surface area in 33 minutes through the formation of pinocytotic vesicles. When phagocytosis begins, there is a rapid increase in the rate of membrane lipid synthesis, but it is not known how the cell restores its membrane proteins, since new synthesis of these components is quite insufficient to account for so rapid a rate of turnover. The engulfed membrane must therefore be rapidly dispersed and components in short suply are probably recycled to the cell surface in a constant flow.

After a bout of pinocytotic activity, a refractory period may be observed, presumably to allow the recycling process to catch up with the intake of membrane material. Cellular functions presumably depend on the speedy relocation of receptor proteins withdrawn from the cell surface as a result of endocytosis. Perhaps alterations of surface proteins predispose to the fusion of macrophages to form giant cells, as happens in certain circumstances in the body when macrophages are activated

New membrane synthesis, as opposed to recycling, does also occur, particularly prior to mitosis. Perhaps this newly synthesised membrane material may originate in the endoplasmic reticulum and may be processed through the Golgi apparatus, like so many other cell products. In general, a membrane can probably expand by the diffuse incorporation, at any point, of additional components, whether recycled or newly synthesised.

Pinocytosis is related to phagocytosis, described many years previously. In phagocytosis, particulate material is engulfed by the cell along with some extracellular fluid. Phagocytosis of small particles proceeds as described for pinocytosis, but larger particles are engulfed more slowly, by an *amoeboid movement* in which the cell throws out cytoplasmic processes which reach round the particle to enclose

it (Plates 63, 67a). The phagocytic vacuole becomes, in effect, an intracellular stomach, into which hydrolytic lysosomal enzymes are discharged to digest the contents, the component chemical parts of which are then absorbed by the cell across the membrane of the phagocytic vacuole. This subject is dealt with in greater detail in Sections 4.6 and 7.1. Endocytosis by specialised cells plays an important part in the defence mechanisms of the body against *foreign materials* and *infecting organisms*.

Although pinocytosis as originally described was observed by light microscopy, an analogous phenomenon can also be recognised at the fine structural level. Appearances taken to be suggestive of *micropinocytosis* include the presence of small flask-shaped invaginations of the cell surface, around 50 nm in diameter. These are known as *caveolae*. In the underlying cytoplasm circular vesicles of a similar size are often seen, formed by detachment of the flask-shaped caveolae from the surface membrane (Plates 52, 57).

Appearances such as these are particularly common in the endothelial lining cells of blood vessels (Fig. 3). Tracer studies have shown that

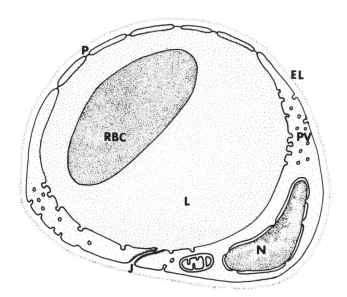

Fig. 3 *Micropinocytosis*
This diagram shows the typical specialisations of a blood capillary. The lumen, L, contains a red blood corpuscle, RBC. The nucleus, N, and an endothelial cell junction, J, are identified. The capillary is surrounded by a continuous external lamina, EL. Two other specialisations are identified. The capillary pores, P, are bridged across by a delicate diaphragm. The second specialisation is the formation of micropinocytotic vesicles, PV, which pinch off from the luminal surface and fuse with the outer surface, providing a pathway for transcapillary fluid movement.

micropinocytosis participates to some extent in the transfer of materials in both directions across the vessel wall. Similar vesicles are seen at the surfaces of peritoneal and pleural mesothelial cells (Plate 77), where a similar transport function can reasonably be assumed, and in smooth muscle cells (Plate 88), where their function is less clear. Micropinocytosis is occasionally observed in the intestinal epithelium, but is not thought to be of quantitative significance in mammals, except in the neonatal period. Newly born suckling infants absorb intact antibodies from their mother's milk by this pathway, gaining *passive immunity* to infection (Plate 4b). The widespread occurrence of micropinocytotic vesicles in cells of various types suggests that this is a commonplace and fundamental part of the cellular economy.

A second variety of micropinocytotic vesicle is recognised in various situations. Its extracellular surface is thickened by an external fuzzy coating and small spikes project into the cytoplasm from its intracellular aspect. Such *coated* or *fuzzy vesicles* are sometimes termed *acanthosomes*, or spiky bodies. They are seen in various cells including macrophages (Plate 32), endothelial cells, renal tubular epithelial cells and red cell precursors. Their distinctive appearance is associated with a receptor mechanism, which is believed to promote the uptake of specific molecules by a *selective* form of micropinocytosis. In red cell precursors it is ferritin which is involved. The characteristic electron-dense ferritin molecules (Plate 32) can be seen attached to the surface receptors at the points of formation of these vesicles. In kidney tubule cells, coated vesicles arise from apical membrane invaginations and are believed to be responsible for the selective reabsorption of plasma proteins from glomerular filtrate. Coated vesicles can be associated with exocytosis as well as endocytosis, since they have, for example, been implicated in the process of casein secretion. Their frequency in the Golgi region and their reported association with contractile proteins suggest that they may play some wider role in translocations within the cell.

Since it is not yet possible to study living cells effectively by electron microscopy, the dynamic process of micropinocytosis cannot be directly observed. The course of events taking place must instead be inferred from the appearance of fixed material. Since each micrograph represents a single moment of time in a single cell, only a tentative composite picture can be reconstructed from different micrographs. The mere presence of vesicles as described above is thus only presumptive evidence for the existence of a dynamic process, giving no indication of direction or total volume of transfer, unless linked with metabolic or tracer studies.

It has been suggested that the different membrane systems in the cell may at times be able to become continuous with one another. The outer membrane of the nuclear envelope at times joins with the endoplasmic reticulum (Plate 12b) and the nuclear envelope appears to re-form from fragments of endoplasmic reticulum following mitosis. There is frequent continuity between granular and smooth endoplasmic reticulum, while in zymogenic cells the membranes of secretion granules fuse with the cell membrane prior to granule release (Fig. 13). Connections have even been claimed between the endoplasmic reticulum and the cell surface, alhough this has not been demonstrated conclusively in animal cells. It is still not clear, in view of the structural and functional differences between different membranes, to what extent membrane flow and interchange normally occur, but dynamic changes of this type are probably a fundamental feature of the living cell.

2.2.3   Cell contact and adhesion

Cells lying in contact tend to adhere to one another. The details of the mechanisms involved in cell adhesion are not yet fully understood, although it is recognised that calcium ions play an important part in the process, since their removal leads to cell separation. Distinctive areas of membrane specialisation are often seen with the electron microscope at points along the contact surfaces between cells. Cell adhesion does not depend entirely on such areas since they may be absent from cultured cells, which are known to stick together.

In the body, however, the presence or absence of specialised cell contacts can be closely correlated with cell adhesion and with other aspects of tissue function such as intercellular communication. There are few specialised contacts in the connective tissues, where the intercellular material serves a mechanical function and where the cells are usually widely separated. In epithelial tissue, however, where the cells are closely packed, intercellular adhesion must often be strong, as in wear and tear surfaces such as skin. Here the specialised cell contacts are particularly large and numerous. In epithelia less subject to injury, contact specialisations are less prominent.

Contact specialisations are best studied in a simple columnar epithelium, such as the intestinal epithelium, sectioned in the long axis of the cells (Fig. 4). The base of each cell thus lies close to the underlying connective tissue and the free apical surface forms the absorptive surface of the intestine. The lateral surface of the cell, extending from the base to the apex, is the contact surface. In stratified epithelium with several layers of cells, such as epidermis, the cells in the middle layers are in contact with their neighbours over their entire surface (Plate 73).

The greater part of the contact surface in most cells is without obvious structural specialisation. The external surface of each cell is defined by its membrane which appears at low resolution as a single thin dense line. At unspecialised areas the membranes of adjacent cells lie parallel to each other, separated by an interspace measuring between 15 and 20 nm. Each of these two cell membranes can be resolved at higher magnification into the typical trilaminar membrane structure already described (Plate 4).

2.2.4 The epithelial junctional complex

An area of membrane specialisation known as the *junctional complex* is always present at the apical or distal end of the contact surface of a columnar epithelium (Fig. 4). It is seen by light microscopy as the terminal bar. This is made up of three areas of distinctive fine structural specialisation, the *zonula occludens*, the *zonula adhaerens* and the *macula adhaerens*. The term zonula, which means girdle, is used in an epithelial situation to emphasise the fact that this specialisation encircles the cell apex, forming an unbroken line of attachment to adjacent cells. The term macula implies a localised spot or disc-shaped area of specialisation.

The most apical part of the epithelial junctional complex is the zonula occludens. Other synony-

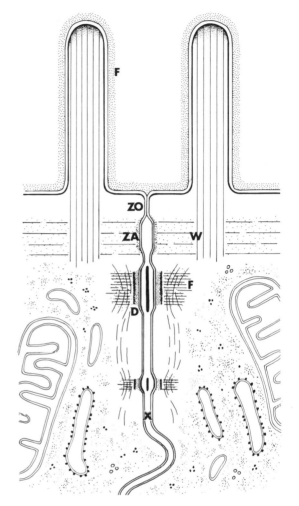

Fig. 4 *Epithelial adhesion specialisations*
This diagram shows the typical contact specialisations found towards the apex of a columnar epithelium such as that of the small intestine. Two microvilli are shown, with the external glycocalyx, or fuzzy coat, F. The zonula occludens, ZO, forms a tight seal by membrane fusion. The zonula adhaerens, ZA, provides an insertion for transverse filaments of the terminal web, W. Beneath these specialisations lie one or more desmosomes, D, characterised by the insertion of cytoplasmic filaments, F, into a dense attachment zone. The desmosome has a central dense linear intercellular component. Elsewhere, the contact surfaces of the two cells shown are unspecialised, as shown at X. The trilaminar pattern of the cellular membranes is indicated. Note that the apical membrane covering the microvilli is the thickest of the membranes shown. The actin filaments which form the core of the microvillus are attached to the tip by an area of diffusely increased density.

mous terms in common use are *close junction* and *tight junction*. Here the normal 15 to 20 nm interspace between adjacent cells is absent and there is fusion between the outer components of the trilaminar membrane of the two cells (Plate 4), resulting in a shared five-layered boundary. The cells do not lose their separate identity despite this close association and can still be separated from one another by chemical and physical means. When studied by freeze-etching, the tight junction is recognised by the occurrence of membrane particles, 3 to 4 nm in diameter, forming linear arrays.

The second specialised area of the junctional complex is the zonula adhaerens. Another synonym commonly used is *intermediate junction*. It lies below or basal to the zonula occludens and again forms a continuous girdle around the cell. At the zonula adhaerens (Plate 4), the membranes of adjacent cells are no longer fused, but are separated by a 20 nm interspace. The cytoplasmic surface of each membrane is associated with a patch of increased density produced by an aggregation of fine filaments which extend from the point of attachment into the adjacent cytoplasm. Linear dense material is not seen between the cells at this point, although there may be some form of mucopolysaccharide-rich intercellular 'cement,' which is not seen with conventional electron microscopic techniques.

The third component of the junctional complex lies at a variable distance beneath these two previous zones. This is the macula adhaerens, or *desmosome*, which forms a discrete plaque or button-like specialisation on the contact surface, as opposed to the zip-fastener configuration of the two zonulae (Plate 6). A desmosome completes the junctional complex, but other desmosomes are found at intervals elsewhere along the contact surfaces of epithelial cells. The detailed structure of the desmosome varies in different species and different cells. There is a 20 to 25 nm interspace, occupied by a median linear condensation of extracellular cement or cell coat material. As in the zonula adhaerens, cytoplasmic filaments are attached to the dense cytoplasmic area which flanks the desmosome (Plate 27).

The junctional complex of an epithelium appears to have several general functions. There is evidence that *permeability to ions* between cells, normally low at unspecialised contact surfaces, is greatly increased at the tight junction. The ready passage of ions and the consequent low electrical resistance

between cells allows a closer functional association between separate cellular units than was once thought possible. It may be that free ionic movements are possible throughout an entire sheet of epithelial cells, as a result of the presence of such junctions. The possibility of communication between cells at the junctional complex may help to explain, for example, the coordinated function of adjacent cells in a ciliated epithelium and may have a wider significance in many other ways. The junctional complex, in particular its occludens specialisation, also acts as a *seal* or gasket at the free margin of the epithelium. This ensures that any substances crossing the epithelial surface in either direction pass through the cells under cellular control, rather than between cells by free diffusion. Different epithelia have been shown to have different degrees of tightness, or resistance to permeation, which may correlate with particular features of their junctional complexes. The zonula adhaerens, as its name suggests, along with the desmosome, is apparently designed to promote *cell adhesion*. This mechanical attachment function is of great importance, since the apical surface of the cell is often exposed to external deforming forces. These are resisted by the junctional complex, thus preserving the coherence and integrity of the epithelial surface.

2.2.5. Desmosomes and gap junctions

The desmosomes are apparently important in cellular adhesion, being particularly numerous and prominent in sites such as epidermis, where *resistance* to mechanical damage is essential (Plates 1, 73). The bundles of filaments associated with the desmosome form the histological image of the tonofibrils. The desmosomes and their associated filament bundles were misinterpreted as 'intercellular bridges', until the true nature of the desmosome was clarified. When the contact surfaces of cells are separated by any unusual increase in tissue fluid, the desmosomes are the last points at which adhesion is maintained.

The attached cytoplasmic filaments are often seen running parallel to the cell surface and linking adjacent desmosomes (Plate 6) within the same cell. This complex of desmosomes and associated filaments probably represents a *structural frame-*

work within the cell which helps to maintain its shape and its relation to its neighbours. The not uncommon occurrence of mitochondria in close proximity to desmosomes raises the possibility that some energy-consuming function is located at these sites.

When epithelial cells fuse to form a syncytium, as in the formation of syncytiotrophoblast from placental cytotrophoblast, the desmosomes which formerly joined adjacent cells may persist within the cytoplasm of the syncytium. Such *intracytoplasmic desmosomes* are seen occasionally in other normal tissues and are quite commonly found in some tumours of squamous cells, such as squamous carcinoma. Their occurrence may be associated with defective cellular adhesion and with the remodelling of contact surfaces.

Finally, a specialisation known as the *hemidesmosome* occurs in squamous epithelium at the interface between the basal layer of epithelial cells and the basal lamina (Plates 5, 8). This consists of half the structure of a typical desmosome, including the dense plaque into which cytoplasmic tonofilaments are inserted and the dense extracellular component, which remains distinct from the basal lamina. It is supposed that the hemidesmosome promotes the adhesion of the epithelium to its basal lamina and so to the underlying stroma.

There are several other named specialisations of cell contact, of which the *gap junction* is perhaps the most important. Junctions of this kind were once classified along with the tight junctions described above. In many cases however, after careful examination, actual membrane fusion was not found to occur and a narrow 2 nm gap was identified between the closely apposed outer membrane lamellae. The gap junction was clearly shown to differ from the tight junction by the use of freeze-etching techiques (Fig. 1), which showed patches of closely packed 2 nm particles, instead of the linear arrays of 3 to 4 nm particles found in tight junctions.

Gap junctions have been found in association with *electrical coupling* between cells. It is thought that they accomplish this through enhanced permeability, allowing a thousand-fold increase in the rate of passage of small molecules from cell to cell by comparison with unspecialised areas of the cell surface. The membrane particles seen on freeze-etching are probably the carrier molecules which provide the channels for this enhanced communication between cells. The gap junction is perhaps better known as the *nexus* specialisation between smooth muscle cells (Plate 88). Similar, but more extensive areas of gap junction specialisation are found in cardiac muscle at the *intercalated disc* (Plate 90). The significance of close contact in cardiac and smooth muscle relates to the need for electrical coupling to ensure the integrated function of the separate cells which make up these tissues. The permeability to ions which characterises such junctions allows the spread of surface excitation from cell to cell throughout the tissue.

2.2.6 The cell coat

In many forms of life an elaborate external coating or shell is built up outside the true cell membrane. In plants, this coating forms the *cell wall* which consists principally of cellulose and provides an important structural scaffolding. A cell wall is also found in bacteria and an additional more diffuse external layer, the *capsule*, is often present. It now seems that all animal cells have some form of external polysaccharide-containing coat, although on electron microscopic examination this is present as a significant structural feature only in certain cases. The term *glycocalyx* has been proposed for this layer, but it is more generally known as the *cell coat*.

A visible cell coat covers the cell membrane at the free surface of various cell types, including the intestinal absorptive cell in certain species including man. This layer has been called the *fuzzy coat* of the microvilli. At high resolution the fuzzy coat appears as fine filaments or 'antennulae microvillares' radiating out from the trilaminar membrane. They bridge the narrow spaces between the microvilli and extend to about a tenth of a micron beyond their tips. The fuzzy coat has been shown to contain an acid mucopolysaccharide component attached to the cell membrane (Plates 7, 33, 115) and is not merely a mucus film on the cell surface. Investigations using autoradiography have revealed that its mucopolysaccharide component is synthesised continuously by the epithelial cells themselves and that there is rapid turnover of this component of the coat.

In the intestine (Plate 7), this fuzzy coat is

visible, on electron microscopy, only at the apical surface of the absorptive cell, but it is thought that the glycocalyx also persists in some tenuous form not demonstrable by the present techniques of electron microscopy at the sides and base of the cell. This may account for the apparently universal 15 nm interspace which is found between adjacent cell membranes at the unspecialised parts of their contact surfaces.

Several functions can be suggested for the intestinal fuzzy coat. It may form a *protective* layer, preventing possible physical and chemical damage to the cell membrane and acting as a buffer against sudden changes in the external environment. The intestinal epithelium is exposed to contact with food particles, digestive enzymes, gastric acid and bacteria. The fuzzy coat might perhaps form a defence against injuries which could lead to cell damage.

More generally, the glycocalyx may present a *selective barrier* to diffusion, allowing some substances and not others to come into direct contact with the cell membrane. This function would have general significance in relation to cell metabolism and special importance at an absorptive surface. The *receptor mechanisms* which control numerous aspects of cellular function, including for example the selective form of pinocytosis, may be situated in the cell coat. Enzymes may be located within the glycocalyx with functions relevant to surface activity. In summary, through its selective permeability and enzyme activity, the cell coat may help to control the composition of the micro-environment which lies immediately external to the true cell membrane, while also playing a vital role, through receptors, in the control of intracellular activity.

Finally, the glycocalyx may influence the relationships between cells. The surface antigens which determine transplantation rejection are located here. Association between cells of similar types may be promoted by the mutual recognition of distinctive surface properties. For example, the normal uptake of lymphocytes by endothelial cells of post-capillary venules in lymph nodes is stopped if the surface coating of the lymphocyte is damaged by the action of glycosidases, which do not harm the cell in other ways. It is possible that some of the disturbances of cell association seen in malignant disease might be related to abnormalities of the cell coat.

2.2.7 'Basement membrane'

With the increase in our knowledge of the biochemistry and fine structure of membranes, the word 'membrane' has acquired a more precise definition than histologists accorded it in the past. The light microscopic 'basement membrane' is not a true membrane in the fine structural sense, but represents the interface at the base of an epithelium where there is an apparent condensation of extracellular components.

When this epithelial 'basement membrane' of the light microscope is examined by electron microscopy, several distinct structures can be identified. First there is the true cell membrane at the base of the epithelial cell, which appears trilaminar on high resolution microscopy. Below the base of the epithelial cell a clear zone 30 nm wide separates the cell from an underlying continuous homogeneous lamina of medium density, composed of diffuse flocculent or perhaps filamentous material with non-organised fine structure. This layer is known as the *lamina densa* or *basal lamina* (Plates 5, 8). It is 30–70 nm thick, much thicker than a true membrane. It is, however, too thin to be visible on light microscopy and cannot alone correspond directly to the 'basement membrane' of the histologist. Deep to this lamina densa lie collagen fibrils and thin process of fibroblasts. The collagen fibrils often run parallel to the basal lamina and may seem at times to merge with it. There are also special anchoring filaments, which seem to tie down the basal lamina to the underlying connective tissue (Plate 74).

Thus the 'basement membrane' as seen by the light microscope represents a fusion of the images of cell membrane, basal lamina, collagen fibrils and connective tissue cell processes, along with any components of the ground substance which may be able to react with the histological stain being used. For this reason the term basement membrane should be reserved for use only in a histological context, to avoid confusion with the ultrastructural entity more correctly termed the basal lamina.

The 'basement membrane' of an epithelium was thought by histologists to arise by condensation of the connective tissue ground substance around the base of the epithelial cells. It has now been shown that there are antigenic cross-reactions demonstrable by the fluorescent antibody technique between epithelial basement membranes in different sites and species, showing the presence of a common epithelial antigen apparently located in the basal lamina. In certain cases it has been shown that the basal lamina is at least partly produced by the adjacent epithelial cells and it seems likely that this may be a general rule. In this and possibly in other respects the basal lamina and the cell coat may be related. In addition, biochemically identifiable collagen components are also present in the basal lamina. This amorphous collagenous component lacks the periodic fibrillar pattern characteristic of collagen in other sites. This is type IV collagen (Section 9.1)

2.2.8. Basal and external laminae

Boundary layers with many of the features of basal laminae are found close to the surface of muscle cells (Plates 85, 87, 88), fat cells (Plate 70), cells of peripheral nerve (Plates 28, 99, 100) and capillary endothelium (Plates 53, 54). When such layers surround cells which, unlike epithelium, have no true base, they are best termed *external laminae*. They appear antigenically distinct from the boundary layers of epithelia.

The basal lamina varies greatly in appearance in different situations. It is sometimes delicate and even apparently discontinuous, at other times thick and coherent. Descemet's membrane in the cornea, among the thickest of basal laminae, is thick enough to be seen distinctly with the light microscope. Although the basal lamina is usually closely applied to the epithelium, it does not follow infoldings at the cell base such as are seen in kidney tubule cells (Plate 51), but usually forms a flat shelf upon which the base of the cell, however complex, can rest as a whole. When cells are shed from the epithelial surface, the gap which they leave is rapidly filled by migration of neighbouring cells. The presence of an even-surfaced basal lamina may be of importance in this process of orderly cell replacement. Since the basal lamina must be crossed by all substances which enter or leave the base of the cell, it has the potential to influence epithelial metabolism.

It is possible that the coherent linear image of the basal lamina is misleading, suggesting as it does the presence of a firm physical barrier. The true physical nature of the basal lamina and the extent to which it restricts free passage are unknown. It can, however, be breached quite easily by migrating cells such as lymphocytes (Plate 68) and polymorphs, which can cross it from the underlying connective tissue. Perhaps, under localised pressure from a migrating cell, the basal lamina undergoes a change of physical state and loses its cohesion, re-forming again spontaneously after the migrating cell has passed through. Rapidly growing embryonic cells may push partly through the basal lamina to the underlying connective tissue while tumour cells can break right through the basal lamina which may surround them, coming to lie in direct contact with the connective tissue stroma.

The widespread occurrence and strategic location of basal laminae in general suggests a fundamental importance which far exceeds our present limited understanding of the significance of these boundaries. In one situation, however, a fundamental functional correlation is recognised. One of the most important of specialised basal laminae is found in the renal glomerulus, separating the endothelial cell of the glomerular capillary from the glomerular epithelial cell, or podocyte (Plates 61, 62). This thick basal lamina is known sometimes as the *glomerular filtration membrane*, since it represents the primary filter in the first stage of the formation of the urine from the circulating blood. The glomerular filtration interface normally retains virtually all of the plasma proteins, the glomerular filtrate being protein-free. It may, however, may be damaged by the deposition from the blood of circulating antigen-antibody complexes (Plates 109, 110). This results in a poorly-understood alteration of its function, characterised by abnormal leakiness with regard to the smaller albumin molecules of the plasma proteins. The occurrence of albuminuria is thus an important sign of renal disease.

3

Structure and function of the nuclear components

3.1 NUCLEUS

3.1.1 Nuclear morphology

It is now possible to investigate the structure of the nucleus using a wide variety of techniques. These range from the simple stains used for light and electron microscopy, to elaborate labelling and analytical methods which can probe the genes themselves.

In fixed and stained preparations for light microscopy, using dyes such as haematoxylin, the cell nucleus shows a characteristic pattern of staining. This is one of the more distinctive features by which different cell types can be recognised by the histologist. The nuclear substance is known as chromatin. Those parts of the nuclear material which stain deeply are termed heterochromatin, while the remaining paler areas are known as euchromatin. The heterochromatin may form patches scattered throughout the nucleus, or may be clumped towards the nuclear envelope, often forming a thin rim on its inner aspect. Small patches of heterochromatin are commonly associated with the nucleolus.

The proportions and patterns of euchromatin and heterochromatin are widely variable, and often reflect cell function. For example, the different stages of maturation of a particular cell type, such as the red cell precursors in the bone marrow, show a characteristic progression from a diffuse pattern with abundant euchromatin to a dense aggregate consisting almost exclusively of heterochromatin. In general, the more restricted the functional role of the cell becomes, the more predominant becomes the heterochromatin component of the nucleus.

The patterns of pale and dense nuclear material

seen in electron micrographs correspond closely to the patterns of histological staining, since heterochromatin takes up the heavy metals used in fixation and staining to a much greater extent than euchromatin (Plates 5, 9, 19). The detailed arrangement of the chromatin is, of course, more clearly seen, as in the case of the thin peripheral rim of heterochromatin which lines the inner aspect of the nuclear envelope in most cell types. When peripheral clumps of heterochromatin are seen, they are separated by pale euchromatin channels (Plates 10, 12, 14), which lead from the centre of the nucleus to the nuclear envelope at points corresponding to the positions of the nuclear pores. The resulting alternation of blocks of heterochromatin and channels of euchromatin may, in an extreme case, give rise to a 'cartwheel' pattern in the nucleus, which is particularly characteristic of the plasma cell (Plates 3, 15, 68).

Various special techniques have helped us to understand the chromatin patterns more fully. Autoradiography with labelled RNA shows that active transcription occurs within the euchromatin areas, although not all the euchromatin is active at any one moment. The RNA is localised to a variety of structures known as perichromatin granules and filbrils, but little is known of their molecular configuration.

In contrast, heterochromatin seems to be the inactive form of the nuclear material. Thick and thin filaments can be identified, the thick type measuring up to 30 nm in diameter, while the thin type is around 10 nm in diameter. The thick filaments may be due to secondary coiling of the thin filaments, although their exact molecular configuration is still uncertain. More detailed study of the 10 nm filaments shows a beaded structure, the individual beads being known as nucleosomes. Each nucleosome has a cylindrical protein core of four different basic, or histone proteins, around which is looped the thread of the nuclear DNA. It is only through the tight coiling and supercoiling of these fundamental nuclear components that the 3 metre length of DNA in each human cell is accommodated within the nucleus only 5 μm in diameter.

In general, it must be acknowledged that the ultrastructure of the nucleus is difficult to interpret in any precise functional terms. The main struc-

tural components, which appear in sections as granules or filaments of various sizes, often seem to be randomly arranged in thin sections. It is difficult to distinguish a discrete granule from a cross-sectioned filament in a thin section. It is also difficult to detect relevant three-dimensional patterns, since these basic structural elements have dimensions in the same order of magnitude as the section thickness. Finally, the conventional electron micrograph provides no basis for the identification of the molecular components of the nucleus. For these reasons, the study of the

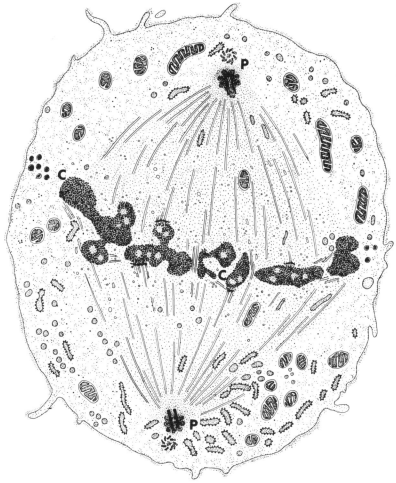

Fig. 5 *Metaphase of mitosis*
The dense masses, C, spread horizontally across this cell are the chromosomes. These form the metaphase plate. The chromosomes are connected by microtubules to the two pairs of centrioles which form the poles, P, of the metaphase spindle. One centriole of each pair is shown cut in longitudinal section, the other in transverse section. During mitosis, the Golgi apparatus and endoplasmic reticulum break up into small fragments which are equally distributed between the two daughter cells.

Diagram by courtesy of R. V. Krstić

intimate nuclear fine structure is likely to remain of interest mainly to the specialist, who can call on parallel studies in the field of molecular biology to assist in the interpretation of these complex images.

The description of nuclear structure has concentrated so far on the resting, or interphase nucleus, in which the individual chromosomes are dispersed and obscured by their particular configuration and by their complex associations with the various proteins of the nucleus. In the dividing nucleus, however, the chromosomes become condensed and can be separately identified. The typical features of mitosis have been well described by the light microscope

The earliest changes of mitosis are metabolic rather than structural. Autoradiography shows that DNA synthesis takes place for about 7 hours before the onset of the structural changes of mitosis. The total nuclear DNA doubles during this synthetic, or S-phase. The initial structural change, known as prophase, involves an alteration of the chromatin pattern, which results in the formation of the discrete chromosomes. Each consists at this time of two equivalent parallel chromatids (Plate 11), each of which contains the full complement of genetic information which is carried by that chromosome. The nuclear envelope then breaks down.

In the next stage of mitosis, known as metaphase, the chromosomes line up together across the centre of the cell (Plate 10b). A refractile double cone called the spindle occupies the greater part of the cytoplasm at this time. This consists of microtubules linked to a pair of centrioles at each pole (Fig. 5). The chromosomes lie across the equator of the spindle and are linked to the microtubules. During anaphase the chromatids of each chromosome separate, and the two daughter chromosomes are pulled apart, under control of the centrioles, to the opposite poles of the spindle.

In the final stage, known as telophase, the two separate chromsome masses coalesce to form two separate daughter nuclei, the chromosomes losing their separate identify once again as the interphase pattern of the nucleus reasserts itself. The nuclear envelope is reconstituted from fragments of endoplasmic reticulum in the cytoplasm. Meanwhile, a progressive constriction appears across the equator of the dividing cell, resulting in the formation of an increasingly slender bridge between the two cells. The main structural link at this time is a residual bundle of overlapping spindle microtubules known as the telophase midbody (Plate 29). The function of this persistent narrow bridge is not clear, but perhaps, by holding the two cells apart, the midbody may minimise the risk of cell fusion at this critical period. Finally, the cytoplasm splits completely, forming two independent nucleated daughter cells.

The highly condensed chromosomes of metaphase have been studied by various techniques. Staining by various dyes can result in the production of a characteristic banding pattern. The study of these patterns can be of major importance in the diagnosis of structural disorders of chromosomes, such as translocations and deletions.

It is difficult to read meaning into the fine structural image of the mitotic chromosome. As in the interphase nucleus, the dense histological staining of the mitotic chromosomes corresponds to areas of equivalent heavy metal staining on electron microscopy. The mitotic chromosome is an irregular mass of dense heterochromatin material which is not partitioned in any way from the surrounding ground substance of the cytoplasm, since the nuclear envelope has already been dispersed during prophase. Different levels of ultrastructural organisation have been described, including dense particles and fine spirals and chains, which may be as small as 2 nm in diameter, which approximates to the molecular diameter of DNA.

In the thin sections used for electron microscopy, only a small portion of each chromosome is normally found lying in any given plane of section (Plate 10). For this reason, the contribution that conventional thin-section techniques can make to cytogenetics is necessarily limited. It will be obvious that the number and form of the chromosomes in any given mitotic figure can only be accurately judged when they are spread out in special preparations, best studied by light microscopy. Thus many important problems concerning chromosome numbers and anomalies are most appropriately studied by the use of the light microscope. Soon, however, it is possible that scanning electron microscopy may play a greater part in cytogenetics (Plate 11).

There is now a growing interest in the use of new techniques. Enzyme digestion of sections embedded in water-soluble media can lead to the identification of material of a particular chemical nature. For example, the comparison of experimental and control sections following digestion with DNase can demonstrate that a particular area contains DNA. Techniques have now been developed to study spread preparations of mitotic chromosomes and of interphase chromatin (Plate 116b). The main disadvantage of some of these techniques is the problem of specimen thickness. This can be overcome in part by the use of high-voltage electron microscopy, which makes possible much greater specimen penetration by the electron beam (Section 14.3.4). Image superimposition in these thick preparations is overcome by tilting the specimen and taking pictures at two different angles to produce a stereoscopic effect, allowing visualisation of structural details in three dimensions. Surface details of appropriate preparations can also be visualised by scanning electron microscopy (Plate 11). Techniques such as these, associated with autoradiography and biochemical investigations, may help to resolve the many problems which still remain concerning the fine structure of the nucleus.

3.1.2 Nuclear function

The essential component of the nucleus is DNA, which exists in association with various protein components, including the enzymes involved in nuclear metabolism. The DNA of the nucleus is the carrier of genetic information, which is the basis of heredity. However, even in cells which do not divide, the nucleus is still essential for the daily control of cell activity. Any cell deprived of its nucleus will soon die. Protozoa so treated lose their motility and their power to feed. Once the red blood corpuscle of the mammal loses its nucleus during maturation it not only becomes unable to reproduce itself, but its potential lifespan becomes limited. It is now clear that the nucleus directs the activity of the cell according to instructions coded in the nuclear DNA.

The molecule of DNA forms a long chain arranged in a double stranded spiral, the two strands of which are complementary. The precise sequence of bases in the nucleotides, the molecular

subunits of the DNA molecule, carries genetic information in the form of a *coded message* in which the information essential for heredity is contained. The molecule of DNA has two important properties which make it particularly suitable as a carrier for the genetic message. The first feature is the ability to *reproduce* itself under suitable conditions. Each molecule of DNA unwinds into two complementary halves. Each half is then a template for the reconstruction of a new molecule identical to the original. The molecule is constructed in such a way that mistakes during replication are extremely rare. The second feature of DNA is its *stability*. The sequence of bases, and thus the coded message, remains exactly the same, irrespective of the number of cell divisions or of the passage of time. The message can, of course, be altered by physical or chemical damage to the DNA molecule, producing a *mutation*, or by a virus infection, which inserts new instructions into the message.

The nucleus exerts control over the day-to-day activities of the cell by determining the nature of every protein molecule synthesised by the cell. Since this includes the enzymes necessary for each step in metabolism, this control by the nucleus extends to all forms of cellular activity. In any cell of the body, however, only an appropriate part of the nuclear DNA is active in the control of cell function at any given time. At active sites located in the euchromatin areas of the nucleus, the message locked in the DNA seems to be first transcribed to very large RNA molecules of a special type, known as heterogeneous nuclear RNA, or hnRNA. This has a short half-life and does not leave the nucleus, so it is presumed that these molecules are degraded to form messenger RNA, or mRNA. Messenger RNA formed in this way is not stable, so it becomes coupled to protein before leaving the nucleus through the nuclear pores. This messenger RNA can now instruct the cytoplasmic ribosomes to make the specific protein for which it codes, and its base sequence mirrors the instructions contained in the base sequence of the coded message carried by the original nuclear DNA. The control of this process of information transfer, known as transcription, is the primary function of the somatic cell nucleus. In this indirect way, the active DNA in the nucleus regulates cell functions. The remaining part of the DNA, which contains genetic information not immediately required by the cell, appears to remain functionally inert. In some cells as little as 5 per cent of the genetic material is in the active form while the remaining 95 per cent is dormant.

The tightly coiled inactive DNA of heterochromatin is associated with nucleohistones which not only repress the DNA, preventing it from taking part in transcription, but which may also offer physical protection to the genetic material. In the diffuse extended DNA of euchromatin there is an accumulation of acidic non-histone proteins, which can specifically inactivate the repression of DNA, making possible the transcription of individual genes. Special techniques have now been developed for the direct observation of the molecular structures involved in transcription. DNA molecules can be isolated in special cases from the nucleus and picked up on thin carbon films for electron microscopic examination. In such preparations, the attached partially synthesised molecules of mRNA can be seen extending at right angles to the thread of DNA.

Different proportions of euchromatin and heterochromatin are found in different cells. Embryonic cells engaged in growth and multiplication, and rapidly dividing cells in bone marrow or intestinal crypts have a larger proportion of their nuclear material in the euchromatin form. This indicates that much of the genetic material is in use, directing the numerous aspects of cell differentiation. Mature cells with more narrowly specialised functions are often distinguished by the predominance of dense heterochromatin. Two examples of this type of cell are the plasma cell, devoted to the manufacture of antibody molecules, and the normoblast, the immediate precursor of the red blood corpuscle, which is engaged in the synthesis of haemoglobin. In these cells, very little of the genetic information is in active use. In an extreme case, the nuclear material of the spermatozoon is virtually all in the protected, metabolically inactive form, appearing very dense in electron micrographs (Plate 95). This underlines the role of this cell as a simple transport vehicle.

The normal human interphase nucleus, with its 23 pairs of chromosomes, contains an amount of DNA which is termed *diploid*, or twofold. Shortly before cell division, the DNA of the nucleus doubles itself in a brief synthetic phase which lasts about seven hours in human cells and which results in a *tetraploid* or fourfold amount of DNA. The subsequent mitotic division redistributes the doubled DNA to provide each of the daughter cells with its full diploid complement. In this way the genetic message is carried intact to every cell of the body, even though any individual cell may only use a small part of the information for its daily functions.

The germ cells, the sperm and ovum, are the only exceptions. They are formed by a specialised type of cell division, called *meiosis*, in which there is halving of the normal genetic information, each daughter cell containing only one member of each of the 23 chromosome pairs. The term *haploid*, or single, describes the DNA complement of the germ cell, corresponding to half of the full genetic material. Germ cells are the only haploid cells in the body. When the germ cells fuse, the *zygote* which is formed once again has the full 46 chromosomes, the diploid amount. The subsequent mitotic divisions of the zygote produce the cells of the new individual, all of which have an identical genetic make-up.

If the sequence of bases is altered in the DNA molecule, a *mutation* is said to have occurred. The alteration may be so slight as to be unnoticed, or so severe as to cause the death of the cell or prevent it from ever dividing successfully. Since the mutation involves an alteration in the DNA molecule, it constitutes a permanent change in the genetic information of the cell, which is transmitted to all the progeny of the mutant cell if mitosis is still possible. There are, however, intricate enzyme mechanisms within the nucleus for the repair of damaged strands of DNA. Because of these controls, only a few of the many episodes of damage which must regularly occur express themselves in a recognisable form.

In the past, the nucleus was often the primary interest of the cellular biologist. This was partly because of the distinctive behaviour of the nucleus during cell division and partly because of the difficulties of resolving the finer cytoplasmic details. To a certain extent the electron microscope has restored the balance. Since all nuclei contain the same essential biochemical blueprint, it is not surprising that study of the nucleus with the electron microscope has been relatively unrewarding. Differences between the nuclei of different cells are most clearly expressed at the molecular rather than the fine structural level, whereas cytoplasmic differentiation is often intimately expressed in the ultrastructure of the cell.

3.2   NUCLEOLUS

3.2.1   Nucleolar morphology

The nucleus normally contains one or more small discrete nucleoli clearly seen on light microscopy. A single nucleolus is, of course, not always seen in thin sections of the nucleus, since the plane of section may not pass through that part of the nucleus in which it lies. The nucleolus is basophilic and has been shown to contain a substantial proportion of RNA, by cytochemical techniques and by biochemical analysis. On electron microscopy the nucleolus is readily identified, if the plane of section permits, by its high electron density (Plates 10, 13, 80). The nucleolus is not separated from the substance of the nucleus by any definite specialisation such as a membrane. Two components of the nucleolus have been described, a coiled coarse skein of granular material called the *nucleolonema*, or granular portion, and a paler substance between the loops of the nucleolonema which is termed the *pars amorpha* or fibrillar portion. Patches of chromatin are often associated with the rim of the nucleolus.

The nucleolonema, the denser part of the nucleolus, may have a coiled structure, or may at times best be represented as a sponge-like component. The pars amorpha, less dense, fills the interstices between the twisted components of the nucleolonema. The main fine structural subunit of the nucleolonema is a slightly angular dense particle, structurally similar to the ribosomes seen in the cytoplasm. Similar particles are found elsewhere in the nucleus. In addition, thin filaments, less than 10 nm in diameter, are described within the nucleolus in association with these granules. The granules often lie in groups, densely packed, but may also be more loosely arranged. The pars amorpha has few distinctive fine structural features, containing only fine granules with few obvious interrelations.

3.2.2   Nucleolar function

The nucleolus is an essential link in the chain of communication between the nucleus and the cytoplasm, taking part as it does in the synthesis and processing of the various species of ribosome RNA. The initial step in the biosynthesis of protein is the formation of messenger RNA, a molecule which contains the detailed information originally locked in the nuclear DNA. This information is necessary for the linking of amino acids in the correct sequence to form individual protein molecules, a process which takes place on the ribosomes in the cytoplasm. It is thought that the nucleolonema may contain newly synthesised ribosome subunits, which pass out of the nucleolus into the nuclear substance, and finally into the cytoplasm, combining at some point with messenger RNA already formed by the DNA of the nucleus. The messenger RNA can then instruct the ribosomes to synthesise a specific protein. In this way the nucleus can control and direct cell function. It is obvious that the passage of material from nucleus to cytoplasm is of great importance in the expression of genetic information. The importance of the nucleolus in protein synthesis is emphasised by its prominence in cells distinguished by activity of this type, such as tumour cells engaged in rapid growth. The structure of the nucleolus can be modified substantially by experiments designed to affect its function, involving dietary and chemical influences.

3.3   NUCLEAR ENVELOPE

3.3.1   Nuclear envelope morphology

The partition which separates the nucleus from the cytoplasm is a complex structure known as the *nuclear envelope*. It is composed of two separate membranes. The inner nuclear membrane forms the limit of the nuclear contents and is separated by a gap of about 50 nm from the outer nuclear membrane. The width of this gap varies in different cells and even at different points around a single nucleus. This narrow cavity surrounding the nucleus is called the *perinuclear cisterna*. On one side, the cisterna is in contact with the heterochromatin clumped on the inner surface of the inner nuclear membrane. On the other side, the outer nuclear membrane lies in contact with the cytoplasm (Plate 12a).

In cells with a prominent granular endoplasmic reticulum, continuity may occasionally be seen between membranes of the cytoplasmic cisternae and the outer nuclear membrane (Plate 12b). The ribosomes, many of which are attached to the cytoplasmic surfaces of the granular endoplasmic reticulum, are also commonly seen on the outer surface of the outer nuclear membrane. There is thus a potential intercommunication between the cavities of the endoplasmic reticulum and the perinuclear cisterna. Since it is known from light microscopic examination of the living cell that its interior is in a state of constant movement, it seems likely that such connections are being constantly broken and re-formed.

Nuclear pores form an interesting and important specialisation of the nuclear envelope. These pores, usually from 50 to 70 nm in diameter, are more numerous in some cells than in others, the area occupied by them ranging from 5 per cent to 30 per cent of the total surface area of the nucleus. At the nuclear pore the inner and outer nuclear membranes meet, producing a circular hole between the nucleus and the principal phase of the cytoplasm (Plates 10a, 12, 13). The pore thus formed is usually bridged by a diaphragm which has no obvious trilaminar structure and is more diffuse in appearance than a membrane (Plate 30). The diaphragm is the only apparent barrier between nucleus and cytoplasm across the nuclear pore. Around the margins of the outer aspect of the pore a cylindrical collar or annulus is often seen projecting into the cytoplasm. When a tangential section of the nucleus is examined, the nuclear pores and their annuli can be seen in surface or face view (Plates 13a, 28a). The annuli can at times be seen to consist of a ring of tiny subunits extending out from the circumference of the pore. Finally, in certain cases, a flange can be seen, extending into the perinuclear cisterna in the plane of the diaphragm. The detailed morphology of the nuclear pores is subject to great variation, for reasons not yet clear.

3.3.2 Nuclear envelope function

Communication between nucleus and cytoplasm is of vital importance in the control of the cell. Since the nuclear envelope must be crossed by molecules taking part in any interchange, it is likely that the nuclear pores play a role in this biochemical *communication*. The extent to which the nuclear pore presents a *barrier* to diffusion is not fully known, but it is the only well defined structural channel between the cytoplasm and the nucleus; it certainly provides a pathway to the nucleus for various substances introduced into the cell by experimental means.

The potential continuity of the perinuclear cisterna and the endoplasmic reticulum is underlined by other observations. During fat absorption by the columnar cells of the intestine (Plate 14), droplets of particulate lipid material are found in the channels provided by the endoplasmic reticulum. On occasion, however, lipid droplets have also been seen within the perinuclear cisterna. In the developing Paneth cell of the intestine, a cell with prominent granular endoplasmic reticulum, the presence of newly synthesised material with a distinctive crystalline appearance has been reported in both the endoplasmic reticulum and the perinuclear cisterna. Such results suggest that these cavities share a common function, emphasising their dynamic unity. The possibility also exists, at least in developing cells, that the endoplasmic reticulum may in part arise from the nuclear envelope.

The behaviour of the nuclear envelope during cell division confirms the view that it is related to the endoplasmic reticulum. In the early stages of mitosis, as the chromosomes form, the nuclear envelope breaks down, becoming dispersed in the cytoplasm in the form of vesicles and membrane profiles similar to the scattered components of the endoplasmic reticulum. During most of mitosis the chromosomes lie in direct contact with the cytoplasm, without any intervening structural barrier. At telophase, however, when the two chromosome masses of the separating daughter cells are coalescing to form the daughter nuclei, each nucleus becomes surrounded once more with its complex envelope, apparently formed by the linking up of the scattered cytoplasmic vesicles and cisternae of the reticulum. The mechanism responsible for the orderly breakdown and reformation of the nuclear envelope is not known.

4

Structure and function of the cytoplasmic components

4.1 RIBOSOMES

4.1.1 Ribosome morphology

Early studies of cytoplasmic fine structure showed the presence of the distinctive small dense granules which have become known as *ribosomes*. The name ribosome reflects the presence in these granules of a significant proportion of ribonucleic acid (RNA) in combination with ribonucleoprotein. They are 15 nm in diameter and have a slightly angular profile when seen at high magnification. Detailed analysis has shown that ribosomes have two subunits of unequal size. These granules, among the most fundamental of subcellular particles, are present in virtually every type of cell. They are found singly and in groups in the cytoplasm, often arranged in clusters or rosettes, sometimes in spirals. Identical granules are attached to the outer surfaces of the cisternae of the granular endoplasmic reticulum, giving this system its characteristic morphology.

When a cell contains many ribosomes, it has a characteristic *cytoplasmic basophilia* by light microscopy. Diffuse basophilia is seen when numerous ribosomes are free in the cytoplasm and when there are few organised cisternae of the endoplasmic reticulum, as in the intestinal crypt cell (Plate 6a). Ribosomes associated with areas of organised granular endoplasmic reticulum show a patchy basophilia which has been given different names in different cell types. Such scattered patches throughout the nerve cell were termed *Nissl bodies* (Plate 98a). In the pancreatic zymogenic cell the basophilic area, often predominantly basal, was termed *ergastoplasm* (Plate 53a). The intensely basophilic plasma cell (Plates 3, 15, 16, 68) is virtually full of granular endoplasmic reticulum, except for the Golgi region near the nucleus, which appears as an isolated, pale, non-basophilic patch on light microscopy.

4.1.2 Ribosome function

The messenger RNA-ribosomal complex is the site of protein synthesis in the cytoplasm. This fundamental activity is ultimately controlled by the genetic message carried in the nuclear DNA. The ribosomal precursors are synthesised in the nucleolus, but they receive their instructions for the manufacture of specific protein molecules from messenger RNA, in which the coded message is transcribed. Each ribosome has two subunits of unequal size, each with a specific function, attached to a thread of messenger RNA. A number of ribosomes concerned in the manufacture of single large protein molecules are arranged together, linked by the thread of messenger RNA into a functional group known as a *polyribosome* or *polysome* (Plates 3, 16). The rosettes and spirals of ribosomes commonly seen in the electron micrograph represent these functional units of protein synthesis.

4.2 ENDOPLASMIC RETICULUM

4.2.1 Endoplasmic reticulum morphology

In the early days of biological electron microscopy, when crude sectioning techniques were inadequate to demonstrate detailed fine structure, cells could be examined whole, spread out on thin plastic films. This method was unsatisfactory in many ways since the specimen was not only thick, but was unsupported by an embedding medium and distorted by dehydration. However, the main cell features could still be seen in outline. The cytoplasm was found to contain a lace-like network extending through the interior of the cell, to which the name *endoplasmic reticulum* was given. With greater refinement of thin sectioning methods, the endoplasmic reticulum appeared as a complex system composed of pairs of membranes enclosing interconnecting cavities or *cisternae*.

Although wide structural variations occur from cell to cell, the endoplasmic reticulum possesses common features which have led to its recognition as a fundamental cytoplasmic organelle. Two patterns of endoplasmic reticulum are recognised on a morphological basis, the *granular* and the *agranular* endoplasmic reticulum. The alternative terms *rough* and *smooth* are often used.

4.2.2 Granular endoplasmic reticulum

In most cells there are are at least some cytoplasmic membranes with ribosomes attached to their outer surfaces, forming a cisternal pattern classified as endoplasmic reticulum. The presence of attached ribosomes suggested the use of the descriptive term granular reticulum. The cisternae of the reticulum form patterns of varying complexity, ranging from a few small isolated profiles (Plates 14, 42) to systems with interconnections and cavities which form an apparently continuous phase of the cytoplasm (Plates 3, 16, 20), limited by an extensive area of granule-covered membrane surface. Individual cisternae usually take the form of flattened envelopes (Fig. 6), but may appear in thin section as tubular or vesicular profiles.

The outer nuclear membrane, which is also studded with ribosomes, appears structurally similar to the membranes of the endoplasmic reticulum, with which it occasionally becomes continuous (Plate 12b). When this occurs, the perinuclear cisterna becomes confluent with the cisternae of the endoplasmic reticulum.

It is now believed that the granular endoplasmic reticulum is a dynamic system. Connections between different cisternae and between the endoplasmic reticulum and the perinuclear cisterna are thought to be continually made and broken as a result of *cytoplasmic streaming* movements, but this activity is halted at the moment of fixation to give the static picture shown in the electron micrograph.

Fig. 6 *Granular endoplasmic reticulum*
This diagram shows the relationships between cisternae of the granular endoplasmic reticulum and a mitochondrion. The membrane-limited cisternae have numerous small dense ribosomes attached to their outer surfaces. Ribosomes are also seen in the cytoplasmic matrix between cisternae. The lumen of the cisterna, L, contains protein material newly synthesised by the ribosomes attached to the membranes. The mitochondrion is enveloped by the endoplasmic reticulum, to which it supplies the energy required for protein synthesis.

The ribosomes, site of cytoplasmic protein synthesis, are attached to the membranes in variable numbers, at times closely packed, at times more widely spaced. The occasional tangential section of a cisterna can reveal the distribution pattern of the ribosomes across the two dimensions of the membrane surface (Plates 3, 16). Clusters and spirals forming polysomes, or interrelated groups, are often found in addition to randomly spaced single ribosomes. It is likely that the members of a group are linked together functionally and structurally to produce molecules of a specific type of protein.

There are many cell types with a well developed granular endoplasmic reticulum. The associated basophilia on light microscopy has long been recognised to be due to the presence of abundant RNA. Typical examples are zymogenic or enzyme-secreting cells, antibody-secreting plasma cells (Plate 15) and fibroblasts which manufacture collagen and connective tissue ground substance (Plate 80b).

Biochemical studies of homogenised pancreatic cells have shown that protein synthesis is carried out by the *microsome* fraction of the cell. This fraction, which remains after heavier components like mitochondria (Plate 23b) have been spun down, is composed largely of ribosomes and fragments of the endoplasmic reticulum. Experiments like this confirm that the presence of an elaborate granular endoplasmic reticulum is an indication of active synthesis of a protein product designed for eventual use outside the cell. In cells which 'export' the protein they make, the membrane system seems essential for the segregation and transport of the finished product through the cytoplasm to the Golgi apparatus in preparation for further processing and discharge. By way of contrast, numerous cytoplasmic ribosomes with few associated membranes suggest significant protein synthesis for internal use, as in the rapidly growing or dividing cell, or in the red cell precursor, in which haemoglobin is retained as it is synthesised.

4.2.3 Smooth endoplasmic reticulum

In many cells there are membrane-bound cisternae which form a complex tubular system of interconnecting cavities, without associated ribosomes. Cytoplasmic membranes of this type are all classed, despite wide morphological variations, as *smooth* or *agranular endoplasmic reticulum*. Although the membranes of the granular reticulum are sometimes seen in continuity with areas free of attached ribosomes, it is more common for agranular and granular reticulum to occupy different areas of the cell. Cells containing significant proportions of agranular reticulum, especially when combined with a prominent mitochondrial population, are generally eosinophilic when examined with the conventional techniques of light microscopy.

The detailed structure of the agranular endoplasmic reticulum is extremely variable and appears to depend to a considerable extent on the method of fixation. With osmium tetroxide alone, the reticulum is discontinuous, consisting mainly of empty-looking smooth-surfaced vesicles. With glutaraldehyde fixation, however, the reticulum

may appear as narrow interconnecting tubules forming a network more like that of the granular endoplasmic reticulum. The vacuole form of the smooth reticulum probably results from fixation damage, so that the tubular form represents a closer approach to the original state of the living cell.

The various cell types characterised by a prominent agranular endoplasmic reticulum have no single common function. They include cells of the apocrine sweat glands and sebaceous glands as well as cells which secrete steroid hormones, such as the interstitial cells of the testis and the cells of the adrenal cortex (Plate 18). Intestinal epithelial cells have a variable amount of agranular endoplasmic reticulum, which appears to be involved in the process of fat absorption (Plate 14). This apparent link with lipid metabolism may also account in part for the presence of agranular reticulum in liver cells (Plates 45, 46, 55a), although here these membranes have additional functions, such as the metabolism of drugs. For example, the extent of the membranes of the agranular reticulum of the normal liver cell can be greatly increased by the administration of barbiturate drugs, which are detoxicated by membrane-associated liver enzymes. The reticulum may also be associated with glycogen metabolism in liver cells, since there is a close topographical relationship between smooth-surfaced membranes and glycogen granules. An extreme specialisation of the agranular endoplasmic reticulum is seen in cardiac and skeletal muscle. Here the smooth surfaced membranes, commonly termed the *sarcoplasmic reticulum* (Plate 85a, 86), appear to be concerned with calcium binding and release, essential to muscle contraction (Fig. 22), as discussed more fully in Section 10.1.1.

The acid-secreting gastric parietal cells of the mammal and the acid-secreting cells of other species present a special problem of nomenclature. These cells contain a prominent system of smooth-surfaced membrane profiles which have in the past been classified as smooth endoplasmic reticulum. These profiles, in the form of tubules, vesicles or vacuoles, undergo changes according to the functional state of the cell and appear to play a fundamental part in acid secretion. The membranes of this system are, however, thicker than the usual agranular reticulum and may be more closely related to the membranes of the cell surface than to those of the endoplasmic reticulum (Plate 40).

4.2.4   Endoplasmic reticulum function

Morphological subdivision of the endoplasmic reticulum into distinct granular and agranular components could be misleading, since it might be thought to imply a single functional distinction. Although granular endoplasmic reticulum is unequivocally concerned with protein synthesis, there is no equivalent specific function which can be assigned to agranular reticulum. Indeed, the wide range of cells in which smooth-surfaced cytoplasmic membranes are prominent makes it very difficult to suggest such a functional link. Perhaps it would be more realistic to regard the membranes of the endoplasmic reticulum, whether granular or agranular, as having a common general function. This can be seen as the formation, at the subcellular level, of a flexible *structural framework* which can be made to serve widely differing biochemical purposes in differently specialised cells.

Any membrane-limited cisternal system has several important functional properties. First, it effectively *partitions* the cytoplasm into two phases, separated by a single membrane surface. In this way ideal conditions are provided for the segregation of material within a closed system in the cell. Secondly, a large area of *membrane surface* can be accommodated within a limited cytoplasmic volume. Surface-limited reactions can therefore be carried out with maximum efficiency in the confines of the system. Thirdly, enzyme systems attached to the membranes can be *organised* to achieve their optimum spacing, and three-dimensional relationships can be established which would not be possible without a structural framework. Thus the ribosome, which is bathed in the cytoplasmic ground substance between cisternae, is attached to the outer surface of the membrane by its large subunit, allowing the forming protein molecule to be extruded into the lumen of the cisterna. Fourthly, such a system of interconnecting cisternae provides channels throughout the cell which can be used for the *transport* of material from place to place.

These properties of a simple cisternal framework can easily be adapted to different functions by the attachment of different enzyme systems to the membrane surfaces. In this way, the endoplasmic reticulum in different cells can adopt functions as different as protein synthesis, steroid metabolism, absorption, detoxication, glycogen metabolism, excitation-contraction coupling and ion transport. In most cases, the attachment of different enzymes to the membranes of the endoplasmic reticulum does not produce any fine structural specialisation of the membrane surface, since the specialisation is at the level of molecular rather than microscopic structure.

There is, however, one biosynthetic system associated with a clearly recognisable structural unit; the apparatus for *protein synthesis* lies in the ribosome which can be identified readily in thin sections. Thus when the basic membrane framework of the endoplasmic reticulum is adapted for the purpose of protein secretion, the membranes become structurally as well as biochemically specialised and can be recognised by electron microscopy as the granular reticulum.

The full significance of the endoplasmic reticulum in cell metabolism is not yet known, with even the extent of its links to other parts of the cell being far from clear. It has been suggested that the endoplasmic reticulum may connect directly with the exterior of the cell, but in the animal cell the evidence for this is so slight that it can occur only rarely, if at all. Evidence for indirect communication between the endoplasmic reticulum and the Golgi apparatus through the formation of transport vesicles is more convincing, indicating the existence of a physiological pathway through the cell, such as is employed in fat absorption (Section 6.1.1).

4.3   ANNULATE LAMELLAE

4.3.1   Annulate lamellar morphology

The annulate lamellae are stacks of membrane-limited cisternae perforated by distinctive ring-shaped *fenestrations*. Each lamella consists of two parallel membranes forming an elongated cisterna from 30 to 50 nm wide. There are frequent interruptions of these cisternae by circular pores or discontinuities (Plate 18c). At times the pores appear bridged by a diaphragm and there is a diffuse ring or collar of dense material forming an annulus around the pore margin. These pores are virtually identical to the nuclear pores, described in a previous section (Section 3.3.1). Continuity can be found between the annulate lamellae and the nuclear envelope, which they resemble and from which they seem to arise. Intranuclear annulate lamellae have also occasionally been seen. In some instances there is continuity between cisternae of the annulate lamellae and adjacent cisternae of granular endoplasmic reticulum, although annulate lamellae themselves are generally free of ribosomes.

4.3.2 Annulate lamellar function

Annulate lamellae are particularly common in the developing oocytes of many species and occur also in embryonic and in tumour cells. These all share the common feature of rapid *growth* or *differentiation*. There are, however, reports of the sporadic occurrence of small aggregates of annulate lamellae in a wide range of adult cells. Such lamellae must therefore be considered as basic cell organelles.

While it is generally accepted that the annulate lamellae arise from the nuclear envelope, several possible modes of formation have been described. The lamellae might arise, for example, by fusion and aggregation of vesicles formed from extensions of the outer nuclear envelope. Alternatively they might peel off into the cytoplasm from areas of repeated folding of surplus lengths of the full thickness of the nuclear envelope. It has been suggested that nucleolar material might be incorporated into the lamellae. The common occurrence of annulate lamellae in cells which are rapidly growing or differentiating implies that they are involved in enhanced nucleo-cytoplasmic exchanges. The lamellae might act as a carrier for information derived from the nucleus for which there is a special cytoplasmic requirement.

4.4 GOLGI APPARATUS

4.4.1 Golgi apparatus morphology

When cells are impregnated under suitable conditions with silver or osmium solutions, a specific structural network is demonstrated by deposition of the metal in parts of the cytoplasm. Since this

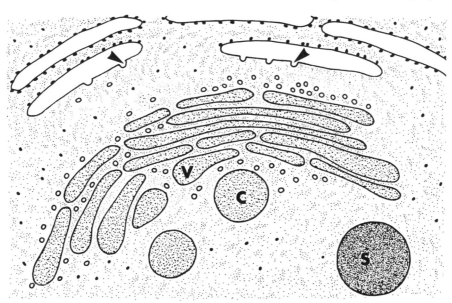

Fig. 7 *Golgi apparatus*
The Golgi apparatus consists of parallel membranes stacked together. Parts of the membrane system are dilated to form vacuoles, V, while numerous small vesicles lie around the apparatus. Some of these are formed by budding from the nearby cisternae of the granular endoplasmic reticulum, as shown by arrows. Secretion granules, S, are released from the apparatus after maturing through a stage of accumulation to form structures known as condensing vacuoles, C.

reticular system was first described by Golgi, it became known as the *Golgi apparatus* or *Golgi complex*. The significance of the Golgi apparatus became the subject of a prolonged cytological dispute. From observations with the electron microscope it is clear that the Golgi apparatus is a distinctive membrane system in which osmium is deposited selectively under the conditions of the classical Golgi impregnation.

In the thin sections used for electron microscopy, the full three dimensional nature of the network (Fig. 8) is not immediately appreciated, since its interconnecting strands appear in section as isolated units at different points in the cell. This is especially true of the neurone, in which the Golgi apparatus often entirely surrounds the nucleus. In contrast, epithelial cells of columnar or cuboidal shape are said to be polarised, having a definite apical and basal surface. In these cells the Golgi apparatus generally lies at the apical pole of the nucleus, occupying an oval or a horseshoe-shaped area of the cytoplasm in thin sections (Plate 14). The position and morphology of the Golgi apparatus is relatively constant in a given cell type. It is small in muscle cells and lymphocytes but large in plasma cells (Plate 15), goblet cells (Plate 20), the secretory cells of the exocrine pancreas and salivary glands and the absorptive cells of the intestine.

The Golgi apparatus can be distinguished from the endoplasmic reticulum by its circumscribed appearance, its position in the cell, the thickness of its component membranes and the distinctive architecture of its different elements. The three principal components forming the electron microscopic Golgi apparatus are membrane *lamellae*, dilated membrane-bound *vacuoles* and small *vesicles* (Fig. 7). The membranes of the Golgi apparatus often appear slightly thicker than those of the endoplasmic reticulum, but the usual trilaminar unit membrane structure can be made out at high resolution. The membrane lamellae which make up most of the apparatus are paired, each pair forming a single closed sac, the walls of which can lie close together or widely separated. The interior of this sac is usually apparently empty in an electron micrograph. The Golgi vacuoles often appear to consist of the dilated ends of such sacs and are thus continuous with the spaces between the paired membrane lamellae. It is usual for the

lamellar structure of the Golgi apparatus to be built up from several parallel Golgi sacs, their separation often being less than the width of each individual sac. Around this main framework there lies a population of small membrane-bound vesicles, each around 30 to 50 nm in diameter. These Golgi vesicles may be very numerous and at times seem to contain material of appreciable density. Two types are recognised, smooth surfaced, which are the more numerous, and coated vesicles, which have a thicker wall lined with a fuzzy layer and with an outer rim of radially projecting spikes.

4.4.2   Golgi apparatus function

Early light microscopic studies linked the Golgi apparatus with the process of *secretion* in certain cells. This is fully supported by electron microscopic evidence which confirms a close association between forming secretion granules and the Golgi membranes.

In the goblet cell and in the zymogenic cells of the digestive glands, the Golgi apparatus is particularly complex. Granules of secretion first appear close to the Golgi apparatus, apparently formed by the accumulation of material within its components (Plate 37). These granules subsequently become detached from the apparatus and pass towards the apex of the cell where maturation takes place, perhaps by withdrawal of fluids, and where the granules are stored prior to release. A similar relationship between secretory granules and the Golgi apparatus is also seen in cells with an endocrine secretory function such as the islet cells of the pancreas, the intestinal endocrine cells (Plate 42a) and the anterior pituitary cells.

The early suggestion that the Golgi apparatus may participate in secretion by concentrating or packaging material made elsewhere is supported by recent work. In protein-secreting cells such as the cells of the exocrine pancreas, the major part of protein synthesis takes place in the endoplasmic reticulum. The newly formed protein is then transported within its cisternae to the region of the Golgi apparatus. Small vesicles filled with the newly synthesised contents of the cisternae then bud off from the membranes adjacent to the Golgi apparatus. These *transport vesicles* appear to carry material across to the apparatus, releasing it into

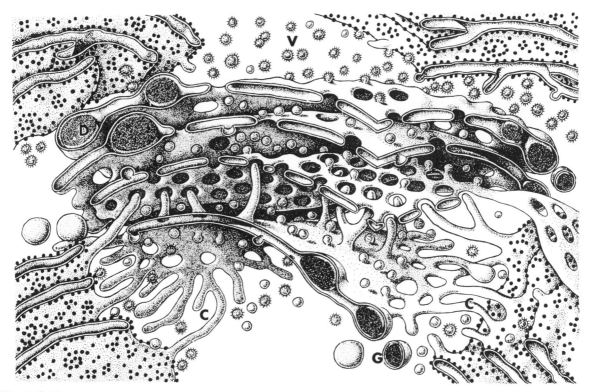

Fig. 8 *Golgi apparatus*
This artist's impression shows the three dimensional relationships between the Golgi apparatus and the endoplasmic reticulum. The upper part of the diagram shows the formation of small vesicles, V, from the endoplasmic reticulum and their fusion with the outer aspect of the Golgi cisternae. Some of these are of the fuzzy or coated type. The lower part of the diagram shows tubular connections, C, between the granular endoplasmic reticulum and the innermost sacs of the Golgi apparatus. This inner area is known as the GERL complex. Dense material, D, forms granules within the Golgi vacuoles, which become detached to form secretory granules, G. The outer cisternae of the Golgi apparatus are rich in thiamine pyrophosphatase, while the innermost GERL cisternae show acid phosphatase activity, suggesting that lysosomes may originate here.

Diagram by courtesy of R. V. Krstić

the cavities of the Golgi sacs, at the so-called *forming face* of the Golgi system.

The secretory substance then undergoes concentration and perhaps chemical alteration, finally accumulating in a secretion granule at the far side, the so-called *mature face* of the apparatus. Since the formed granule is surrounded by a membrane derived from the Golgi apparatus, it is possible that enzymes attached to the membrane may continue to act on the granule after its release from the apparatus. In the goblet cell it has been shown that this sequence of granule formation and release is continuous, secretion granules leaving the apparatus and passing to the cell apex at the rate of about one every two minutes (Plates 20b, 38, 39b).

In addition to the segregation, concentration and packaging of secretions synthesised elsewhere, the Golgi apparatus has independent *synthetic functions* of its own. There is evidence that the polysaccharide components of certain types of secretion are manufactured by the Golgi apparatus and then become conjugated with a protein moiety synthesised in the endoplasmic reticulum. In the intestinal cell, the Golgi apparatus has been shown to be the site of synthesis of the carbohydrate component of the fuzzy coat of the microvilli. It also takes part in the process of *absorption*. In certain protozoa the Golgi apparatus may be concerned with the regulation of the *fluid balance* of the cell; this represents a function sufficiently fundamental to account for the almost universal occurrence of this organelle in plants and animals, although there is no evidence for such a role in higher species.

Finally, the innermost part of the Golgi apparatus, in association with adjacent portions of the endoplasmic reticulum, appears to be involved in the formation of lysosomes. This association has been formalised in the *GERL* (Golgi-Endoplasmic Reticulum-Lysosome) concept, which draws attention to the apparent functional co-ordination between these cellular components.

4.5 MITOCHONDRIA

4.5.1 Mitochondrial morphology

Small cytoplasmic structures in the shape of threads and granules were described by cytologists before the start of this century. These structures, thought to be intracellular organelles, were called *mitochondria*. Rod-shaped, spherical and sinuous mitochondria were described and wide variations in their numbers, sizes and shapes were noted in different cells. For many years, however, the significance and function of the mitochondria were disputed.

Electron microscopy has shown that mitochondria are cytoplasmic organelles with a distinctive fine structure. All mitochondria are constructed of membranes arranged in a similar organised pattern, shown diagrammatically in Figure 9. The membranes of the mitochondrion are thinner than the cell surface membrane. The mitochondrion is limited by the *outer* mitochondrial membrane, forming an unbroken boundary with the cytoplasm. Within this limiting membrane is the *inner* mitochondrial membrane, separated from the

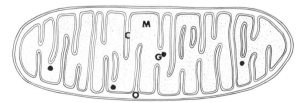

Fig. 9 *Mitochondrion*
The outer mitochondrial membrane and inner mitochondrial membrane are both shown as trilaminar structures. They are separated by the outer mitochondrial space, O. The interior of the mitochondrion is known as the matrix, M. Intramitochondrial granules, G, are located here. The inner mitochondrial membrane forms folds or shelves known as cristae, C.

outer membrane by a constant narrow gap, the *outer mitochondrial space*, which measures approximately 8 nm in width. There are distinct chemical and functional differences between the inner and the outer mitochondrial membrane. The inner membrane is unusual in its particularly high protein content and lower lipid content than other biological membranes. This membrane appears to function as a significant *permeability barrier*.

The inner mitochondrial membrane is thrown into shelves or *cristae*, which extend from the side wall of the mitochondrion into its centre. Each of the cristae consists of two parallel layers of the inner mitochondrial membrane, separated by an extension of the outer mitochondrial space. The cristae are very variable in morphology, at times extending only a short distance into the centre, at times bridging the mitochondrion from side to side (Plate 22b), forming virtual partitions. The space within the mitochondrion into which the cristae project is completely enclosed by the inner mitochondrial membrane. This is known as the *inner mitochondrial space*, or the *matrix* of the mitochondrion. The matrix is often finely granular but varies in density, being occasionally so condensed that the outer mitochondrial space, at the rim of the mitochondrion and within the centres of the cristae, stands out in negative contrast (Plate 95c). Variations such as this can be induced in isolated mitochondria (Plate 23b) by varying the composition of the suspending fluid. In particular, the condensed configuration can be induced by the addition of ADP.

Some mitochondria contain in their matrix irregular dense *intramitochondrial granules*, around 50 nm in diameter. These granules are believed to be binding sites for mitochondrial calcium. In the adrenal cortex, dense intramitochondrial lipid droplets may appear. Other mitochondrial inclusions may be seen in special sites. In some cells, threads and bar-shaped inclusions within the matrix have been identified as *mitochondrial DNA*, now known to be a normal constituent of all mitochondria.

The variability in the numbers and outlines of mitochondria reported by light microscopy has been borne out by electron microscopy. Although the general architecture of the mitochondrion is essentially similar in different cells, there are detailed variations in mitochondrial fine structure. Cells with high energy needs met by *oxidative phosphorylation* have large mitochondria with closely packed cristae. The red blood corpuscle, by contrast, which has no mitochondria at all, fulfils all of its energy needs by *glycolysis*. In the intestinal epithelium the mitochondria are usually thin and elongated (Plate 14); in the liver they are rounded and quite large but with sparse and poorly organised cristae (Plate 22a). Cells with a significant role in steroid hormone metabolism, such as adrenal cortical cells and the interstitial cells of the testis, commonly have cristae which are tubular rather than shelf-like (Plate 23a). The significance of these variations is unknown, although structural variations may reflect distinctive spatial arrangements of the mitochondrial enzymes.

The variations in the shape and size of mitochondria are often exaggerated by the thin sections used for electron microscopy. An elongated, sinuous structure may be represented in an electron micrograph by different profiles depending on whether the plane of section cuts it obliquely, transversely or longitudinally. Mitochondria may present confusing images (Plate 42) on account of the failure of tangential sections to resolve the mitochondrial membranes as distinct structures (Section 15.1.5.).

4.5.2 Mitochondrial function

The technique of differential centrifugation of cell homogenates made it possible to isolate mitochondria for biochemical investigations (Plate 23b). In this way it was shown that the mitochondrial matrix contains most of the enzymes of the *citric acid cycle*. This is a sequence of biochemical reactions concerned with the breakdown of many of the simple molecules from which energy is obtained by the cell. The citric acid cycle is the final common pathway of catabolism for carbohydrates, fatty acids and many of the amino acids. At each step in the cycle, hydrogen atoms are removed from the substrate molecule by enzyme action and are eventually combined with oxygen to form water. During these oxidative reactions, energy stored in the chemical structure of the molecule is released for use by the cell.

The energy is made available for use by a further sequence of enzyme-controlled reactions. The hydrogen atoms removed in the citric acid cycle are passed along a series of carriers which are located on the inner mitochondrial membrane and which bring about their oxidation step by step through a process of *electron transfer*. In this way the energy of oxidation is released in small amounts, which can either be used directly to fuel the functions of the cell, or can be trapped and converted into a form which can be stored for later use. The energy, for example, can be used to produce molecules of *adenosine triphosphate* (ATP) by the phosphorylation of adenosine diphosphate (ADP). The ATP molecule contains two high energy phosphate bonds which provide a source of stored energy for many of the activities of the cell. In muscle contraction, for example, when the ATP is broken down to ADP, this energy is released to produce movement. Numerous aspects of cellular metabolism, synthesis and transport depend equally on the constant availability of this adaptable energy source.

The conversion of ADP to ATP through the simultaneous oxidation of simple molecules is called *oxidative phosphorylation*. This is the main biochemical function of the mitochondria. While much is known about mitochondrial function, some aspects of energy production are still unclear. One possible model of mitochondrial function is expressed in the chemi-osmotic hypothesis, which proposes that hydrogen ions are pumped across a gradient into the space between the mitochondrial membranes during the oxidation of citric acid. These hydrogen ions could then move back across the inner membrane at certain sites, releasing energy to nearby ADP and PO_4, making possible the formation of ATP. The trapping of energy, however, is not entirely efficient: in fact, only about 45 per cent of the energy of glucose oxidation is trapped as ATP. The remaining energy is dissipated as the heat of metabolism, the basic source of heat for the maintenance of body temperature.

The process of electron transport is an essential factor in oxidative phosphorylation. Structural subunits of the mitochondrial membrane have been isolated from disrupted mitochondria; these subunits apparently contain components of the electron transport chain. Such enzyme assemblies have been called *electron transport particles* (ETPs) and it is estimated that from 10 000 to 50 000 of them may be present in a single mitochondrion. Negative staining techniques have shown that the surfaces of the mitochondrial cristae are covered by closely packed 9 nm subunits, or *stalked particles*, which might represent the ETPs. On theoretical grounds, the more ETPs present in a mitochondrion, the greater its capacity to produce energy from simple molecules.

In cells which produce large quantities of energy by oxidative phosphorylation the high metabolic rate of the cell is reflected in the number, size and complexity of the mitochondria. Among the types of cells which have prominent mitochondria are the acid-secreting gastric parietal cell (Plate 21), the cardiac muscle cell (Plate 89) and the brown fat cell (Plate 70c).

Since gastric juice contains hydrochloric acid at pH 1 and the tissue fluid surrounding the base of the gastric gland cell is at a pH level greater than 7, it follows that the acid-secreting gastric parietal cell can concentrate hydrogen ions (H^+) by a factor of 1 000 000. The *secretion of electrolytes* against such a concentration gradient is a remarkable biochemical activity, accomplished only with the assistance of large amounts of energy produced by numerous large mitochondria.

The heart provides the motive force for the circulation from the beginning to the end of life, with a reliability greater than that of any mechanical pump. The muscle cells are in the phase of active contraction for nearly half of their life span and may be called on to increase their output of work by a factor of ten for prolonged periods during exercise. The energy supply for this *constant activity* must come from the continued oxidation of simple molecules in the mitochondria.

Brown fat is particularly prominent in animals which hibernate and in the very young of warm-blooded species. In both instances, the function of brown fat seems to be the *generation of heat*, important for the restoration of body temperature after hibernation, and for combating lack of insulation at a critical period in the life of the neonate. Brown fat seems to act as an auxiliary heat source, to make good the deficiencies of general metabolic heat generation at particular times of need (Section 8.2).

Brown fat cells seem to accomplish this by *uncoupling* the processes of mitochondrial oxidation and phosphorylation. A much greater proportion of the stored energy of the fat can thus be released as heat than in other cell types. This can be seen as an example of the constructive use of inefficiency, in a biological metabolic system designed primarily for the opposite purpose. The generation of metabolic heat, usually a waste product of oxidation, has become the primary role of the brown fat cell. Recent work suggests that brown fat may have an additional continuing role in adult life, involving the *regulation of body weight*. By burning off surplus calories as heat, excess storage of depot fat may be avoided. Brown fat is under direct autonomic nervous system control (Plate 70c).

In all of these cells the mitochondria are numerous and very large. In addition, their parallel cristae are closely packed and often cross from wall to wall. These structural features provide an extensive surface area on which the electron transport particles and other enzyme systems can be assembled in large numbers.

The mitochondria are often found gathered in the particular region of the cell in which energy is being consumed to produce metabolic work. There are many illustrations of this tendency to minimise the gap between energy source and energy consumer, across which the ultimate fuel, ATP, must diffuse.

In striated muscle the mitochondria lie in direct contact with the myofibrils, the contracting units of the muscle cells (Plates 85, 86). This may have a wider significance in some types of muscle, since mitochondria have biochemical mechanisms for the uptake of calcium. In some circumstances the mitochondria may take part in the regulation of calcium levels in the muscle cell cytoplasm, the vital trigger for muscle contraction. In the kidney tubule cells and striated duct cells of salivary glands (Plates 41, 51) the mitochondria lie sandwiched between complex basal infoldings of the cell membrane, where large-scale transfer of water and solutes takes place (Fig. 10). The mid-piece of the spermatozoon consists of a tightly-wound spiral of mitochondria around a central core of functional units which give the sperm its motility (Plates 95c, 96a). Subapical aggregates of mitochondria supply energy for constant ciliary activity in respiratory

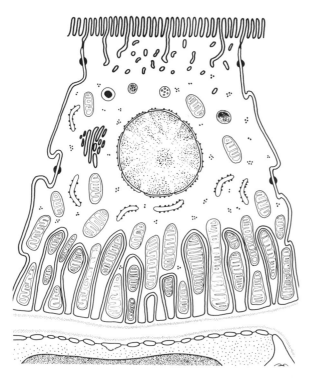

Fig. 10 *Mitochondria and basal infoldings*
This diagram of a kidney tubule cell illustrates the importance of associations between mitochondria and infoldings of the membrane at the base of the cell. The mitochondria provide energy for the active transport of ions, while the extensive membrane surface produced by the infolding provides the necessary area for this important membrane function. An underlying capillary vessel is shown at the base of the epithelium.

epithelium. Mitochondria are often closely related to the granular endoplasmic reticulum and to desmosomes. In some cells there is an equally close association between mitochondria and lipid droplets, a depot form of raw material suitable for mitochondrial oxidation. Cardiac muscle provides an example of this juxtaposition of fuel store and energy generator.

The biology of mitochondria ranges much more widely than oxidative metabolism. They play an important part in the *biosynthesis* of *haem*, an essential component of the blood and a vital part of other enzyme systems in the cell. In steroid-producing cells, some of the enzymes responsible for the early stages of the conversion of cholesterol to the *steroid* ring system are located within the mitochondria, as reflected in their distinctive tubular cristae.

There are other less easily explained features of mitochondrial biology. Mitochondria contain their own DNA, in a form quite unlike that of the chromosomal DNA of the animal cell. Mitochondrial DNA, like that of bacteria, is a *circular molecule*. Alongside this source of genetic information there exists the apparatus for transcription, including RNA in the form of ribosomes similar to those of bacteria. The presence of the machinery for protein synthesis is confirmed by the ability of mitochondria to incorporate amino acids into peptide chains, although it appears that less than 10 per cent of mitochondrial protein can be coded by the limited mitochondrial DNA. It may be that these are proteins which cannot easily be transported into the mitochondrion from without and are better synthesised within the mitochondrial matrix. In any event, it is clear that mitochondria have a significant capacity for *self-replication*. There is a continuous process of replacement of worn-out mitochondria, removed from active service through autophagocytosis (Section 4.6.2).

The effects of drugs on mitochondria have proved to be of great interest. There are certain antibiotics which have no effect on general cytoplasmic protein synthesis, but which disrupt the synthesis of bacterial protein. It has been shown that such antibiotics produce similar disruption of mitochondrial protein synthesis. Other curious similarities include the absence of cholesterol from the inner mitochondrial membrane and from bacterial cell membrane.

4.5.3 Mitochondrial origins

Metabolic similarities such as these have led to speculation on the *biological origins* of mitochondria. It has been suggested that mitochondria might be the simplified remnants of some primitive aerobic prokaryotic cell, the present day descendants of which could be identified as the aerobic bacteria. In the early stages of cellular evolution, primitive cells may have ingested, and then developed in *symbiosis* with captured aerobic organisms; indeed, endosymbiotic relationships of this kind are known to exist today between certain species of protozoa and bacteria. These hypothetical primitive endosymbiotic organisms might be regarded as promitochondria.

In return for a secure and protected environment, these endosymbiotic organisms would contribute to the economy of the cell an entirely new oxidative pathway. Their newly acquired energy surplus could have made possible for these early cells a whole new range of activity and potential development, perhaps providing the critical stimulus for the evolution of the definitive eukaryotic cell.

Whatever the origin of the mitochondrion, its fine structure illustrates the uses of membranes in the cell and provides an example of the link between morphology, molecular biology and cellular function.

4.6 LYSOSOMES

The term *lysosome* was introduced to describe a distinctive group of subcellular particles first isolated by de Duve using the technique of cell fractionation and biochemical analysis. These particles were of the same general size as mitochondria, but could be separated from the larger mitochondrial fraction of the cell by differential centrifugation. The small light fraction obtained in this way was shown to contain powerful *hydrolytic enzymes*. Since the enzyme activity was fully demonstrated only after a period of storage, or disruptive treatment of the particles, it was deduced that they were surrounded by a membrane which normally prevented leakage of their active contents. The name lysosome was intended to convey the biochemical concept of a discrete cytoplasmic particle containing hydrolytic enzymes.

Over forty lysosomal enzymes have now been identified, most of which are active at an *acid pH*. Among the characteristic enzymes of the lysosome are acid phosphatase, beta glucuronidase, beta galactosidase, various cathepsins, ribonuclease and deoxyribonuclease. These and other lysosomal enzymes provide a comprehensive system capable of handling the intracellular digestion of most of the components of biological material.

4.6.1   Lysosome morphology

Since lysosomes were first discovered and characterised by biochemical techniques it follows that their morphological recognition should rely on *cytochemical methods* which demonstrate the presence of their characteristic enzymes. A cell component can be confidently identified as a lysosome if it can be shown to contain at least two known lysosomal enzymes, but the ultrastructural demonstration of *acid phosphatase* activity alone is often taken as presumptive evidence of lysosomal identity (Plate 26b).

Cytoplasmic structures of widely differing appearance have now been shown to have the cytochemical features of lysosomes. A common structural link between them is the presence of a single intact trilaminar limiting membrane, which isolates the substance of the lysosome from the rest of the cytoplasm. The presence of this resistant membrane is responsible for the biochemical *latency* of the lysosomal fraction of the cell: the chemical activity of the enzymes can be demonstrated only after the membrane is damaged. The lysosomal membrane appears normally to prevent leakage of enzymes into the cytoplasm, where their uninhibited action might injure the cell. Substrates undergoing digestion within the lysosome are similarly confined until they have been fragmented into small simple molecules which can diffuse across the membrane.

Some uncertainty exists concerning the simplest form of lysosome, the newly formed *primary lysosome*, which has not yet been involved in digestive activities. The enzyme molecules themselves are manufactured, like other proteins, by the granular endoplasmic reticulum and are passed to the Golgi apparatus, where the primary lysosomes originate as small vesicles. In macrophages small homogeneous structures of medium density have been identified as primary lysosomes (Plates 24a, 65).

More complex *secondary lysosomes*, which contain the typical enzymes along with other material, provide evidence of intracellular digestive activity. The secondary lysosome can arise in two ways. It can result from the uptake, by phagocytosis, of extracellular material, in which case it may also be termed a *heterolysosome*. Alternatively, a secondary lysosome can indicate *autophagocytosis*, a process by which worn out or damaged cytoplasmic organelles such as mitochondria or unwanted secretion granules are isolated and broken down within a lysosome. Secondary lysosomes of this kind are also known as *autolysosomes*, *cytolysosomes*, or *autophagic* vacuoles, while lysosomes involved in the recycling of secretion granules are termed *crinophagic* vacuoles. Such lysosomes may contain assorted material including dense granular debris and closely packed lamellae resembling myelin, thought to arise from the partial breakdown of phospholipid-containing cell components. Lipid materials are poorly handled by lysosomes.

Eventually the enzymatic breakdown within the secondary lysosome comes to an end. The organelle, still limited by its membrane, now contains undigestible and often pigmented lipid-rich residues and is known as a *residual body* or *telolysosome* (Plate 24). Such structures accumulate within cells with age, appearing as a form of pigmentation known to histologists as *lipofuscin.* Their composition presumably reflects the absence of an appropriate lysosomal lipase capacity.

In general, the class of biochemically identified lysosomes includes a wide range of structures which have often been described in the past by electron microscopists as *pleomorphic dense bodies*. In addition, however, the vesicle-containing or *multivesicular body*, which is a relatively common cell component, is now considered to be related to the lysosome (Plates 40, 76a). It consists of a single large vacuole, surrounded by a membrane, within which are found a number of smaller membrane-limited vesicles. The origin and function of these bodies remains uncertain.

The commonest source of a pure sample of isolated lysosomes is the liver. Here the lysosomes have been identified as the peribiliary or *pericanalicular dense bodies* (Plate 45), usually situated close to the bile canaliculus. The granules of macrophages and some of polymorphonuclear leucocytes have been shown to be lysosomes, in keeping with their phagocytic role (Plates 19, 63, 64a, 65). In view of the structural variability of recognised lysosomes, it is desirable when possible to use cytochemical criteria for lysosome identification. In purely morphological terms, lysosomes can be difficult to distinguish from secretion granules in some cell types.

4.6.2 Lysosome function

In general it seems that the lysosome represents a safe and convenient store of potent destructive enzymes which can be used by the cell in different ways. The process of *phagocytosis* undertaken by the macrophage and the polymorphonuclear leucocyte can be only one aspect of the function of the lysosome, since this organelle can be identified in many other cells which have no recognised phagocytic properties. It is now known that lysosomes are involved in many other aspects of normal cell function.

For example, lysosomes are responsible for the *disposal* of worn-out or damaged cell components, such as mitochondria, which have a limited life span in the cell. This is an essential part of the normal process of cytoplasmic maintenance and turnover. Lysosomes in kidney tubule cells play an important part in the *digestion* of protein reabsorbed from the glomerular filtrate (Plates 51, 60). In secretory cells, unused granules are taken up and destroyed by lysosomes. The lysosomes in endocrine gland cells, through this process of *crinophagocytosis*, may play an important part in the regulation of secretory activity. In the thyroid gland, lysosomal action is involved in the splitting of thyroxine from thyroglobulin, when stored colloid is mobilised under the stimulus of TSH. In bone, lysosomes are involved in the resorption of old matrix during the constant metabolic turnover in this tissue. The *acrosome* of the spermatozoon can be regarded as a large single lysosome (Plates 26a, 95), the contents of which are released on contact with the ovum, to assist penetration of the zona pellucida.

In *injured cells*, the involvement of lysosomes in the turnover of cell organelles becomes more apparent. In the cells of the intestinal crypt after exposure to radiation, prominent lysosomes containing damaged cytoplasmic components and nuclear fragments appear after a few hours and may develop to a remarkable size (Plate 25). Similar effects can be produced in many cell types by hypoxia and toxic chemicals. These complex structures are often found in unhealthy, degenerating or dying cells but are also a feature of physiological processes such as the post-partum involution of the uterus, the remodelling of an embryonic limb bud, or the resorption of the amphibian tail.

In some instances the activation of lysosomes leads to the destruction of the cell, as in the case of the polymorphonuclear leucocyte which ingests bacteria. The explosive release of lysosomal enzymes from the leukocyte granules into the cytoplasmic vacuole containing the organisms (Plate 63), leads eventually to the death of the cell, as well as to the destruction of the bacteria. In experimental studies in which lysosomes are primed by the uptake of certain photosensitive dyes and are subsequently ruptured on exposure of the cell to light, the resultant release of enzymes into the cytoplasm can kill the cell. At one time it was thought that the death of cells in the course of physiological cell turnover or during morphogenesis might result from the programmed rupture of lysosomes. This idea led to the term 'suicide sacs' being applied to the lysosomes. There is, however, no convincing example of cell death resulting from spontaneous lysosome rupture in the normal body and a programmed 'suicidal' function of lysosomes is no longer regarded with favour.

Lysosomes are now known to play a part in *cell injury* in a number of diseases. In *silicosis*, for example, macrophage lysosomes are damaged by ingested silica particles and lysosomal enzymes leak out into the cytoplasm with damaging results. Exposure of cells to excess amounts of vitamin A leads to instability of lysosome membranes, with similar consequences. In contrast, the beneficial effects of cortisone and aspirin in the treatment of some chronic diseases such as *rheumatoid arthritis* may in part be due to their ability to stabilise the lysosome membrane, reducing the tendency to abnormal enzyme release. It has been suggested that lysosome rupture might act as an abnormal stimulus to mitosis and that during this process the exposed chromosomes might be damaged by free lysosomal enzymes. If this were the case, such injuries might lead to somatic mutations and even to the development of tumours.

Finally, many *storage diseases* have now been shown to result from the genetically determined absence of particular lysosomal enzymes. This leads to abnormal accumulation of metabolites, such as glycogen or complex lipid, within the defective lysosomes in cells throughout the body, resulting in widespread functional abnormalities. A classical example is *Pompe's disease*, a variety of glycogen storage disease, caused by the absence of a lysosomal glucosidase. The normal cellular glycogen, particularly plentiful in heart and liver, finds its way into lysosomes, where it cannot be metabolised. It accumulates progressively, causing heart and liver enlargement and finally death, by interfering with myocardial function. It is remotely possible that ways may be found in the future to replace absent enzymes with exogenous substitutes, making good the genetically determined defect.

4.7   MICROBODIES

4.7.1   Microbody morphology

The *microbody*, or *peroxisome*, is a small intracytoplasmic organelle limited by a single membrane thinner than that of the lysosome. Microbodies occur only in certain cell types, being most common in liver and kidney. In the liver they measure around 0.6 μm in diameter and are only one quarter as numerous as mitochondria (Plate 26c). They are believed to arise directly from the endoplasmic reticulum and are usually located close to agranular cisternae. The *matrix* of the microbody is finely granular and in many instances it contains a dense core or *nucleoid*, with a degree of internal structure which may differ from species to species and organ to organ. Some nucleoids are amorphous, others contain bundles of fine tubular structures and others again have a highly organised crystalline pattern. A peripheral dense aggregate known as the *marginal plate* is found in certain microbodies.

4.7.2   Microbody function

The microbody takes no part in the segregation of ingested material and has no acid phosphatase activity, both of which distinguish it from the lysosome. It has, however, an *oxidative* activity. The matrix of all microbodies contains large amounts of *catalase*, the enzyme typical of the organelle. Catalase is involved in two principal reactions. The first reaction, from which it takes its name, involves the combination of two molecules of hydrogen peroxide to produce oxygen and water. The second reaction, known as the *peroxidase* reaction of catalase, involves the combination of hydrogen peroxide with a hydrogen donor substrate, resulting in the oxidation of the donor and the formation of water. The recognition of peroxidase activity in the microbody suggested the now popular alternative name *peroxisome*. The peroxisome concept holds that the principal significance of catalase lies in its peroxidase reactions with various substrates, including perhaps carbohydrates, lipids and bile acids. Its catalase potential is seen mainly as a safety precaution against the risk of the accumulation of potentially harmful concentrations of hydrogen peroxide.

Microbodies usually contain several other oxidative enzymes, active at slightly alkaline pH. These include urate oxidase and D-amino acid oxidase. The presence of urate oxidase is usually correlated with the occurrence of a highly structured microbody nucleoid. This enzyme appears to participate in uric acid metabolism. The function of D-amino acid oxidase is not clear, since most naturally occurring amino acids are in the L configuration, the main exceptions being found in some bacterial cells. Perhaps the prominence of liver microbodies relates to the handling of D-amino acids absorbed from the intestinal flora.

Although the significance of the microbody remains to be established, it is thought that it may play some part in carbohydrate or lipid metabolism. Perhaps the microbody might represent the remnants of the original energy-producing oxidative system of the primitive cell which became largely supplanted by the more efficient mitochondria later in the course of evolution. The persistence of certain parts of its original oxidative pathways could be explained by their value for cell survival and adaptation during the difficult evolutionary period when free oxygen began to appear in the atmosphere in large amounts as a result of photosynthesis.

4.8   CYTOPLASMIC FILAMENTS

4.8.1   Actin and myosin morphology

Most cells contain demonstrable cytoplasmic microfilaments of various types (Plates 20a, 30). Morphology alone is a poor guide to the nature and function of these filamentous protein aggregates, but various functional studies have now helped to clarify the appearance seen by electron microscopy. Filament diameter is a useful start in classification.

Two fundamental filamentous structures, seen to best advantage in striated muscle, are the contractile proteins *actin* and *myosin*. Actin filaments are around 7 nm in diameter, while myosin filaments are 12 nm thick. It is now clear that actin and myosin are not confined to muscle cells, but are much more widely distributed. Actin, in particular, is now recognised as one of the commonest of cell proteins, constituting from 10 to 15 per cent of total protein in various non-muscle cell types, while myosin is found only in much smaller amounts. In many cells, particularly in tissue culture, the actin filaments form sub-surface aggregates parallel to the cell membrane and extending along the line of cell projections. These are sometimes termed *stress fibres*. The filaments appear to be attached to the cell surface at certain points, as in smooth muscle.

As in skeletal muscle (Plates 85, 86, 87), such actin filaments can be identified provisionally by their diameter. Their chemical identity has been proved by immunofluorescence studies and by the observation by electron microscopy of their ability to bind a myosin subunit, *heavy meromyosin*, giving rise to a distinctive arrowhead pattern. The myosin component of non-muscle cells is present in much smaller amounts.

Cytoplasmic actin, however, may not always be in the recognisable filamentous form. Depolymerisation can be induced experimentally by exposure of cells to the substance known as *cytochalasin B*, which has been widely used by experimentalists to investigate the cellular role of actin filaments. It is probable that actin in most cells is capable of rapid polymerisation and depolymerisation under as yet unrecognised stimuli. Thus at a given time, much of the actin in a cell may be in the invisible disaggregated form, as in the red blood corpuscle, rather than being present as discrete filaments.

4.8.2   Actin and myosin function

It is obvious that the actin and myosin of muscle cells are related to their function of *contraction*. It now seems likely that the widely distributed actin and rather less abundant myosin of non-muscle cells have a similar role. Actin has been isolated from the cytoplasm of amoebae, where it is associated with *motility*. Actin has also been implicated in the active cytoplasmic movements of macrophages during *phagocytosis*. In general, cytoplasmic actin may well form the basis for all cell motility and cytoplasmic contractility as well as for the *streaming* which is involved in the distribution of cellular metabolites. The identification of actin in the cores of intestinal microvilli (Plate 27a) provides a mechanism by which these surface projections could shorten and lengthen, perhaps with significance in the process of absorption.

4.8.3   Intermediate filament morphology

The presence in many cells of a family of filaments of intermediate diameter, distinct from actin and myosin, has been recognised for some time by electron microscopists. Such *intermediate filaments*, typified by the *keratin* tonofilaments of epithelial cells, are from 8 to 10 nm in diameter (Plates 5, 6, 27b). Keratin filaments are attached to the inner aspects of the desmosomes and are believed to form a three-dimensional network throughout the cytoplasm, the so-called cell web (Plates 5, 73). Aggregates of such filaments in epidermal keratinocytes may be prominent enough to be visible, on histological examination, as tonofibrils.

Several other classes of intermediate filaments have now been recognised. In muscle cells, especially smooth muscle, a network of 10 nm filaments with a presumed cytoskeletal role has been identified. The name *desmin* has been applied to these filaments. Actin and desmin are tightly bound at certain sites, such as the membrane attached and cytoplasmic dense bodies of the smooth muscle cell, as well as in the Z line of striated muscle, now seen as their homologue. In striated muscle the Z line desmin filaments may form lateral links between adjacent myofibrils, maintaining the orderly registration of the sarcomere structures across the muscle cell.

A further 10 nm diameter filamentous system now recognised in some mesenchymal cells is the perinuclear network of 'wavy' filaments, identified as *vimentin*. These filaments are seen as having a structural role in relation to the positioning of the interphase nucleus and of the mitotic apparatus. Two other intermediate filament types are the *neurofilaments*, found within nerve processes and the *glial filaments*, composed largely of glial fibrillary acidic (GFA) protein. These various types of intermediate filament can only be confidently identified by the use of immunocytochemical techniques, since they are morphologically similar.

4.8.4   Intermediate filament function

More than one of these various types of intermediate filament can coexist in the same cell, indicating that they may serve different purposes. The details of their subcellular role, however, are as yet poorly understood. The evidence suggests that they play a fundamental role in the mechanical integration of the various components of the cell, perhaps with the power to regulate or modify cell shape, or, as in the case of the keratinocyte, to preserve epithelial integrity in the face of mechanical stress.

Emphasis has been placed on the three-dimensional relationships of the cytoplasmic filament networks. With the help of high voltage microscopy and stereoscopic photography, an elaborate three-dimensional microtrabecular network has been postulated by Porter, incorporating the cytoplasmic filaments and microtubules and linking the polyribosomes, the endoplasmic reticulum and other cytoplasmic inclusions. There can be no doubt that the purposeful organisation of the cytoplasmic matrix relies largely on the presence of these *cystoskeletal* filamentous components. Recent evidence also suggests that the cytoskeleton may have important functional links with cell surface receptors, thus providing a pathway for the integration of many aspects of cell response and cell function.

4.9 MICROTUBULES

4.9.1 Microtubule morphology

With the introduction of glutaraldehyde fixation for electron microscopy, fine tubules 23 nm in diameter have been identified as an almost universal component of the cytoplasmic matrix. These *microtubules* are less apparent with osmium fixation alone and are liable to dissociate when fixed in cold solutions. They are not composed of membranes, since they lack the characteristic trilaminar structure. At high resolution, a 4 nm periodic substructure can sometimes be distinguished, while on cross-section the tubule can be shown to be composed of some thirteen subunits. The whole thickness of the tubule is often contained within the thickness of the thin tissue section. When orientated in the correct plane, individual microtubules can be followed for long distances in the cytoplasm (Plates 93, 98b).

Microtubules are composed of polymerised macromolecular subunits known as *tubulin*. The formation of the tubules occurs through a rapid process of self-assembly of these tubulin subunits. Unlike the self-assembly process which gives rise to the permanent structure of the collagen fibre from tropocollagen subunits, microtubular self-assembly is fully and rapidly *reversible*. Microtubules, in other words, can depolymerise, vanishing as quickly as they can appear.

The *mitotic spindle* is the most prominent cell component in which microtubules are involved (Plate 10b), giving the spindle its birefringent double cone appearance on light microscopy. The spindle microtubules appear to originate on or near the centrioles (Plates 30, 31), which form the *poles* of the spindle and act as microtubule organisers. Some microtubules link the poles to the individual chromatids, to which the microtubule is attached at the kinetochore. Other microtubules extend from pole to pole without contacting the chromosomes.

Microtubules make up the central cores of *cilia* and *flagella*, where they are organised into particularly complex and well-defined structural patterns. Elsewhere, microtubules are plentiful in situations where there are extended or asymmetrically arranged cytoplasmic processes, such as in the podocyte of the renal glomerulus and in the axons and dendrites of nerve cells. Elliptical cells, such as the nucleated erythrocytes of fish and birds and the disc-shaped platelets in man and other species have well-defined marginal bands of microtubules around the rim of the disc, close to the cell surface.

4.9.2 Microtubule function

The wide distribution of microtubules makes any generalisation about their function particularly difficult. It has been suggested, however, that they may be concerned in some way with the maintenance of *cell shape*, as in the case of the asymmetrical structures mentioned above. It has also been suggested that microtubules form cross-bridges with other cellular organelles, raising the possibility of a ratchet-like mechanism for some types of intracellular *organelle movement*. Perhaps also they might play a passive role in the establishment of pathways for intracellular *diffusion*. Any or all of these suggestions could account for the presence of particularly large numbers of microtubules in nerve processes, where axoplasmic flow is prominent, and for the involvement of tubules in the basic cytoskeletal network proposed for cells in general. Finally, the existence of close connections between cytoskeletal components and the cell surface has raised speculation that the cytoskeleton and its associated microtubules could play some role in communicating changes in activity at the cell surface to the cytoplasm as a whole.

The involvement of microtubules in *mitosis* clearly indicates some important role in co-ordinated cellular movement, but again it is not clear whether this is an active or a passive role. The rate of chromosome movements corresponds exactly to the rate of growth of microtubules by assembly from their tubulin subunits, a process more complex than a simple polymerisation reaction. It may be, therefore, that it is the controlled growth and dissociation of the spindle microtubules that directs movement in mitosis. There remains a possibility that actin may be involved, but this is controversial. Actin, however, is certainly involved in the process of cleavage of the daughter cells at telophase.

The presence of microtubules in *cilia* and *flagella* shows their potential for the generation of movement. This function, however, requires the attachment of a further protein to the microtubules. This component, known as *dynein*, forms the 'arms' seen in cross-sections of the peripheral doublets of the axial complex. The ATP-ase function of dynein allows it to act as a kind of *ratchet mechanism*, producing sliding between microtubules which causes bending of the structure as a whole. Cilia which lack dynein molecules, as in certain mutant protozoa and in the human disease known as Kartagener's syndrome, retain their structural integrity but are non-motile. It is this same mechanism that is proposed to be involved in the suggested capacity of cytoskeletal microtubules to move cytoplasmic organelles, through the establishment of cross-bridges between them.

Many experimental tricks can be played on microtubules. Cytoplasmic microtubules, although not those in cilia and flagella, can be disaggregated by various external influences, including *cold*, raised hydrostatic *pressure*, certain *anaesthetics* and various drugs, including *colchicine* and *vincristine*. The well-known ability of colchicine to arrest mitosis in metaphase is due to this action on spindle microtubules.

4.10 CENTRIOLES

4.10.1 Centriole morphology

On light microscopy of many cells, an area of the cytoplasm can be recognised which appears more homogeneous than its surroundings, is free from mitochondria and has two tiny spots at its centre. This area of the cell, often associated with the Golgi apparatus, is known as the *centrosome* and its two central spots are the *centrioles* (Plates 30, 31, 32b). On electron microscopy, each centriole is a cylindrical structure usually around 0.5 µm in length and 0.15 µm in diameter.

The two centrioles normally seen in the resting cell lie commonly with their long axes at right angles to each other. The fine structural appearance of a single centriole in a thin section depends on whether the cylinder is cut in its long axis, obliquely, or transversely. In transverse section the wall of the cylinder is found to be composed of *nine sets* of microtubular structures which run the length of the centriole in a gentle spiral. Since

transverse section shows these structures 'end on', their detailed organisation can be distinguished (Fig. 11). Each set consists of three *subunits* which are termed *a*, *b* and *c*, arranged in a characteristic pattern. Subunit *a* of each triplet has two short arms, one connecting to subunit c of the adjacent triplet, the other reaching in to the centre of the centriole. When the cylinder is cut in its long axis, the parallel 'walls' are seen but there is little indication of their detailed structure. The centre of the cylinder may be occupied by dense material and one end often appears closed. Dense *satellites* which may be attached to the outside of the centrioles serve as points of insertion for the microtubules of the mitotic spindle. The centriole appears to control the formation of these microtubules.

4.10.2 Centriole function

The centrioles are best known to cytologists through two aspects of their behaviour. Before *cell division*, the centrioles reproduce themselves and the resultant two pairs take up positions at opposite sides of the nucleus, where they form the two poles of the mitotic spindle (Fig. 5). In *ciliated cells*, on the other hand, the centrioles divide repeatedly and form the *basal bodies* which give rise to the cilia, acting as their anchorages in the cell and perhaps also as centres for their control. When centriolar division takes place, the daughter centriole grows from the satellite region of the parent at right angles to its long axis, perhaps explaining the consistent orientation of the two centrioles of each pair.

The mode of replication of the centrioles, the significance of their internal fine structure, their association with the chromosomes at mitosis and their relation to the microtubules which comprise the spindle are still not fully understood. The association of centrioles with chromosomal movement and their role as basal bodies of cilia and flagella indicate that the centrioles are concerned with certain aspects of the organisation of movement at the subcellular level.

The reason for the universal occurrence of nine components in centrioles is unknown. The constancy of this pattern may reflect the preservation throughout evolution of a satisfactory structural

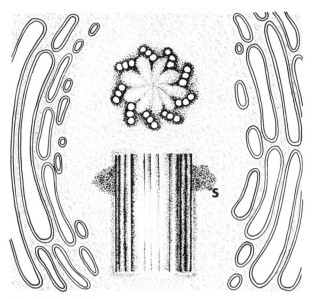

Fig. 11 *Centrioles*

This diagram shows a pair of centrioles, one of which is cut in transverse section, the other in longitudinal section. In transverse section, each subunit of the centriole is seen to consist of three microtubular components, the innermost being designated component A. Note the absence of any central component, unlike the cross-sectioned axoneme of the cilium. In longitudinal section, details of the microtubular arrangement are obscured by overlapping planc of section effects. Areas of diffuse density, S, are known as the centriolar satellites. The commonly seen relationship between the centrioles and the Golgi apparatus is indicated.

formula developed for an essential function at an early stage. As with the mitochondria, the suggestion has been made that the centriole may represent the remnant of some self-replicating and formerly free-living entity, which on capture adopted a symbiotic relationship with a primitive cell early in the course of evolution. A self-replicating structure which became associated successfully with as fundamental a function as cell division might be expected to remain substantially unchanged in its essentials throughout the subsequent course of evolution.

4.11 CELLULAR METABOLITES

There are numerous chemically defined substances in the metabolic systems of the cell, but there are few which have a clearly defined ultrastructural identity. Three of these are worth description, glycogen, ferritin and lipid.

Glycogen is a long chain polysaccharide molecule, which acts as a storage form of carbohydrate in the cell. The carbohydrate content can be demonstrated by the reactivity of glycogen to variants of the histological PAS stain (Plate 33b). Glycogen is present in two main identifiable configurations, alpha glycogen and beta glycogen. *Alpha glycogen*, the form typically present in liver cells, consists of aggregates of dense particles forming rosettes. These are often gathered in areas of the cell in which smooth or agranular endoplasmic reticulum predominates (Plates 18b, 45, 46). The particles are distinguished from ribosomes on two counts; they are not attached to the membranes and the individual subunits of the rosettes are of larger diameter than ribosomes. *Beta glycogen* is found in single particles rather than rosettes, each particle measuring about 30 nm in diameter, approximately twice the size of a ribosome (Plates 19, 33, 85a). This configuration is found in many cell types including, in particular, polymorphonuclear leucocytes and muscle cells. Once again, these particles are not attached to membranes. In some cases, particularly when it is present in large aggregates, glycogen is partially or totally dissolved in the course of processing, leaving areas of apparently empty cytoplasm.

Ferritin is the common storage form of iron. It consists of an iron-rich core surrounded by a protein carrier molecule. The iron core is electron dense, its electron scattering power being related to the localised content of relatively high atomic number. These small dense particles are about 9 nm in diameter, substantially smaller than ribosomes. They are often gathered in iron-rich dense bodies, or telolysosomes, corresponding to the granular iron-rich brown pigment known to histologists as haemosiderin (Plate 32). Ferritin molecules are found in macrophages of spleen and bone marrow, which are involved in the turnover of red blood corpuscles. The presence of ferritin indicates the role of these cells in the recycling and storage of the iron of haemoglobin. Ferritin-laden telolysosomes are sometimes known as *siderosomes*.

Lipid or fatty material may have many different ultrastructural appearances, depending largely on its chemical composition. Depot fat, consisting largely of triglyceride, is usually dissolved out in processing, leaving empty spaces or fat ghosts

(Plates 70a, 70b, 71). Some lipid, stabilised rather more by fixation, remains visible as homogeneous droplets of uniform texture and greyish appearance (Plates 14, 70c). Some lipid-rich material is intensely osmiophilic, as in residual bodies or telolysosomes (Plate 24), while cholesterol, itself soluble, may leave flattened slit-like spaces within cells, where its crystals have been dissolved away. Complex lipids like gangliosides and some lipid-rich membrane derivatives such as myelin and phospholipids, typically display dense lamination (Plates 97, 101, 112a).

CELLS WITH SPECIALISED FUNCTIONS

All cells are built from the structural units described in Section I. The following section will describe the patterns of cell structure which are formed when this machinery of metabolism is assembled in the various factory floors which make up the specialised cells and tissues of the body. The examples have been chosen to illustrate a number of basic relationships between cell structure and function, but without seeking to make a comprehensive survey of the numerous organs and tissues of the body.

5

Secretion

The process of *secretion* is defined as the elaboration by a cell, from precursor materials, of a new substance which is then released from the cell. Special groups of cells gathered together for the purpose of secretion are known as *glands*, but although gland cells are particularly concerned with secretion, many other types of cell also show this activity. The formation of collagen by connective tissue cells is as much a secretory process as the production of digestive enzymes by the pancreas. A secretory product may be stored in the cell for some time in the form of secretory granules, which often have a characteristic morphology. These are eventually discharged from the cell for use elsewhere. The fine structure of cells engaged in secretion has been found to vary in characteristic ways according to the nature of the secretion.

Glands fall into two groups, *exocrine* and *endocrine*. The exocrine gland passes its secretion through a duct on to a surface or into a hollow organ in the body. The salivary glands and the exocrine pancreas which secrete digestive enzymes are examples of exocrine function. Exocrine *zymogenic* secretion such as this is taken as the model for protein secretion in this chapter. Other exocrine glands can form secretions of different chemical natures, such as gastric acid and simple sweat. The endocrine gland, by contrast, has no duct and secretes a chemical messenger or *hormone* which passes from the gland cell into the tissue fluids. The pituitary and adrenal are examples of endocrine glands. Hormones are usually carried to their destination in the blood stream. The products of the various endocrine glands have different chemical compositions and produce different specific biological effects. The ultrastructure of some typical endocrine glands is outlined below.

Secretory cells *release* their products in different ways. In some cases the product is not formed into visible aggregates such as secretion granules but is released from the cell in molecular or near-molecular form. There is often no ultrastructural indication of secretory release in cells of this type, which include the fibroblast and the plasma cell.

Histologists described three mechanisms for the release of aggregated cell products. These are known as merocrine, apocrine and holocrine secretion. In zymogenic cells, prominent granules, each surrounded by a limiting membrane, are stored for a while (Plates 36a, 39a) and are finally released one by one at the cell apex. The membrane of the zymogen granule fuses with the cell membrane, allowing the contents to be released without loss of the structural integrity of the cell. This process is known as *merocrine* or *eccrine* secretion. The goblet cell shares this mechanism (Plates 38, 39b), but may on occasion discharge its stored mucus in a more explosive fashion, losing in addition some of the apical cytoplasmic organelles which are caught up in the mass of granules. When part of the cytoplasm is shed in this way along with the secretion product, the process is described as *apocrine* secretion, but this mechanism is probably rather rarer than was previously believed. It occurs in the mammary gland as the method for secretion of fat droplets. The so-called apocrine sweat glands, however, do not show true cytoplasmic shedding when examined by electron microscopy. The third secretion mechanism occurs in the cells of the sebaceous glands of the skin, which undergo a process of sequential maturation from relatively undifferentiated precursor cells at the gland margin. These cells accumulate their fatty product within the cytoplasm and finally disintegrate entirely, pouring out their residual cytoplasmic organelles along with their secretory products. This is the process known as *holocrine* secretion.

5.1   SECRETION OF PROTEIN

5.1.1   Zymogenic cells

Perhaps the most extensively studied protein-secreting cell is the *zymogenic cell* of the exocrine pancreas, taken here as the model, (Fig. 12) but similar features are seen in other zymogenic cells such as those of the salivary glands and the stomach. The essential feature of this type of protein-secreting cell is the presence of an elaborate system of *granular endoplasmic reticulum* which may often fill most of the cytoplasm (Plate 52). The cisternae are long, often extensively interconnected and at times so closely packed that the ribosome-filled cytoplasmic space between adjacent cisternae is narrower than the width of individual cisternae. Interconnections between the cisternae may allow widespread continuity throughout the system, while the connections seen at times between the cytoplasmic cisternae and the perinuclear cisterna emphasise the dynamic unity of these components.

The contents of the cisternae vary in appearance in different cells, but are usually of low density in comparison with the adjacent cytoplasm with its numerous ribosomes. The material contained in the cisternae may be flocculent and the cavities dilated, suggesting that the protein synthesised by the cell can accumulate and possibly be stored within the cisternae. In the pancreatic zymogenic cells of the guinea pig, *intracisternal* zymogenic granules are present (Plate 17), but in general the

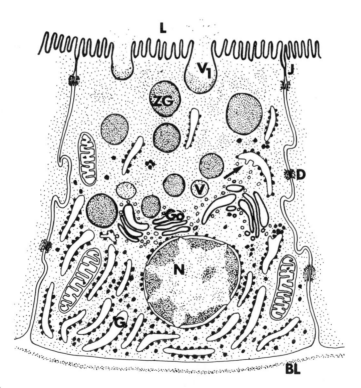

Fig. 12 *Zymogenic secretion*
This diagram shows the principal features of a zymogenic cell. The basally situated nucleus, N, is surrounded by granular endoplasmic reticulum, G. The Golgi apparatus (Go) lies above the nucleus. Small vesicles form from the adjacent endoplasmic reticulum as indicated by the arrow. These carry material to the Golgi apparatus for processing. Immature granules form as condensing vacuoles, V, and pass towards the apex of the cell where they are stored as mature zymogen granules, ZG, until they are released by fusing with the membrane of the cell surface as shown at V_1. The epithelial cells rest on a basal lamina, BL, and are attached to each other by junctional complexes, J, and desmosomes, D.

material in the endoplasmic reticulum is not organised and the only granules are those formed in the Golgi apparatus.

The membranes of the cisternae of the endoplasmic reticulum effectively partition the cytoplasm into two phases. The interconnecting cavities of the reticulum are related to the inner nuclear membrane through the perinuclear cisterna. The ground substance, on the other hand, which contains the ribosomes and other cytoplasmic components, is related to the nucleus through the nuclear pores.

In cells of this type the *Golgi apparatus* is prominent (Plates 36a, 37). It usually lies between the nucleus and the cell apex, where it forms a complex cup-shaped pattern not always apparent in thin sections. The lamellae of the apparatus are long and the vacuoles often prominent. The Golgi apparatus is generally surrounded by a shell of endoplasmic reticulum, the inner cisternae of which lie close to the outer aspect of the apparatus. Transfer of newly synthesised protein material from the cavities of the cisternae to the Golgi apparatus takes place here by the production of small blebs or buds similar to micropinocytotic vesicles, which form from the membrane of the cisterna, fill with its contents and finally pinch off to form a free vesicle (Plate 20a). This transport vesicle then passes to the Golgi apparatus and presumably fuses with its outer aspect, releasing its contents into the Golgi sacs (Fig. 5).

The components of the Golgi apparatus are structurally and functionally polarised, since the formation and output of the secretion granules occur on the side of the apparatus opposite the input of material from the endoplasmic reticulum. The input side is known as the immature, or forming face, while the output side is termed the mature face. In cells where intracisternal granules appear in the endoplasmic reticulum, their presence seems to reflect a transient holdup in the transfer of protein to the Golgi apparatus. Intracisternal granules are not themselves a mature secretory product and are not intended, nor are they suitable, for secretory release.

The final *secretion granule* forms by the accumulation of material of appreciable density within a Golgi sac. The formed granules which lie closest to the Golgi apparatus are often paler and more mottled in texture than those which have been released from the apparatus and have passed to their storage area in the cell apex. Pale inclusions of this type near the Golgi apparatus are often termed condensing vacuoles, implying maturation to form the final granule, perhaps by withdrawal of fluid or perhaps by the continuing action of enzyme systems associated with the Golgi membranes from which their limiting membrane originated.

The Golgi apparatus clearly plays a central role in the secretory process. Not only does it receive and package proteins manufactured in the granular endoplasmic reticulum, but in addition it contributes synthetic functions of its own. The apparatus has been shown by autoradiography to be the site of formation of the carbohydrate component of secretory granules. It is here also that the sulphate component is added to mucus in the goblet cell (Plate 20b).

The dense mature secretion granules are released without damage to the cell surface by the process of *merocrine secretion*, as illustrated in Figure 13. The limiting membrane of the granule is thinner than the apical cell membrane, but although the granules rarely coalesce with one another within the cell, their membranes can readily fuse with this thicker cell membrane, leading to release of their contents. On exposure to the extracellular fluid, the dense substance of the secretion granule becomes pale and flocculent and is rapidly dispersed, leaving a flask-shaped concavity at the cell apex (Plate 39a). Once fusion has occurred, the membrane of the now empty granule adopts the thicker pattern and takes on the functional properties of the cell surface, since further granules may now be released by fusion in turn with this altered membrane. At no time does the continuity of the cell surface become broken during the secretory process. As in most epithelial cells, the apical surface bears a few irregularly disposed microvilli which are of uncertain function, since here they have no obvious association with absorptive activity.

The zymogenic cell provides a model for the investigation of the functional aspects of protein secretion, using a combination of biochemical and ultrastructural techniques involving labelled cell fractions and autoradiography. There are two main

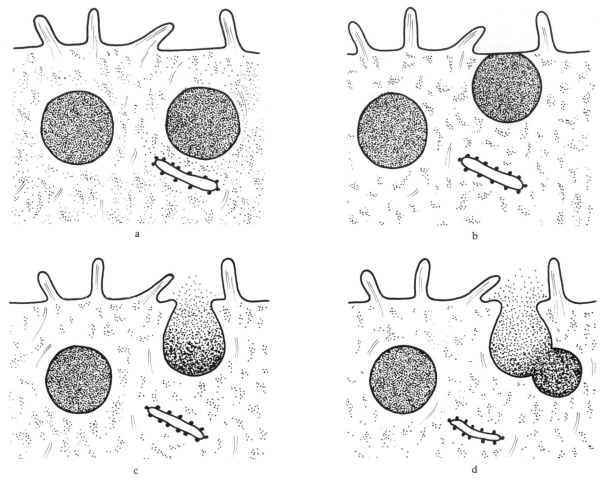

Fig. 13 *Zymogen granule release*
These four figures show the sequence of events occurring during zymogen granule release. In (a), two intracytoplasmic zymogen granules are shown. In (b), one of these granules has moved to the cell surface where its surrounding membrane has fused with the surface membrane between two microvilli. In (c), the fused membrane has broken down allowing the granule contents to be discharged into the lumen. A flask-shaped concavity remains for some time as the granule contents disperse. In (d), this concavity has acted as a further area of cell surface with which yet another granule has fused, resulting in discharge of its contents through the remaining shell of the original granule.

steps, the initial synthesis of the protein molecules and their subsequent movement through the cell to the point of discharge. Protein synthesis is now known to take place in the ribosome of the granular endoplasmic reticulum, as shown by the incorporation of labelled amino acids into protein molecules by cell fractions containing ribosomes. The subsequent transport of this newly formed protein can be broken down into several stages. These include the entry of protein from the ribosome into the cisternae of the granular reticulum, the journey within the endoplasmic reticulum to the Golgi region, the movement, in transport vesicles, from the endoplasmic reticulum to the Golgi apparatus, the passage through the apparatus to the condensing vacuoles and the subsequent journey of secretion granules to the cell surface prior to discharge. Only one of these steps appears to require substantial energy expenditure; this step is the movement of transport vesicles from the granular reticulum to the Golgi apparatus.

Once a protein molecule has been synthesised by the ribosomes, its subsequent movement does not depend upon continuing protein synthesis. Granules can still be formed and released from the cell using formed protein 'in the pipeline' despite the arrest of further protein synthesis by specific metabolic inhibitors. In other words, the movement of secretory material through the cell is not dependent on pressure from the upstream accumulation of further products of ribosomal synthesis.

5.1.2 Protein secretion by other cells

The function of protein secretion is not confined to the exocrine glands, which are mainly concerned with the production of digestive enzymes, since protein molecules have innumerable metabolic and mechanical functions in the body. The thyroid gland manufactures thyroglobulin, a protein which is conjugated with the iodine-containing hormone, thyroxine, and released into the storage or colloid vesicles of the gland. The beta cells of the pancreatic islets (Plate 42b) secrete the relatively simple protein hormone, insulin. Other cells secrete protein but are not contained in glands. These include the plasma cell (Plates 3, 15, 68), which produces antibody protein as part of the immunological defences of the body and the fibroblast (Plate 80b), which manufactures the structural proteins of connective tissue.

In all of these cell types there are found the hallmarks of protein secretion; a prominent granular endoplasmic reticulum arranged around an elaborate Golgi region. The extent of these specialisations can be related to the functional state of the cell. In the plasma cell, the granular reticulum may become progressively dilated owing to the accumulation of antibody, leading finally to the formation of Russell bodies. The active fibroblast in healing wounds has particularly elaborate membrane systems, whereas in the mature fibrocyte, a much less active cell, these features are less prominent.

5.2 SECRETION AND TRANSPORT OF IONS

5.2.1 Acid secretion

Acid secretion is among the most remarkable of cellular functions, since the physical nature of the secretion is so far removed from that of the normal body constituents. The *parietal cells* of the mammalian gastric gland secrete hydrochloric acid at pH 1, a level of acidity which could cause damage and death to many cells. The essential function of the acid-secreting cell is the concentration and secretion of *hydrogen ions*. Since the pH of blood and tissue fluids is maintained at a level slightly higher than pH 7, it appears that the parietal cell can concentrate hydrogen ions by a factor of 10^6 — a million times. The gastric parietal cell has several distinctive structural adaptations related to this function, as shown in Figure 14.

Surface specialisations are important in the parietal cell (Plates 21, 35, 40), since the rate of transport of ions appears to be in proportion to the available area presented by the surface membrane of the cell. The *intracellular canaliculus*, a tubular invagination of the apical surface of the cell, is its most characteristic feature. Branches of this canaliculus ramify in the cytoplasm around the nucleus and often reach close to the base of the cell, forming an elaborate system which greatly increases the potential secretory surface of the cell and ensures that no part of the cytoplasm is far removed from the effective 'apex' of the cell. The surface area of the canalicular system itself is increased by the presence of numerous bulbous or club-shaped microvilli which project into the canalicular lumen and may at times almost fill it with their close-packed profiles. Through these specialisations of the apical surface of the cell, an adequate area for secretory transfer is attained despite the limited portion of each cell apex which is able to reach the narrow lumen of the gastric gland. Since the lumen of the canalicular system is in continuity with the lumen of the gastric gland (Plate 35), acid secretion passed into any part of the system from the cytoplasm is discharged finally into the gland. Without this canalicular specialisation it would be impossible to accommodate the full acid-secreting potential of the highly evolved mammalian stomach within the restricted space available in the gastric glands.

The base of the parietal cell presents a further specialisation which increases the available membrane surface. *Basal infoldings* of the cell membrane occur at several points (Plate 21). In certain species, the extent and the complexity of these

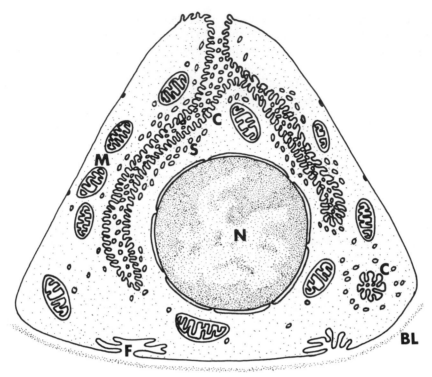

Fig. 14 *Acid secretion*
This diagram shows the main distinctive features of the acid-secreting gastric parietal cell. The intracellular canaliculus, C, represents a tubular extension of the cell surface, lined by microvilli. The many large mitochondria, M, indicate a high level of oxidative metabolism. The smooth-surfaced tubulovesicle system, S, is also involved in acid secretion. The morphology of the canalicular microvilli and the tubulovesicles alters dramatically during active acid secretion. Basal infoldings, F, increase the surface area of the base of the cell, but the basal lamina, BL, does not follow the cell membrane at these points.

folds varies with the functional state of the cell, the folds becoming more elaborate during active acid secretion. The basal lamina does not follow these labile infoldings, but remains as a simple flat platform beneath the cell base.

Within the cytoplasm of the acid-secreting cell there are two major specialisations of fine structure. Firstly, the *mitochondria* are large and numerous (Plate 22b), occupying a high proportion of the cytoplasmic volume. This indicates the considerable demands which acid secretion places on the energy supply of the cell. Each mitochondrion is elaborately organised, with cristae which often extend across its full width and are packed closely together. The rate of oxidative activity in the parietal cell is among the highest measured in the body.

The second specialisation of the cytoplasm of the parietal cell is an extensive system of *smooth-surfaced membranes*, formerly regarded as agranular endoplasmic reticulum. These membranes, however, are thicker than those of typical agranular reticulum, as seen, for example, in liver and adrenal cortex. The tubulo-vesicular cavities formed by these membranes (Plate 40b) apparently communicate, at least intermittently, with the lumen of the gastric gland and in general are found close to the intracellular canaliculus. This system of membranes may therefore be more closely related in functional terms to the cell surface than to the agranular endoplasmic reticulum as recognised elsewhere. In some acid-secreting cells these *tubulovesicles* accumulate in the cytoplasm during the resting phase and are reduced in amount after active acid secretion, suggesting that they play a fundamental part in this process. Their prominence contrasts with the poor development of the Golgi apparatus and the granular endoplasmic reticulum, which are presumably of little importance in acid secretion. What little granular reticulum there is may be related to the synthesis and secretion of intrinsic factor by the parietal cell.

The pattern described above relates to the *mammalian* stomach, in which acid secretion and enzyme secretion are undertaken by different cells, the parietal cell and chief cell. Certain species such as amphibia and birds, however, combine the two activities in a single type of gland cell. In the stomach of birds, the gastric gland cell has typical zymogenic specialisations, such as a well developed granular endoplasmic reticulum, a prominent Golgi apparatus and membrane-bound secretion granules. In addition there is a well developed system of cytoplasmic tubules and vesicles, the mitochondria are characteristic of an acid-secreting cell and surface membrane specialisations such as basal infoldings are prominent. Instead of an intracellular canalicular system to increase the surface area available for secretion, there are deep clefts between adjacent cells, which serve a similar function by allowing the sides as well as the apex of the cell to be used for ion transfer. These clefts are made possible by displacement of the junctional complex from the conventional location at the cell apex, towards the base of the cell. During active secretion, the cell surface, including the clefts between the cells, becomes covered with the typical club-shaped microvilli of the acid-secreting cell, the basal infoldings become more elaborate and there is a reduction in the number of tubulovesicles in the cytoplasm. These changes can be produced by administration of *histamine* or *gastrin*, which stimulate acid secretion. The structural differences between the highly specialised and separate mammalian gastric parietal and chief cells and the more primitive but dual function acid and enzyme secreting cell of the avian stomach suggest an evolutionary pressure towards increasing specialisation and efficiency of function.

5.2.2 Ion transportation

In several situations, highly specialised cells can be recognised whose function is the large scale *transport of ions* other than the hydrogen ion. A striking example of this is the cell of the avian *salt gland*, a nasal gland which can secrete concentrated NaCl as part of the regulation of electrolyte balance. A similar cell type is present in the gills of some fish, where the term *chloride cell* is applied. In the brain the *choroid plexus*, which secretes the cerebrospinal fluid, has some parallel features. In all of these cells some combination of basal infoldings, club-shaped microvilli, prominent mitochondria and smooth-surfaced cytoplasmic membranes is encountered, underlying the similarity, in cellular terms, of any ion transporting function.

A final example of a similar specialisation is found in the salivary tissue. Certain portions of the duct system of the major salivary glands have a distinctive tall columnar pattern with a histological appearance of basal vertical striations. These *striated ducts* are engaged in the modification of the primary salivary secretions through the active transport of ions. The basal striation is due to elaborate infoldings of the basal and lateral cytoplasm of these cells (Plate 41), associated with aggregates of closely-packed mitochondria, which have highly organised cristae.

5.3 SECRETION IN ENDOCRINE GLANDS

5.3.1 The APUD cells

Many of the major endocrine glands were thought by Pearse to share a common embryological origin from the neural crest, and to retain certain common biological features despite their functional diversity. The list included the anterior pituitary, the adrenal medulla, the pancreatic islets, the intestinal endocrine cells and the C cells of the thyroid. The common functional link, summarised under the acronym APUD, is Amine Precursor Uptake and Decarboxylation, often associated with the production of various polypeptide hormones. This hypothesis has undergone various modifications from its original form, particularly after it was shown that the pancreatic and intestinal endocrine cells were clearly of endodermal rather than neural crest origin. From the electron microscopist's viewpoint, however, there is certainly a common structural link between these various cells. In all cases the secretory product is formed into granules (Plate 36b, 42, 69). The *small dense granules*, surrounded by limiting membranes derived as usual from the Golgi apparatus, accumulate in the cytoplasm. In some cases such as in the enterochromaffin or argentaffin cells of the intestine, these granules, which are responsible for the specific histochemical reactions of the cell, show a predominantly basal distribution in the cytoplasm (Fig. 15). Granules may occasionally be seen as they are discharged singly at the cell base, in much the same way as individual secretion granules leave the apex of the exocrine zymogenic cell.

By careful study of their ultrastructure, the various types of endocrine cell in this group can be separately identified. The different cells of the anterior pituitary have distinctive patterns of granulation. In the human, among other species, the beta cells of the pancreatic islet have granules which contain *crystals* with species-specific morphology (Plate 42b). In the adrenal medulla, the granules which contain adrenaline can be distinguished, following glutaraldehyde fixation, from those which contain noradrenaline. In the intestine, similar cells are believed to be the source of the local hormones of the gut. The various morphological types of intestinal endocrine cell which can now be recognised reflect the different products stored. In certain cases, when secretion is stimulated, degranulation of the cells has been shown to be accompanied by a rise in the level of the particular hormone in the bloodstream.

Immunocytochemical methods have made it possible to identify specific products within particular cells, confirming, for example, the identity of the gastrin-secreting cell of the stomach and the insulin and glucagon secreting cells of the pancreatic islets. It is now possible to recognise tumours originating from cells of this type by the presence of their characteristic small dense granules, and to confirm the nature of their secretion by immunocytochemical staining both at the histological and the ultrastructural level.

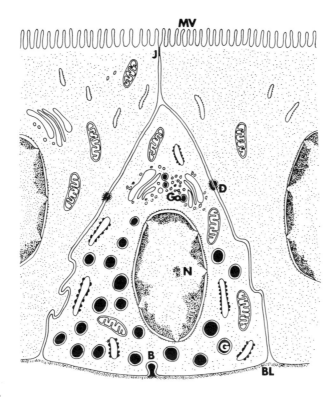

Fig. 15 *Endocrine secretion*
This diagram shows an intestinal endocrine cell between two absorptive cells which are joined by a junctional complex, J, and have numerous surface microvilli, MV. The endocrine cell has basal granules, G. Its apex rarely reaches the lumen. Its nucleus, N, is identified. The secretion granules are formed in the Golgi apparatus, Go, and are eventually discharged at the cell base, B. The presence of desmosomes, D, emphasises that the endocrine cell is an integral part of the epithelium, separated by a basal lamina, BL, from the connective tissue below.

5.3.2 Other endocrine glands

The secretion of *steroid hormones*, by the cells of the adrenal cortex and by the interstitial cells of the testis, is associated with several of the distinctive fine structural features connected with lipid metabolism. The mitochondria of these cells have tubular cristae instead of the more typical shelf configuration (Plate 23a) and there is a moderately well developed agranular endoplasmic reticulum (Plate 18a). Accumulations of lipid material in the form of irregularly shaped droplets are commonly encountered.

The cells of the thyroid are unusual, since they combine some of the features of an exocrine cell with others of an endocrine cell. The initial product of the cell is *thyroglobulin*, a protein conjugated with the thyroid hormone. Thyroglobulin, however, is not released into the circulation, but is stored instead in the thyroid follicle, a blind vesicle lined by thyroid gland cells. The cytoplasm of these cells contains a prominent granular endoplasmic reticulum with dilated cisternae and a large Golgi apparatus. The thyrogobulin is released into the follicle by a process morphologically identical to exocrine secretion, but the follicles are not drained by ducts and simply act as storage space for hormone reserves. At the same time, the cell also functions as a true endocrine gland. Thyroglobulin is removed from the follicle by pinocytosis on the part of thyroid gland cells, which then detach the hormone component and discharge it from the base of the cell. The final secretory activity of the gland at any time is the resultant of these two opposing aspects of secretory function.

The typical endocrine gland, as mentioned before, has no lumen or duct and *discharges* its secretion into its rich supply of capillary *blood vessels*, which come into close relationship with every endocrine secretory cell. Each cell in such a gland has at least one and often two surfaces within close reach of a capillary. The capillary endothelial walls are thin and are full of tiny holes or *fenestrations*. The vessels are separated from the secretory surface of the cell by the minimum of connective tissue space. And yet, despite the close relationship between cells and circulation, several barriers still remain. After discharge from the cell, the hormone molecules must diffuse across the epithelial basal lamina, the connective tissue space, the capillary basal lamina and the capillary endothelium, although the presence of fenestrations may ease passage across this barrier. The hormone, dissolved in the tissue fluids after its release from the cell, makes this journey by diffusion, assisted by the concentration gradient from the cell base to the capillary lumen.

Electron microscopy can make an interesting contribution to the study of *tumours* originating from endocrine glands (Plate 108). The tumour cells may secrete hormones in excess of normal needs, giving rise to symptoms of disease. In many cases these tumour cells retain structural features which are closely related to those of the normal cell. The ultrastructural pathology of the endocrine tumours is dealt with in greater detail in Chapter 13.

5.4 SECRETION BY LIVER CELLS

The liver cell or *hepatocyte* is difficult to classify, since it has numerous functions. It is involved in carbohydrate, protein and lipid *metabolism* and plays an important role in drug metabolism and *detoxication*. In addition, the liver is engaged in *excretory* activities as well as both exocrine and endocrine *secretory* functions. Since the basic pattern of the liver is that of a gland, the hepatocyte is perhaps most conveniently described under the heading of secretion.

Liver cells are polygonal in outline (Plate 45), with some faces in contact with neighbouring cells while others are exposed to the circulation. The blood is carried through the liver in the *sinusoids* (Plate 55a). These are loosely arranged blood channels, lined partly by endothelial cells and partly by Kupffer cells (Plate 24a), which are fixed macrophages, capable of engulfing foreign particles from the circulation. The sinusoidal wall is widely fenestrated and has no basal lamina over much of its extent, so that the plasma freely bathes the irregular microfolds and microvilli of the sinusoidal surface of the liver cell. The space between the sinusoidal lining cells and the hepatocyte, into which these cell processes project, is known as the *space of Disse*. Between adjacent liver cells run the *bile canaliculi*. These are narrow tubular channels formed from matching grooves on the surfaces of neighbouring cells (Plate 46). The lining of the canaliculus is, therefore, the surface of the liver cell itself, although specialised by the presence of short irregularly disposed microvilli. These narrow bile channels communicate with the smallest bile ducts at the edge of the liver lobule and provide the pathway for bile excretion. The lumen of the canaliculus is sealed off by junctional complexes linking the adjacent liver cells, requiring that interchange between the blood stream and the bile be normally channelled through the liver cell and not between cells.

The hepatocyte is often taken as an example of the 'typical' cell, since it contains all of the principal cytoplasmic organelles. The mitochondria are large and numerous, although their cristae are typically rather sparse and irregularly disposed (Plate 22a). Patches of organised granular endoplasmic reticulum, with parallel stacks of cisternae, alternate with areas of agranular reticulum, which have a tortuous tubular configuration. There is a considerable quantity of particulate glycogen in the liver, much of it scattered between the tubular components of the agranular reticulum in clusters or rosettes (Plate 45). At first sight this may lead the inexperienced to confuse these membranes with granular endoplasmic reticulum, but careful examination shows that the glycogen particles are not actually attached to the membranes and that they are larger than the ribosomes which are the single distinctive feature of the granular reticulum (Plate 18b).

The Golgi apparatus, which has the usual pattern of membranes, exists as discrete units scattered throughout the cytoplasm. Liver cells contain numerous lysosomes, most of which lie surrounding the biliary canaliculi. Before their biochemical identity was defined, these structures were known as pericanalicular dense bodies. The liver cell homogenate is the classical source of isolated lysosomes for biochemical studies. Microbodies are plentiful in liver (Plate 26c), from which they also can be isolated.

The elaborate one-sided structural specialisations seen in other cell types which are more narrowly committed in function are inappropriate to a biochemical workshop as complex as the liver. Nevertheless, certain functions of the liver cell can be related to some extent to its ultrastructural features. The brisk *oxidative metabolism* of the liver is expressed in the size and number of the mitochondria, up to 1000 per cell. The presence of glycogen in large amounts and of occasional lipid droplets points to an involvement of these cells in *carbohydrate* and *fat* metabolism. The well-developed smooth reticulum is linked partly to lipid metabolism and partly to *detoxication* of drugs; the agranular reticulum becomes greatly hypertrophied under the stimulus of chronic barbiturate administration, which induces a parallel increase in the concentration of the membrane-associated drug-metabolising enzymes. The prominent granular endoplasmic reticulum is in keeping with the known hepatocyte function of plasma *protein synthesis*. The pattern of the liver lobule and the associated biliary passages points to the *secretory* and *excretory* functions of this organ. Thus the complex metabolic functions of the liver cell call for a remarkably wide range of cytoplasmic specialisation for their successful execution.

Absorption and permeability

All cells take up substances from their surrounding environment, such as the metabolites required for their function. Some cells, however, are particularly specialised in this direction. They may be adapted, as in the gut, to *absorb* the numerous dietary metabolites required for the functions of the body as a whole; they may have a fairly simple role, as in the gall bladder, *concentrating* secretions by the removal of water; or they may have a complex function, as in the kidney, involving the conservation and control of metabolites through selective *reabsorption* and *excretion*. Even more striking is the wholesale uptake shown by *phagocytes*; this is dealt with in greater detail in Chapter 7. In these examples, the cell is an active participant in the uptake process, expending energy in transport functions.

The second phenomenon dealt with in the present chapter is *permeability*. This describes the property of a boundary layer of cells across which there is relatively free diffusion of various substances. Different tissues are specialised to allow rapid interchange of metabolites; these include the capillary vessels and those special anatomical regions where capillaries are particularly important, such as the lung and the renal glomerulus. One of the factors governing the rate of diffusion is the thickness of the barrier. In general, therefore, such specialisations involve *attenuated cell layers,* the structural details of which are only visible to the electron microscopist. The role of ultrastructural study in the investigation of their normal and abnormal function is correspondingly important.

6.1 ABSORPTION BY INTESTINAL EPITHELIUM

6.1.1 Enterocytes

The intestinal epithelial cells show variations of fine structure which may depend partly on the functional state of the cell and partly on its maturity, as determined by its *position* on the intestinal villus. The cells of the villus are replaced constantly by *division* of precursor cells in the crypts and by *migration* of these newly-formed cells to the tip of the villus, where *extrusion* occurs after a working life of only two or three days. Thus the cells at the tip are more mature than those in the crypt and at the base of the villus. The crypt cell cytoplasm (Plate 6a) has numerous free ribosomes and little formed endoplasmic reticulum; these are characteristic features of a stem cell. In other respects it appears immature by comparison with the cells of the villus. As the crypt cell passes on to the villus it begins to acquire the features of the mature cell while losing the potential for continued division. This structural maturation is accompanied by a development of enzyme systems and an increase in absorptive capacity.

By light microscopy the distinctive feature of the columnar intestinal epithelial cell is the presence of the *striated border*, a refractile zone between 1 and 2 μm thick, at the apical surface of the cell (Plates 14, 48, 113). In view of its critical position, forming the interface between the cells and the contents of the lumen, the striated border was extensively studied, but the details of its structure were beyond the limit of resolution of the light microscope.

The electron microscope showed that the surface of the intestinal cell has a highly organised covering of parallel finger-like projections or *microvilli*, extending into the lumen from the surface of the cell. Each microvillus measures about 1 μm in length and 0.1 to 0.2 μm in diameter and consists of a cytoplasmic core over which is stretched the surface membrane of the cell (Fig. 4). As many as 1700 of these processes may be present on a single cell, increasing the area of its apical absorbing surface more than twentyfold. The microvilli are longer and more numerous towards the tip of the intestinal villus, where absorption is most rapid.

When the microvillus is seen in longitudinal section, a central core of filaments can be seen extending from an attachment at the inner surface of the membrane on the tip of the microvillus, to the cytoplasm at its root. On cross-section, about 40 of these central filaments are found randomly spaced within the core (Plate 27a), quite unlike the organised pattern of clearly-defined microtubules seen in the motile cilia (Plate 94). These features allow cilia and microvilli to be clearly distinguished. The cores of the microvilli are linked together at their bases by a transverse network of filaments extending across the cell apex between the junctional complexes, forming a region of the cytoplasm, the *terminal web,* which contains none of the more elaborate cell organelles. The microvillous border and the terminal web together form a structural unit which can be isolated by fractionation of intestinal homogenates. The filament network of the terminal web gives rigidity to the apex of the cell and prevents deformation of the outlines both of the individual cell and of the epithelial surface as a whole, in this otherwise delicate mucosa.

It is now recognised that the longitudinal filaments of the microvillus are composed of *actin,* the almost universal contractile protein of the cell. It is thought that some *myosin* may also be located in the terminal web, raising the possibility that the microvillus can contract and relax, driven by a tiny 'muscle', perhaps thus aiding the processes of absorption.

The most significant part of the microvillous border is the apical *surface membrane* of the intestinal cell. Through this all absorbed material must pass, since the junctional complex maintains firm adhesion between cells, presenting an effective barrier. The membrane shows the typical trilaminar structure (Plate 48b), but at 10.5 nm in thickness it is markedly thicker than most other biological membranes, such as those of the cytoplasmic organelles. The apical membrane of the intestinal cell is thicker than that which forms its lateral and basal surface (Fig. 4).

It is known that many foodstuffs are broken down by the action of digestive enzymes secreted from the stomach and pancreas into the intestinal lumen. The columnar cells of the intestinal villus were once thought merely to absorb the simple molecules resulting from this entirely extracellular, intraluminal hydrolysis. The structural specialisations of the intestinal cell and its distinctive surface were therefore interpreted as being purely absorptive in nature. It is, however, now known that many of the enzymes such as the disaccharidases and peptidases, thought to conduct hydrolysis in the lumen and supposed to be secreted by the crypt cells as the succus entericus, are not present in the lumen in concentrations sufficient to account for the rapid disappearance of their substrates during the normal course of digestion and absorption. These important digestive enzymes are in fact located in or on the surface of the cells, appearing free in the lumen only as a result of the normal physiological desquamation of cells from the villus.

Biochemical investigation suggests that the membrane covering the microvilli has a *mosaic of enzymes* built into its structure, perhaps in the form of structural assemblies. The presence of subunits of around 7 nm in diameter is suspected from negative staining studies of isolated microvilli. The increased membrane surface provided by the microvilli makes available a greater area to accommodate these digestive enzyme assemblies at the absorptive surface between cell and lumen. We can now look on the fine structural specialisation of the intestinal striated border as being both *absorptive and digestive* in nature, with digestive hydrolysis taking place at the surface of the cell as a necessary preliminary to the absorption of the products of this hydrolysis.

The intestinal absorption of *fat* has often been

studied by electron microscopy since fat droplets, when fixed, are clearly visible. Early work suggested that fat droplets could be taken up directly from the lumen in micropinocytic vesicles which formed at the roots of microvilli, but it is now accepted that this pathway is not quantitatively significant. Fat absorption occurs to a much greater extent in the form of lipid micelles, less than 10 nm in diameter, held in suspension by bile salts and fatty acids in the intestinal lumen. The lipid is probably broken down as it enters the cell and resynthesised in the endoplasmic reticulum, which contains the necessary enzymes. The agranular endoplasmic reticulum is quite well developed in the apical half of the enterocyte, in keeping with this role in lipid processing (Plate 14). Lipid droplets thus appear within vesicles and cisternae, through which they are transported to the Golgi apparatus, which lies above the nucleus. This organelle is quite prominent in the enterocyte, forming a complex system of membranes in which dilated Golgi vacuoles are plentiful. Fat droplets accumulate here as the lipid is processed prior to release from the cell in the form of a discrete droplet, or *chylomicron*. An essential step in this process is the synthesis by the cell of a *lipoprotein envelope* for the lipid droplet, without which it cannot be released from the cell. The chylomicrons pass out through the lateral surface of the enterocyte and percolate down to the lamina propria, where they enter the lacteal in the core of the villus.

Although *pinocytosis* is now not thought to be important in absorption in adults, it does form a pathway for the *uptake of antibodies* by young suckling mammals. The antibodies taken up from the maternal milk are transferred unaltered to the infant circulation and give passive immunity to disease (Plate 4b).

6.1.2 M cells

A distinctive modification of the intestinal epithelium occurs over the lymphoid aggregates in the ileum known as Peyer's patches. There are cells here, now known as *M cells*, which do not have the usual well organised surface microvilli of the enterocyte. The M cell, as seen by the scanning electron microscope, is characterised instead by low anastomosing surface microfolds.

The M cells have two important properties. The first is a distinctive physical relationship with intraepithelial lymphocytes. Such wandering lymphocytes are a normal feature of all epithelial surfaces (Plates 10, 68), probably presenting the process of *immunological surveillance* of these vulnerable tissues. In the ileum over Peyer's patches, however, the lymphocytes gather in groups surrounded and enveloped by M cell cytoplasm. While the lymphocytes approach close to the lumen, they are still separated from it by a thin shell of apical M cell cytoplasm, surmounted by the surface microfolds, and firmly attached to the adjacent enterocytes by apical junctional complexes. The second property of the M cell is its capacity to take up native protein molecules from the gut lumen and transport them in micropinocytotic vesicles to the underlying lymphocytes. The sequence of uptake and transport can be studied in dynamic terms by the use of tracer molecules such as horseradish peroxidase.

These two characteristics suggest that the M cell acts as an immunological 'nose' for the digestive tract, sampling the passing contents for antigens and referring them to the lymphocytes for assessment. In this way the lymphoid system may become alerted to the presence of foreign antigens, and may thus be prepared for any necessary defensive response.

6.2 ABSORPTION BY GALL BLADDER MUCOSA

The gall bladder mucosa modifies the bile secreted by the liver, by adding mucus and reabsorbing water. The *secretory function* is indicated by the presence in the columnar epithelial cells of strands of endoplasmic reticulum and a moderately elaborate Golgi apparatus, in which small mucoid granules are formed. These granules are discharged into the bile at the cell apex. The high concentration of cholesterol in the bile is reflected in the presence of dense residual bodies, containing lipid droplets and lamellar areas, along with typical rhomboid and flattened empty angular spaces, where cholesterol crystals have been dissolved away.

The expression of the *absorptive function* of the gall bladder involves a strikingly labile specialisation of the lateral contact surfaces of the columnar epithelial cells. In the resting state, when the fluid transport across the mucosa is minimal, the cells lie in close contact with one another and the intercellular space is narrow. There are, however, elaborate *flaps of cytoplasm* at the lateral surface of each cell, which interdigitate with each other to form a tightly woven, complex network of processes (Plate 50). In the active state, when water is being transported from the bile to the delicate thin-walled fenestrated capillaries beneath the mucosa, these lateral intercellular spaces open up widely. The interwoven cytoplasmic processes retain contact, but slide over each other as the cells separate; they now come to form a network of delicate partitions across the intercellular space, dividing it up into a labyrinthine channel which extends virtually from the apex to the base of the epithelial cell. The feet of epithelial cells, however, retain close contact along the plane of the basal lamina, however distended the intercellular space may be above the basal region. This restricts the outflow from the epithelial labyrinth.

This layout has been shown to correspond well to a theoretical model of *fluid transport* known as a standing gradient osmotic flow system. This involves the active transport of sodium chloride into the lateral intercellular space, which thus becomes hypertonic, with a standing osmotic gradient decreasing from the lumen to the base. Osmotic equilibration then takes place along the length of the channel, so that the fluid emerging from the base of the epithelium is isotonic.

6.3　ABSORPTION BY RENAL TUBULES

The formation of urine begins in the glomerulus with *filtration* of the plasma, resulting in the production of a dilute urine which collects in the urinary space of Bowman's capsule (Plates 60, 61, 62) and passes into the first part of the tubular nephron. In the human, 120 ml of glomerular filtrate are formed per minute, of which 99 per cent is *reabsorbed,* along with many of the solutes it contains, by the renal tubular epithelium. Much of this reabsorption takes place in the convoluted tubules, the cells of which are highly specialised in fine structural terms (Fig. 10).

The epithelium lining the proximal convoluted tubule is columnar or cuboidal in shape and the cells have a *striated border* on light microscopic examination (Plate 60). As in the intestinal epithelium, the striated border consists of numerous closely packed microvilli, each one too thin to be observed individually by light microscopy. These microvilli are embedded in a dense surrounding glycoprotein coat, reminiscent of the apical surface coat of the intestinal cell. *Tubular invaginations* of the cell surface between the microvilli are common in this type of cell. These have been shown to provide channels for the uptake from the nephron of electron-dense materials such as ferritin and haemoglobin aggregates. The pinching off from these channels of fuzzy or coated vesicles suggests that *selective* pinocytotic uptake of materials may operate at this site. The morphology of these apical channels can be affected by osmotic variations.

The transport of water and ions across the cell, so important a part of the absorptive function of the proximal convoluted tubule, is reflected in the prominent *infolding* of the cell base, consisting of invaginated membranes extending deeply into the cytoplasm (Plate 51). Marked ATP-ase activity has been demonstrated here, suggesting the presence in the membrane of an energy-consuming active transport system. The large and numerous mitochondria of these cells are packed closely between the infolded membranes, with an orientation predominantly parallel to the invaginations of the cell base. These specialisations of mitochondria and of the cell base are seen elsewhere in association with a fluid or ion transport function.

The formation of the definitive urine relies on the integration of numerous separate tubular functions, involving absorption, secretion and diffusion. Different segments of the nephron handle different parts of this work, the regional variations in function being reflected in a wide range of ultrastructural variation on the above themes in the different lining cells.

6.4　PERMEABILITY IN VASCULAR ENDOTHELIUM

The blood *capillaries,* the smallest vessels of the circulation, form the main barrier throughout the body between the circulating blood and the cells of the various tissues. The oxygen and essential metabolites without which the cell cannot survive must pass across the capillary wall to reach the cells, while waste products of metabolism such as CO_2 must pass in the opposite direction to be removed from the tissues.

The capillary blood vessel consists of a delicate tubule (Plates 52, 53), composed of flattened endothelial cells joined together by focal adhesion specialisations, usually of the gap junction pattern, along their contact or meeting edges. Each cell has a central thickened area which contains the nucleus, along with a few of the common cytoplasmic components such as a small packet of Golgi membranes, a few vesicles, occasional mitochondria and some perinuclear filaments. The endothelial tube of the capillary is surrounded by a closely applied *external lamina,* around which are arranged connective tissue elements, such as collagen fibrils, which tend to fuse with the lamina. There is therefore a narrow connective tissue space around the capillary, which separates it from the component cells of the organ to which it is supplying blood.

There are two endothelial specialisations which may aid the passage of materials across the capillary wall. The first of these takes the form of small flask-shaped invaginations of the inner and outer surface of the endothelial cell, known as *caveolae* (Plates 53, 55, 57). These vesicles, containing small quantities of intravascular fluid, pinch off from the cell surface to form isolated spherical *micropinocytotic vesicles* within the endothelial cytoplasm. They pass across the cell and fuse once again with the opposite surface membrane, discharging their contents on the other side. Thick-walled vesicles, suggestive of selective uptake, are sometimes seen as well as thin-walled caveolae. In this small way the endothelial cell participates in transport across the capillary wall.

The second specialisation of capillary endothelium is seen in some tissues but not in others. This is the presence of capillary *pores* or *fenestrations,* which appear as 70 nm diameter circular discontinuities in the endothelial lining (Plates 51, 53, 54). These pores are usually bridged by a tenuous diaphragm, thinner than the surface membrane of the endothelial cell, but still a significant structural barrier. Pores may appear at frequent intervals in the capillary wall, sometimes forming closely arranged aggregates know as sieve plates, which may account for a substantial proportion of the total surface area of the endothelial cell (Plate 61).

Although it is assumed that the caveolae and endothelial pores are of importance in relation to vascular permeability, the extent to which they give either active or passive assistance to the passage of different substances is not clear. Normal physiological vascular permeability probably relies largely on leakage at the points of junction between endothelial cells, aided perhaps by a combination of these other factors mentioned above. It has been clearly shown by tracer experiments that small molecules can pass between the cells, suggesting that their contact specialisations are not designed to seal off the lumen of the vessel completely. During inflammation, when there is increased vascular permeability, the gaps between endothelial cells are increased in width.

Endothelial cells in various sites are characterised by the presence of distinctive cytoplasmic *granules.* These are usually cylindrical, rod shaped or elliptical, measuring up to 0.5 µm and surrounded by a limiting membrane. Their internal structure consists of parallel tubules, closely packed, showing a moderate electron density (Plate 55b). These endothelial granules are generally termed *Weibel-Palade granules,* after the authors of the original description of their typical features. Their function is not yet clearly understood, but some involvement in fibrinolytic mechanisms has been suggested.

Capillary endothelium varies in structure in different sites. In the *brain,* capillary endothelial fenestrations are absent (Plate 99) and there are perivascular neuroglial cell foot processes closely applied to the outside of the vessel, perhaps contributing in some way to the well known restriction of permeability termed the blood-brain barrier. In *skeletal muscle* the vessels are non-fenestrated and of intermediate permeability (Plates 53b, 84). In the *endocrine glands* and the *intestine,* where highly permeable capillaries are receiving hormone secretions or absorbed food materials in addition to allowing normal metabolic interchange, fenestrations are present in large numbers.

Even more permeable are the *hepatic sinusoids* (Plates 45, 55a). Here the endothelium has extensive sieve plates as well as wide discontinuities, measuring up to a micrometre across, features well seen in scanning electron micrographs. Moreover, the basal lamina seen in the conventional capillary is scarcely visible in the sinusoid, being present at best in a thin and discontinuous layer. This exceptional structural deficiency of the sinusoid wall can be related to liver function. The many metabolic activities of the hepatocyte call for the most intimate contact possible between the portal venous blood and the surface of the cell.

The capillary wall is not the only significant barrier to diffusion between tissues and blood. Between the circulating blood and the cytoplasm of the cells within the tissues lie not only the capillary endothelial cell, but also the endothelial external lamina, the connective tissue ground substance, the epithelial basal lamina and the membrane of the epithelial cell itself. Diffusion may be influenced at any of these points both by nonspecific and by selective processes.

6.5   PERMEABILITY IN PULMONARY ALVEOLI

The *lung* is designed to provide an extensive interface between the alveolar air and the circulating blood. Across this interface, in opposite directions, pass oxygen, essential for cellular respiration and carbon dioxide, its end-product. The interface must be thin enough to allow gas exchange to continue at a rate sufficient to meet the needs of the body, but it must be strong enough to resist leakage of plasma from vessels which are poorly supported and to withstand mechanical stress during respiration and coughing. The barrier in the lung between air and blood consists of the alveolar lining epithelium, the capillary endothelium and the intervening connective tissue layer.

The thin-walled pulmonary capillaries (Plates 57, 58) are free from fenestrations and display the flask-shaped caveolae of micropinocytosis. The endothelial cells have the usual closely applied external lamina with a surrounding narrow connective tissue space, in which pulmonary macrophages occur (Plate 24). The capillaries, reinforced by a delicate framework of collagen and elastic fibrils, lie within the cavity of the common wall between adjoining alveoli. The pulmonary alveolar air spaces are lined by a continuous layer of epithelial cells. The nuclei of the main cell type, the flat *type 1 pneumocytes*, or alveolar lining cells, tend to lie in the corners and angles of the alveoli leaving wide areas of alveolar wall covered by their attenuated and often featureless cytoplasm. Since this lining, despite its thinness, represents a continuous epithelial sheet, it lies on a typical continuous epithelial basal lamina.

Thus the entire barrier between blood and air, often less than $0.2\,\mu m$ thick, consists of several distinct anatomical components (Plate 58). These are the alveolar lining epithelium, its basal lamina, the connective tissue space, the endothelial external lamina and the endothelial cell. The epithelial lining of the alveoli is often so closely applied to the capillary that the epithelial basal lamina and the external lamina of the endothelium may apparently fuse, eliminating even the narrowest connective tissue space.

The alveolar construction of the lung is attended by certain physical problems. The moist epithelial surfaces of the alveoli must be separated from each other to ensure aeration, yet the considerable *surface tension forces* acting at the alveolar level tend to cause adhesion between alveolar walls and collapse of the alveoli. Surface tension forces in the alveoli are reduced by the secretion of *surfactant,* a phospholipid substance, by a second type of cell in the alveolar epithelium, the great alveolar cell or *type 2 pneumocyte* (Plate 57). These cells, interspersed between the type 1 alveolar lining cells, are joined to them by junctional complexes and share the same epithelial basal lamina. They are of cuboidal shape, with dense cytoplasm in which are found lamellated secretory inclusions representing stored surfactant. Discharge of these granules at the cell surface between short irregular microvilli leads to dispersal of the surfactant in a molecular layer across the entire moist air-tissue interface.

6.6   PERMEABILITY IN RENAL GLOMERULI

In the renal glomerulus the function of controlled permeability relies on another complex relationship between endothelial and epithelial cells. The *renal glomerulus* is a filter, which consists of a tuft of capillaries projecting into *Bowman's capsule*, the dilated blind end of the nephron. The flat cells of the parietal epithelium lining Bowman's capsule become continuous with the visceral epithelium of Bowman's capsule at the point where the capillaries are invaginated into the capsule. The visceral epithelium of Bowman's capsule is a system of cells forming an investment around the individual capillaries of the glomerular tuft. Between the visceral and the parietal epithelial cells of Bowman's capsule lies a narrow cleft, the capsular space or *urinary space*, which is continuous with the lumen of the nephron and drains ultimately through the ureter into the bladder. The fluid which passes from the capillary through the filtration barrier into the urinary space is the *glomerular filtrate*, a dilute form of urine which is then altered and concentrated during its passage along the nephron.

There is a fundamental difference between the permeability functions of the lung and the glomerulus. In lung, fluid must be retained within the vascular lumen and rigorously excluded from the air space to avoid pulmonary oedema, while gaseous and volatile substances are freely exchanged. The glomerulus, in contrast, is essentially a filter, freely permeable to water and all but the largest of molecules in the plasma, the proteins of molecular weight over 45 000. This explains the contrast between the extensively fenestrated glomerular capillary endothelium and the delicate but entirely non-fenestrated pulmonary capillaries. The pores in the endothelial cells of the glomerulus appear even to lack the tenuous diaphragm which closes such pores elsewhere in the body. The *basal lamina* of the capillary, a vital part of the glomerular filter, is described below.

Each capillary of the glomerulus is invested or surrounded by a sheath of *podocytes*, the visceral epithelial cells of Bowman's capsule. The epithelium forms not a continuous cytoplasmic layer, but a complex interdigitating system of closely packed *foot processes*, or *pedicels*, sent out in different directions by the podocytes which lie between the capillary loops (Plate 59). A single podocyte may have foot processes applied to the surfaces of several capillaries, while its nucleus and surround-

ing cytoplasm lie in a cleft between the vessels. Adjacent foot processes are joined by a delicate connection, the filtration *slit membrane,* which is the main barrier to free passage between the processes, through the clefts known as the *slit pore* system (Plate 62).

Since the podocytes are epithelial cells, they lie in contact with a continuous *basal lamina,* fused with the capillary basal lamina. This vital boundary layer thus intervenes between the foot processes and the capillary endothelium. This single shared 330 nm thick basal lamina, the filtration membrane, is known by histologists as the *glomerular basement membrane.* Although in theory the filtration membrane represents the fusion of an endothelial and an epithelial component, two separate laminae are not seen and it is likely that the bulk of the substance of this thick layer is normally produced by the podocyte (Plates 60, 61, 61).

The basal lamina forms a selective barrier of the greatest importance in glomerular filtration. Its vital role is best appreciated through ultrastructural studies of the glomerulus in some types of kidney disease, when there is deposition of *immune complexes* in the basal lamina (Plates 109, 110). The filtering function of the basal lamina becomes disturbed, resulting in its failure to retain the plasma proteins, causing the heavy proteinuria which characterises such disease. The podocyte, for its part, is probably more than just the mechanical support for the glomerular filter, but the extent of its involvement in glomerular function is still unclear. There are substantial subpodocytic spaces through which the glomerular filtrate must trickle before it reaches the freedom of the open urinary space of Bowman's capsule. The podocyte would certainly be well placed to influence the composition of the glomerular filtrate, but there is little evidence to suggest that it does.

A further important component of the glomerulus is the *mesangium*. This is the connective tissue core of the vascular pedicle, on which are hung the glomerular capillary loops. The mesangium consists of small dense stellate *mesangial cells* set in a *matrix* of material continuous with and of similar appearance to the basal lamina of the glomerular capillaries. The functions of the mesangium are not fully understood, but it seems to play a part in the normal turnover of the basal lamina and is involved in the reactions of the glomerulus to disease. The mesangial cells are to some extent capable of active phagocytosis of particulate material trapped by the glomerular basement membrane.

The ultrastructural study of the glomerulus now forms an important part of the investigation of various kidney diseases, making possible a more rational approach to the detailed interpretation of pathological changes previously barely recognised by light microscopy (Ch. 13).

Defence

The mechanisms of defence deployed by the body range from the cough reflex, through ciliary motion, to the synthesis of immunoglobulin molecules. This chapter, however, will concentrate on the cellular mechanisms of defence and in particular on the phagocytic and inflammatory processes which form the main protection against infection.

7.1 THE PHAGOCYTES

7.1.1 Phagocytosis

This term describes the active uptake of particulate material by specialised cells. The general features of the process have already been outlined in Section 4.6.2, but in this section we will look at the phagocytic mechanism in greater detail.

The amoeba feeds by phagocytosis, but in more complex animals phagocytosis has become the special function of a limited number of cell types known as *phagocytes*. The two important groups of phagocytic cells are the *polymorphonuclear leucocytes* or white blood cells and the mononuclear phagocytes or *macrophages*. Both cell types are found throughout the body.

Phagocytes ingest a wide range of unwanted material ranging from bacteria which have invaded the tissues to debris from damaged cells. Any foreign substance, such as particulate mineral dust, which may gain access to the body, will also be dealt with by phagocytes.

Digestion follows ingestion. The smaller particles are directly incorporated into a lysosome while the larger particles, such as bacteria, are isolated in a phagocytic vacuole into which the lysosomal enzymes are released, thus forming a kind of isolated intracellular stomach (Plate 63). Poly-morphs involved in this activity sometimes die, but in the macrophage (Plates 24, 65), digestion usually proceeds without risk to the cell unless the ingested material is itself toxic. If, however, the ingested material damages the lysosomal membrane, the macrophage can die as a result, through the release of lysosomal enzymes into the cytoplasm.

The products of digestion are absorbed from the secondary lysosome and any indigestible residue persists as a *residual body,* or *telolysosome.* Residual bodies are often pigmented, due to the presence of lipid-rich lipofuscin, reflecting the relative inefficiency of lysosomes in dealing with lipid residues. The wide range of material ingested from time to time by the longer-lived phagocytes leads to wide variation in telolysosome structure (Plate 24b). In cells involved in erythrocyte destruction, iron-rich deposits of haemosiderin are found, termed *siderosomes* (Plate 32).

The phagocytes throughout the body react to the presence of foreign or unwanted material, accomplishing its removal and destruction as far as may be possible. Their scavenging function forms an important part of the mechanisms of defence.

7.1.2 Polymorphonuclear leucocytes

The commonest of these, the *neutrophil polymorph* (Plate 19), is the first line of defence against bacterial invasion. The neutrophils gather where there is cell or tissue injury of any type, such as areas of infection or physical trauma or in areas damaged through being deprived of oxygenated blood. The neutrophil polymorph is actively motile. To reach the damaged area, it adheres to the endothelial lining of a nearby small vessel and migrates through the wall by pushing between the endothelial cells. It then sets off into the interstitial tissues, guided by the concentration gradients of *chemotactic substances* released from the organisms or the injured cells. The granules of the neutrophil polymorph are of several types. They contain various active antibacterial substances including lysosomal hydrolases, lysozyme and peroxidase. Some of these granules are typical lysosomes. Polymorphs are generally effective in killing ingested bacterial cells (Plate 63), but when the organisms are particularly virulent, the polymorphs themselves may be killed in the course of phagocytosis. The resultant accumulation of cellular debris from disrupted polymorphs and injured tissues, along with a fibrin-rich fluid exudate and numerous still viable polymorphs, is known as *pus*. This is a typical finding in acute inflammation.

The *eosinophil polymorphs* with their distinctive granules behave in a rather different fashion. Their elliptical granules are larger than neutrophil granules and have an amorphous matrix surrounding an internal crystalline component (Plate 64a). When crystals form in cells, they indicate the presence of ordered stacking of a pure substance in high concentration. In this case, the crystals are thought to represent the enzyme peroxidase, while the granular matrix contains acid phosphatase.

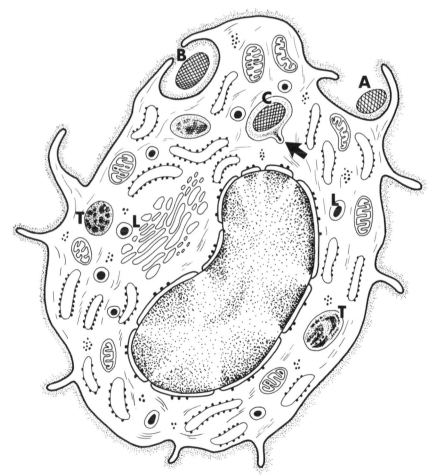

Fig. 16 *The macrophage*
This diagram shows the sequence of events in the process of phagocytosis. Foreign material, A, comes in contact with the surface coat or glycocalyx of the macrophage and excites the engulfment mechanism. The material is enveloped within a phagocytic vacuole, B, and is then interiorised within the cytoplasm, C. Primary lysosomes, L, are formed in the Golgi apparatus, stored in the cytoplasm, and discharged into phagocytic vacuoles, arrow, forming a secondary lysosome. The continuing process of intracellular digestion results in the formation of residual bodies, or telolysosomes, T.

The functions of the eosinophil are still obscure, although it is believed to be involved in the uptake of antigen-antibody complexes which might otherwise cause harmful effects, such as kidney and vascular disease. The numbers of these cells increase in certain *allergic conditions* such as asthma, and decrease in response to corticosteroids, which are known inhibitors of the immune response.

7.1.3 Macrophages

These are the large mononuclear phagocytes. Members of this class of cell can be found throughout the body in the form of wandering or *free macrophages*. Others are present as *fixed macrophages* in particular organs and tissues, such as the Kupffer cells of liver and the macrophages of spleen, lymph nodes and bone marrow. The circulating blood *monocyte* is a known precursor of macrophages. The monocytes can migrate freely into the peripheral tissues from the bloodstream to replenish or reinforce the free macrophage population (Plate 65a), particularly when there is tissue damage and inflammation, which release chemotactic substances which attract macrophages. The location and distribution of macrophages can be demonstrated by injecting carbon particles, dyes or colloidal gold into an experimental animal. Subsequent light or electron microscopy allows the identification of macrophages by the presence of the tracer substance in their cytoplasm (Plate 115a).

The characteristic features of the macrophage (Fig. 16) include the presence of abundant cytoplasm containing numerous lysosomes (Plates 24a, 65b). Their active synthesis is indicated by the presence of granular endoplasmic reticulum and a prominent Golgi apparatus, in which the active enzymes are packed into primary lysosomes. The ruffled and invaginated surface of the macrophage displays the different varieties of pinocytosis. The contours of the cell become more elaborate on stimulation, using injected substances such as lipopolysaccharide or glycerol trioleate, which enhance phagocytic activity. The resulting surface projections appear on electron microscopy as irregular ruffles and flaps limited by the cell membrane. The injected material adheres to the cell surface, which then becomes invaginated to form a cytoplasmic vacuole or phagosome. There is a prominent cell coat, visualised by special staining procedures (Plate 115c). It may play a part in the recognition of foreign material, in the process of phagocytosis and in the mechanism of macrophage adhesion. Within the cytoplasm, fine filaments of actin form the basis for the active motility of the macrophage and its surface projections.

The process of phagocytosis by macrophages forms a fundamental, if at times unsuccessful, defence against infections of various kinds and against foreign materials such as inhaled dust. An aggregate of macrophages and other inflammatory cells formed in response to such a stimulus is known as a granuloma. Macrophages in such circumstances may fuse together to form multinucleated giant cells, and stimulate local collagen deposition, or fibrosis. These are characteristic features of chronic inflammation.

Along with the phagocytosis of foreign material, macrophages in spleen and bone marrow participate in the breakdown of worn-out red corpuscles, and hence play a central role in *iron metabolism* and the recycling of iron stores. Macrophages ingest and remove cellular debris in areas of damaged tissue (Plates 65, 67a), allowing *healing and repair* to progress. They play a part in the process of *involution*, such as in the placental site in the uterus or in the mammary gland following lactation. Macrophages play some part, as yet imperfectly understood, in the *immunological responses* of the body, particularly in antigen uptake and in functional associations with lymphoid cells. Macrophages may secrete *interferon*, the natural anti-viral agent and a fever-producing substance, or *pyrogen*. Macrophages can attack and kill tumour cells in experimental situations and may play some part in the defences against the development and spread of *tumours*. This cell is clearly among the most important of the specialised cells of the body.

7.2 THE IMMUNOCYTES

The lymphocytes and the plasma cells are the two main cells involved in the mechanisms of immunological defence. The *lymphocyte* (Fig. 17) is a small round inconspicuous cell with a compact

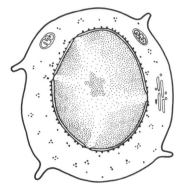

Fig. 17 *Small lymphocyte*
The small lymphocyte has little cytoplasm and shows no distinctive structural specialisations. Scattered free ribosomes, a small Golgi apparatus, and an occasional mitochondrion are the only organelles normally encountered.

nucleus and few organelles. The cytoplasm contains free ribosomes, mostly single rather than in groups, with virtually no organised granular endoplasmic reticulum. The mitochondria are small and scanty and the Golgi apparatus is diminutive.

A second cell, the *immunoblast*, is also recognised (Fig. 18). This is a transformed lymphocyte, stimulated to activity and proliferation by experimental exposure to non-specific mitogens or by natural immunological arousal. It differs from the small lymphocyte in having more abundant cytoplasm and a paler nucleus with prominent nucleoli. The characteristic pyroninophilic histological staining of the cytoplasm is accounted for by a great increase in RNA, as shown by the numerous polyribosomes.

Some immunoblasts, the B cell line, mature into plasma cells committed to the manufacture of specific antibodies against the arousing antigen. Others, the T cell line, give rise to specifically sensitised lymphocytes capable of producing delayed hypersensitivity reactions such as graft rejection. In both the B and T cell lines, some of the daughter cells of immunoblast proliferation revert once more, when immunological stimulation stops, to the appearance of small lymphocytes, distinguished now only by the possession of a specific *immunological memory* towards the stimulating antigen. Neither resting nor stimulated T lymphocytes can be distinguished by morphological means alone from their B cell counterparts.

This distinction can only be made by functional tests which recognise the presence of specific surface receptors in T cells and of surface immunoglobulin in B cells.

The contrast between lymphocytes and plasma cells could not be greater. The *plasma cell* (Fig. 19) is much larger, with bulky cytoplasm and a distinctive 'cart-wheel' nuclear pattern, formed by blocks of densely stained heterochromatin. The cytoplasm is virtually filled with closely-packed cisternae of granular endoplasmic reticulum, with the exception of a pale perinuclear area which contains the elaborate Golgi system.

The insignificant fine structure of the small lymphocyte greatly understates its importance in *defence*. Having access to every part of the body, these are the cells which provide new recruits for the immunological barricades and which carry in their collective memory a comprehensive record of close encounters of the antigenic kind dating back

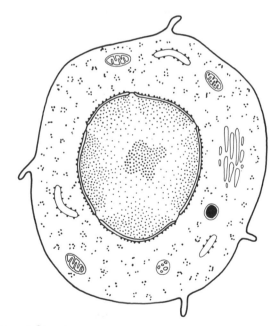

Fig. 18 *Transformed lymphocyte*
On exposure to an antigen to which the cell is able to respond, or on stimulation by certain chemicals known as mitogens, the lymphocyte enlarges and proliferates. This process of transformation is accompanied by an enlargement of the nucleus and nucleolus, along with a greater development of the cytoplasmic membrane systems and an increase in ribosomes.

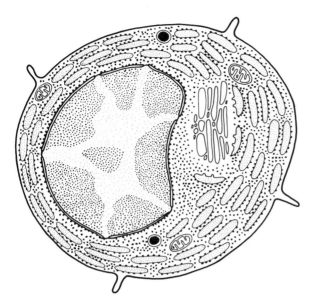

Fig. 19 *Plasma cell*
When an antigen stimulates an immune response in B cells,
some of the proliferating lymphocytes develop an abundant
granular endoplasmic reticulum and a prominent Golgi
apparatus. These cells synthesise and release
immunoglobulins.

to infancy. The lymphocyte, once sensitised to a
specific antigen, is capable of reacting against
subsequent exposure to that antigen at any time
and in almost any location. The specificity of its
immunological memory lies unseen, rather than
being expressed in terms of cytoplasmic organisation. The plasma cell, on the other hand, is a
production line dedicated to the manufacture of a
single type of antibody protein. Its fine structure
is typical of a protein-secreting cell undertaking
the export of its product.

The importance of the various cells involved in
the immunological defences cannot be overstated.
The new discipline of immunology has fundamental implications for science and medicine. The
narrow idea of defence against infection has given
way to the more fundamental concept of the
recognition of 'self' and the rejection of 'non-self'.

8

Storage and protection

The specialised function of certain cells is expressed in the *storage* of materials for widely differing purposes. Such cells may play a part in the metabolic functions of certain systems, as in the case of the *red blood corpuscle* and the *fat cell*. Alternatively, the stored material may have a mechanical or structural significance, often involving a *protective* role, as in the squamous cell and the melanocyte. In such cells the stored material may have its own specific fine structural appearances, while its formation and the metabolic activity of the cell in which it occurs are often reflected in the structure of the other cytoplasmic organelles. Protective specialisations, however, may take less obvious forms, as in the case of *transitional epithelium*, the *mesothelial cell* and the *platelet*.

8.1 RED BLOOD CORPUSCLES

A major function of the blood is the carriage of oxygen to the tissues. The *red corpuscles* of the blood are specialised to assist this function by storing large quantities of haemoglobin, which can enter into a loose and readily reversible chemical association with oxygen. In this way the oxygen-carrying capacity of the blood is greatly increased. In the mammal the red corpuscles carry this specialisation to the extreme, by losing their nucleus as they mature. They cannot divide and have a limited life span. The circulating red corpuscle is hardly a true cell, since it consists of little more than a membrane surrounding a concentrated solution of haemoglobin, without any of the usual cytoplasmic organelles (Plate 69). It is dense and homogeneous on electron microscopy, without internal structure (Plate 58), apart

from the trilaminar surface membrane. The dense texture of the haemoglobin tends to cause sectioning artefacts, leading to faint scoring.

The precursors of the red corpuscle carry in their cytoplasm the biochemical apparatus for haemoglobin synthesis. Since the haemoglobin is not intended for use outside the cell, but simply for *storage* in the cytoplasm, there is no elaborate granular endoplasmic reticulum such as is seen, for example, in the protein-secreting plasma cell (Section 5.1). Instead there are numerous free ribosomes which give the early red cell precursors their characteristic cytoplasmic basophilia. The use of iron in the manufacture of haemoglobin is reflected in the uptake of *ferritin molecules* by a form of selective micropinocytosis directly visible by electron microscopy. As the haemoglobin accumulates in the maturing cell it replaces the cytoplasmic organelles, its concentration finally reaching 33 per cent. This results in an efficient vehicle for oxygen carriage but with the loss of the essential cellular identity of the carrier.

8.2 FAT CELLS

Triglyceride is normally stored as droplets in connective tissue cells and is characterised in the electron micrograph by its homogeneous consistency and by the absence of a limiting membrane. Although droplets of fat may appear as spaces after processing (Plates 70a, 70b, 71), they may at times have a moderate electron density depending on the degree of saturation (Plate 70c, 14). The typical mature depot fat cell contains a single fat droplet which compresses the other cytoplasmic components to an extreme degree. Free ribosomes and mitochondria may be identified. The cell surface shows two main structural features, *micropinocytotic caveolae* and a surrounding *external lamina*. Stored fat is important not only as insulation but as a fuel rich in energy. The oxidation of fat produces over twice as much energy as the oxidation of the same weight of carbohydrate or protein.

Brown fat is a specialised form of adipose tissue, particularly prominent in very young mammals and in animals which hibernate (Section 4.5.2). In contrast to the usual type of fat cell described above, the brown fat cell contains a number of smaller droplets (Plate 70c). Brown fat cells are characterised by the presence of numerous large and well organised mitochondria indicating an unusual rate of oxidative metabolism. The presence of autonomic nerve endings indicates an unusual level of central control of cellular function. Brown fat cells are believed to oxidise stored fat within their own cytoplasm rather than acting simply as a storage depot. This oxidation is uncoupled from ATP production, permitting the *release of heat*, following autonomic stimulation, in order to maintain body heat in the young, poorly insulated animal or to restore the normal body temperature of the hibernating animal prior to its re-awakening. Attention has recently centred on a possible role for brown fat in the regulation of *body weight*.

8.3 SQUAMOUS CELLS

In sites exposed to wear and tear, a stratified *squamous epithelium* is commonly found. Its flattened surface cells form a cushion against mechanical injury and are constantly shedding or desquamating from the free border. They are replaced by the steady proliferation of the progenitor cells in the basal layers.

Particular physiological needs at different sites are met by different types of structural and functional epithelial specialisation. In human *buccal mucosa* and *oesophagus*, for example, the epithelium is soft and moist, lubricated by salivary secretion and by mucus produced by numerous small glands. The *skin*, by contrast, undergoes full keratinisation, which provides extra resistance to injury. A mucosa which is normally non-keratin-

ised is capable of undergoing keratinisation in response to altered conditions involving an unusual level of physical injury over a period of time.

The principal feature of any squamous epithelium lies in its *adhesion specialisations* (Plates 1, 5, 6b, 6c). All cells are linked to their neighbours by means of well-formed *desmosomes*, the number and size of which point to their importance in giving coherence and strength to the epithelium as a whole (Plates 27b, 73). The basal aspect of the deepest cells has related specialisations known as *hemidesmosomes* (Plates 5, 8a), which promote the adhesion of the basal cells, not to each other, but to the underlying connective tissues. Both desmosomes and hemidesmosomes are linked to a complex network of bundles of cytoplasmic keratin filaments, the *tonofilaments*. They appear to form, in three dimensions, an elaborate network of guy-ropes which distribute mechanical forces evenly throughout the epithelium. These bundles are so coarse, in the case of skin, that they can be seen as light microscopic tonofibrils connecting with the 'prickles' of the keratinocyte, which represent the desmosomes.

In other respects the basal cells of squamous epithelium have relatively unspecialised cytoplasm, with free ribosomes and occasional mitochondria, in keeping with their stem cell role. In the mid-zone or prickle cell layer, however, the cells become broader, accumulating the dense tonofilament-desmosome complex described above. Close to the margins of the cell one can now find distinctive small membrane-limited granules (Plate 1) being released into the intercellular space. The contents of these granules, commonly termed *membrane-coating* granules, have a fine periodic lamination. They appear to contribute to the permeability barrier which exists at this level of the epithelium, providing an important protection against unchecked diffusion in either direction. A second type of granule, the large dense *keratohyaline* granule, appears in cells in the upper part of this region of the epithelium (Plate 75a).

The most superficial cell layers become progressively flattened (Plates 33b, 72) and, in keratinised epithelium, extremely dense and difficult to section. Their nuclei, their cytoplasmic organelles and their granules largely disappear, while the desmosomes become modified to *squamosomes* prior to desquamation. The density is due to the keratin mass which remains as the predominant feature of the cytoplasm, an inert accumulation of a cytoplasmic product with an essentially physical protective function. The cell membrane of these surface cells becomes altered. Through these various structural adaptations, the superficial cell becomes converted to an envelope containing little other than structural proteins (Plate 76b). Layers of such cells contribute to the resistance to wear and tear. They cushion impact and desquamate on friction only to be quickly replaced by maturation of the underlying strata to the same end stage.

Thus several integrated functional properties can be correlated with the structure of squamous epithelium. The deepest cells have the features expected in a *dividing* population engaged in the synthesis of new cytoplasmic components. The desmosomes and tonofilaments work together to retain *adhesion* and to distribute mechanical stresses. The keratin produced by the cell gives it physical *resilience*. The membrane-coating granules correlate with the presence of a *permeability barrier*, which protects against the environment as well as restricting fluid loss. Finally, the inert surface cells form an expendable buffer layer, responding to wear by controlled *attrition*.

8.4   MELANOCYTES

The *melanocyte*, originally derived from the embryonic neural crest, is found in the epidermis, where it can be identified as a basally placed clear cell after conventional stains. Silver stains, however, show that it has actually a dendritic shape, with elongated fine processes (Plate 8a).

The characteristic inclusion of the melanocyte is the *immature melanin granule*, or *premelanosome*. This membrane-limited granule, formed in the Golgi apparatus, contains the enzymes required for melanin synthesis but does not, initially, contain melanin. Thus the cell 'product', made in the granular endoplasmic reticulum and packaged in the Golgi apparatus, is not melanin itself. These premelanosomes (Plate 74) range from small vesicles located in the Golgi region to large elliptical inclusions up to 1 μm in length and 0.5 μm in

diameter. Their contents, scanty at first, gradually become aggregated into roughly parallel cross-linked filaments lying in the long axis of the granule (Plate 75b), with a cross-striation showing a repeating period of approximately 8 to 9 nm.

The premelanosome becomes converted into a *mature melanin granule* or *melanosome* by the accumulation of melanin within its matrix. Melanin is an insoluble brown or black pigment formed from tyrosine through oxidation by tyrosinase, a copper-protein enzyme complex located in the premelanosome. As the melanin is laid down the internal periodic detail of the premelanosome becomes obscured and the active sites of melanin synthesis are progressively blocked. The final mature melanin granule, encrusted with dense pigment, has lost both its periodic substructure and its synthetic activity. The amount of melanin laid down and the size and shape of the melanin granules depend on genetic factors. The melanin granule of red hair, for example, is oblong rather than elliptical. It becomes less heavily pigmented than that of black hair, retaining some detail within its matrix.

The histochemical marker of the melanocyte is the *DOPA reaction* in which 3,4-dihydroxyphenylalanine (DOPA), an intermediate product in melanin synthesis, is converted to melanin. The DOPA-oxidase action of tyrosinase is found only in melanocytes, but melanocytes are not the only cells to contain melanin. The melanocyte appears to transfer melanin to *basal cells* in the epidermis (Plates 74a, 75c), by injecting mature melanin granules through a 'cellular syringe' mechanism. Melanin from damaged cells may also be ingested by dermal macrophages, particularly following injury to the skin. Melanin granules may even be found within phagocytes in regional lymph nodes draining areas of damaged skin. Only the melanocyte, however, can convert tyrosine through DOPA to melanin.

Melanin pigmentation provides *protection* against damaging ultraviolet *radiation* from the sun and gives *adaptive colouration* to many species. The importance of this protective function of melanin is shown by the abnormal photosensitivity of albino skin and by the high incidence of malignant tumours in Caucasian skin in sunny countries such as Australia.

8.5 TRANSITIONAL EPITHELIAL CELLS

Transitional epithelium lines the bladder, ureter and renal pelvis. It has several layers of cells which alter in configuration on stretching, the cells being short and plump in the empty bladder but progressively elongating and flattening as the bladder distends. This mechanical adaptability is valuable, but the protective waterproofing function of transitional epithelium is its most important property.

The most superficial layer of transitional epithelium, which lies in direct contact with the urine, has special ultrastructural features (Plate 76a). Its free surface appears gently scalloped and undulating in the distended bladder, but becomes irregular, re-entrant and angulated when the cells retract as the bladder empties. The *surface membrane* of these cells is apparently characterised by an unusual degree of rigidity. The apical cytoplasm contains distinctive *fusiform* or *elliptical vesicles*, which appear as clefts within the cell. These apparently empty membrane-lined vesicles can open at times to the lumen by fusing with the cell surface. These vesicles can be regarded as a *reservoir* of membrane material, making possible the necessary increase in the cell surface as the bladder distends.

The cell membrane at the apical surface of transitional epithelium is around 12 nm in thickness, considerably thicker than the apical membrane in most other epithelial cells. The trilaminar pattern of this membrane is particularly prominent, the external dense lamina being slightly thicker than the inner, or cytoplasmic lamina. The membranes limiting the closed fusiform vesicles are identical in pattern, the thicker lamina facing the lumen of the vesicle, which corresponds, from the cell's point of view, to the lumen of the bladder. In the Golgi apparatus of the transitional epithelial cell, areas of similar membrane thickening can be seen. Elsewhere, secondary lysosomes appear to form from some of the fusiform vesicles, leading to breakdown of the specialised membrane material. It is thought from this that the thick membranes typical of transitional epithelium may be manufactured in the Golgi apparatus, while old or damaged segments of membrane, withdrawn from the surface as part of the lining of a fusiform vesicle, can be *recycled* by lysosomal enzyme action.

The lining cells of the urinary tract form an essential *barrier* to the free osmotic movement of water. Without such a barrier, the presence of hypertonic urine in the bladder would lead to an inflow of water from underlying capillaries until osmotic balance was reached with the blood, thus negating the regulatory function of the kidneys. This relative impermeability of the bladder to water and electrolytes is destroyed by agents known to damage the lipoprotein structure of membranes and is thought to be a function of the unusually thickened surface membrane of the transitional epithelial cell. Since the bladder retains its impermeability in the absence of oxygen and nutrients, this would appear to be a *passive* function of the membrane, not linked to energy-consuming active transport mechanisms. It is already known from biochemical analysis that the thick membrane fraction of a bladder mucosal homogenate has a distinctively high concentration of proline and cerebroside. In time it may prove possible to explain the unusual protective functions of these membranes in terms of their more detailed molecular structure.

8.6 MESOTHELIAL CELLS

The serous cavities of the body are lined by a thin layer of inconspicuous flattened cells, which are often overlooked or neglected. The functional importance of the *mesothelial* layer is, however, considerable, since it provides the smooth slippery surface which protects the mobile viscera, suspended within the cavity, from *mechanical* problems.

The mesothelial cells (Plate 77), form a flattened simple layer which lies on a continuous basal lamina, underlain by a reinforcement of collagen fibrils. Their basal surfaces are quite flat, but the apical aspect of the serosa, the shiny surface, is distinguished by numerous long *microvilli*, particularly well visualised by scanning electron microscopy. These projections have been shown to be most numerous on the surfaces of the most mobile of the viscera, thus correlating well with their proposed mechanical function. The microvilli have a prominent external fuzzy coat or *glycocalyx*, often appearing as strands bridging between microvilli. It is thought that this elaborate surface layer may trap water, preserving lubricant properties without the use of mucinous secretion material.

On transmission electron microscopy, these microvilli are clearly seen as slightly crooked, irregularly orientated projections. Occasional single cilia project from the cell surface, their function being unknown. Between the microvilli, the cell surface displays numerous caveolae, indicating an active *micropinocytotic* function. Vesicles are also present at the cell base and within the cytoplasm. The cells are held together by well-defined desmosomes. The internal organelles are not notably specialised; ribosomes, filaments, some scattered membrane systems and mitochondria are present. The prominence of micropinocytosis underlines the high degree of *permeability* to fluids and solutes which the mesothelial layer exhibits.

The normal free mobility of the abdominal viscera will go unnoticed until abdominal inflammation or a surgical operation damages the serosa and replaces it by fibrous tissue *adhesions*. This is a common source of surgical complications; the loops of bowel, now tethered at some point and no longer freely mobile as before, may twist around the area of adhesion, producing mechanical problems such as obstruction and interruption of blood supply. The serosa, in summary, can be compared to the thin film of protective lubrication between the moving parts of an engine. It is barely detectable and may go unnoticed for long enough, except through the complications which are occasioned by its absence.

8.7 PLATELETS

The protective function of the *platelet* lies in the day-to-day maintenance of the integrity of the circulation. The platelet, or *thrombocyte*, is a cytoplasmic fragment originally derived from a megakaryocyte in bone marrow. It is elliptical in shape (Plate 78), measuring around 2.5 μm, with a limiting membrane. Around the margin of the disc runs a band of *microtubules*, credited with the maintenance of the shape of the platelet. The cytoplasm contains mitochondria, ribosomes and various small granular inclusions, some of which contain 5HT and phospholipids. The platelet has no nucleus and survives in the circulation (Plate 69) for about a week.

The protective role of the platelet is best appreciated by observation of patients who have an abnormally low platelet count. In this condition, known as *thrombocytopenia*, multiple bruises follow trivial injuries and serious bleeding from internal surfaces can readily occur. To understand this protective role, it must be realised that the vascular system is under constant stress. Trivial *trauma*, unnoticed in the day's activity, threatens to breach the continuity of the microvascular endothelium. It is the platelets which plug every tiny gap, adhering firmly to the injured area. Indeed, the underlying basal lamina, which contains type IV collagen, has the property of platelet aggregation, which can be seen as a specific protective adaptation. As the platelets stick to the injured surface and to each other, they lose their regular elliptical configuration and pack closely together or aggregate in an irregular jig-saw pattern (Plate 111). Degranulation of the platelets follows, the clotting mechanisms are triggered and there is deposition of fibrin around the platelet plug. The fibrin takes the form of bundles of electron-dense fibrillar material with a faint and not always detectable cross-striation, the periodicity of which is around 22 nm (Plate 78b). This aggregate of platelets and fibrin, often associated with trapped red and white cells, is termed a *thrombus*.

Many thrombi are probably soon broken down by natural fibrinolysin mechanisms, but if they remain to form an obstruction to the flow of blood they are eventually replaced by fibrous tissue and at least partially recanalised by the process of organisation.

Mechanical support

9.1 FIBROUS TISSUE

Those cells and their products which have a mainly mechanical function are known as *connective tissue*. In connective tissue the cells are relatively sparse while the intercellular materials, both fibres and surrounding matrix, are plentiful. The mechanical strength resides in the fibres produced by the connective tissue cells and in certain cases in the matrix which surrounds them. This is in contrast with epithelial tissues, in which the cells lie close together with little extracellular material.

9.1 FIBROUS TISSUE

The fibrous tissues consist largely of *collagen fibres* in a *mucopolysaccharide matrix*, surrounding the rather scattered cells which are responsible for their production. The *fibroblast*, engaged in the active synthesis of the components of fibrous tissue, is best seen in situations where new connective tissue is being laid down, such as in embryonic tissues or in healing wounds (Plate 80b). In mature fibrous tissue the synthetic function of the cells is reduced and they have a much less active maintenance role, restricted to gradual tissue turnover. These less active cells of mature connective tissue are termed *fibrocytes*. Adjacent cells separated by the substantial intercellular material rarely come into contact with one another.

The fibroblast (Fig. 20) is an elongated cell with irregular outlines and processes. Since these often extend far from the main part of the cell, they may leave the plane of a thin section. Irregular processes of this type are often obliquely sectioned, leading to an appearance of blurring of the cell membrane in an electron micrograph. In this way parts of the cell surface may seem to be deficient, leaving no

clear boundary between the cytoplasm and the intercellular space. This appearance is, of course, an artefact induced by the plane of section.

In keeping with its protein-secreting function, the fibroblast cytoplasm (Plate 80b) contains an extensive well-organised granular endoplasmic reticulum and a prominent Golgi apparatus. The fibroblast manufactures the collagen and elastic fibres of connective tissue and produces the mucopolysaccharide or proteoglycan *matrix* which forms the ground substance between cells and fibres. Although the matrix is not usually visible in conventional electron micrographs, it is important from the functional point of view, since it probably determines many of the physical and mechanical properties of connective tissue and may regulate diffusion between the formed elements. In addition to this synthetic function, the fibroblast is probably also involved in the breakdown of connective tissue, which takes place during normal turnover and re-modelling.

It is not clear how the fibroblast releases its synthetic products. Fibrils which lie close to the cell surface but are in fact external to the cell, have at times been thought to be in the course of extrusion from the cell through a 'gap' in the membrane, but appearances such as these can again be explained as effects of an oblique plane of section. Fibrillar collagen is not normally found in the cytoplasm of the fibroblast. The molecular precursors of the fibres and the components of the ground substance may be released as aggregates in vacuoles at the cell surface or may pass directly through the cell membrane in molecular form leaving no visible evidence of their release.

The main fibre of connective tissue is *collagen*, which appears by light microscopy as dense

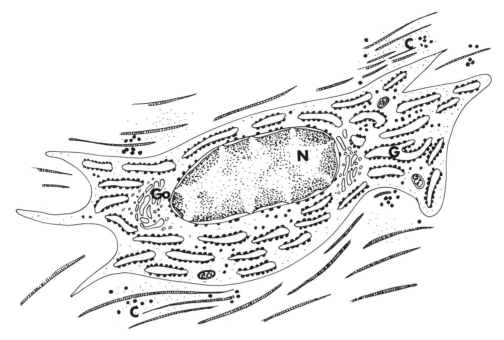

Fig. 20 *Fibroblast*
The fibroblast of connective tissue has irregular outlines and is surrounded by collagen fibrils, C, sectioned in various planes. The cytoplasm contains a well developed Golgi apparatus, Go, and plentiful granular endoplasmic reticulum, G. The nucleus is identified, N. The fibroblast is responsible for the synthesis not only of the fibres of connective tissue, but also of the ground substance which fills the interstitial space.

eosinophilic bundles. These collagen bundles are resolved by electron microscopy into aggregates of individual collagen fibril units of indeterminate length, which may be from 75 to 130 nm in thickness (Plate 79). Histologists distinguish dense eosinophilic mature collagen from the more delicate framework of 'reticular' fibres around the tissue components. The *reticulin* framework is seen only following special silver impregnations. On electron microscopy, however, the individual fibrils of collagen and reticulin share a similar repeating pattern. In relation to their bulk, the scattered individual fibrils of reticulin have a much greater surface area than the closely packed mature collagen bundles, a factor which may in part determine their different staining properties. In addition, there are minor differences between the molecular structure of type I, conventional collagen, and reticulin, known as type III collagen.

Each collagen fibril has a characteristic pattern of *cross-striation*. There is a major periodicity, usually around 64 nm in electron micrographs of sectioned tissue, with up to 13 separate minor repeating components of which four or five are easily seen (Plate 79d). A similar pattern can be recognised in numerous species, indicating that collagen is a fundamental component of animal connective tissues, both vertebrate and invertebrate.

The building-block of the collagen fibre is the *tropocollagen* molecule, an elongated macromolecule 280 nm in length and 1.4 nm in thickness, formed by the aggregation of three spiral chains. Extracellular tropocollagen molecules, present in high concentration near the active fibroblast, line up together through a process of irreversible self-assembly in regularly overlapping chains, forming thin precursor fibrils which may have a modified collagen periodicity. Further aggregation and *cross-linking*, with the lining up of corresponding densities in adjacent molecules, gives rise finally to the full repeating periodicity of the mature collagen fibril.

In chemical terms collagen is distinguished by its high concentration of the amino acids glycine, proline and hydroxyproline. There are now known to be at least five different types of collagen, distinguished by biochemical variations in the subchains of the tropocollagen molecule. Apart from the common type of adult collagen, known as type I, there is type III, known as reticulin, type II specific to cartilage and two types associated with basement membranes, types IV and V. These last two do not normally form recognisable cross-banded fibrils.

The fibroblast is also responsible for the formation of the *elastic fibres*, which give resilience to the connective tissues. The elastic fibre is a coarse, irregular, thick structure which usually appears relatively pale on section (Plate 81). Individual elastic fibres are up to 44 μm in length and 3 μm in thickness, although much smaller ones may be seen. The margins of the elastic fibres are poorly defined and often scalloped. Within and around the amorphous background of the peripheral part of the elastic fibre, dense 10 nm in diameter elongated glycoprotein *skeleton fibrils* may be detected.

The irregular non-periodic structure of the elastic fibre provides a striking contrast to the organised appearance of collagen. These differences are matched by distinctive biochemical features in the molecular architecture of elastin. Instead of the regular periodicity of closely stacked tropocollagen molecules, the structure of elastin is random, with stable crosslinkages between lysine-rich desmosine and isodesmosine subunits. These structural differences no doubt underlie the quite different mechanical properties of these two fibres of connective tissue.

A further fibril found in connective tissue, the *aperiodic* or *anchoring fibril*, may be related to collagen (Plate 74b). This short broad fibril is made up of aggregates of thinner filaments and has a rather irregular pattern of cross-striation, which differs in detail from that of collagen. The function of the anchoring fibril is not clear, but it often appears to be related to basal laminae, perhaps providing additional mechanical strength to the boundary layer between the basal lamina and the connective tissue.

Although the fibroblast or fibrocyte is the principal cell of loose connective tissue, other cell types are commonly found. Most of these are *wandering cells*, originating in many cases from the bloodstream. The histiocytes or macrophages have already been dealt with (Section 7.1). Another more enigmatic cell is the *mast cell*, distinguished by its large metachromatic granules. These vary in detailed morphology, but in the human species they contain distinctive laminated scroll-like inclusions (Plates 64b, 64c). These granules contain histamine and heparin, which are discharged from the cell during certain types of immunological reaction. In sensitive individuals, following an initial exposure to certain antigens, immuno-globulin molecules of the IgE class become attached to the surfaces of mast cells. On a second exposure to the antigen the mast cells may undergo an explosive *degranulation*, flooding the circulation with histamine and causing the dramatic systemic collapse known as *anaphylactic shock*. Presumably the mast cell has some physiological function related to the regulation of vascular permeability, but this remains open to speculation.

The various cellular components of connective tissue are separated from each other to a greater or lesser extent by collagen fibres and connective tissue matrix (Plate 9). Since connective tissue cells do not tend to form closely related groups as epithelial cells do, it is easy to tell at a glance from a low-magnification micrograph whether a tissue is epithelial or connective in type.

9.2 CARTILAGE AND BONE

These fill a vital structural role in the body by providing support and protection for the other tissues. Bone, in addition, forms a large readily available mineral reserve which is in dynamic equilibrium with the tissue fluids. The distinctive mechanical properties of the hard tissues are due mainly to specialisations of their matrix. The cells of bone and cartilage, essentially modified fibroblasts, are widely separated by intercellular material, including substantial amounts of collagen and a predominantly mucopolysaccharide or proteoglycan ground substance.

Over 80 per cent of the matrix of *cartilage* tissue consists of water. *Chondroitin sulphate* is the characteristic mucopolysaccharide, while collagen forms the fibrillar background around which the organic matrix is deposited. The numerous fine fibrils seen in electron micrographs are at times faintly cross-striated but are often free of obvious periodicity, unlike the typical mature type I collagen of most adult connective tissue. This is *type II collagen*, in which all three subchains of the tropocollagen molecule are identical. These fibrils are widely spread throughout the matrix except in the lacunae immediately around the cells (Plate 82). Thick dense irregular elastic fibres lie between the cells of elastic cartilage.

The *chondrocyte*, which secretes these various constituents of the cartilage matrix, is an oval cell with small irregular surface projections. A moderately elaborate granular endoplasmic reticulum and Golgi apparatus are in keeping with its active protein and proteoglycan secretory function. There is often fine fibrillar material in the peripheral cytoplasm and there are numerous glycogen particles and lipid droplets, which may serve as fuel reserves.

In *bone*, as in cartilage, collagen forms the structural background of the tissue, forming up to 40 per cent of the total volume of newly synthesised bone (Plate 83). The collagen is of the common type I pattern; the typical organised periodicity is well seen in bone prior to full mineralisation. The striking rigidity of the bone matrix is due to the deposition of fine needle-shaped crystals of calcium hydroxyapatite, ranging in size from around 10 to 150 nm in length and from 1.5 to 7.5 nm in diameter. The precipitation of crystalline calcium salts is thought to be brought about through the action of alkaline phosphatase produced by the bone cells. The crystals, which lie roughly parallel to the long axis of the collagen fibril, are arranged at intervals corresponding to, and thus accentuating, its major spacing. Eventually the dense sheaves of crystals largely obscure the underlying collagen fibril framework.

The cells responsible for bone formation are the *osteoblasts*, small basophilic cells applied to the forming surface of the bone. These cells, which synthesise the collagen and mucopolysaccharide components of bone, share the complex endoplasmic reticulum and Golgi apparatus of the fibroblast and the chondrocyte. The cell outlines are irregular, with fine projections which may contact adjacent cell processes but which do not form specialised adhesion zones. As the newly formed matrix accumulates and calcifies, thus trapping the osteoblast, these projections come to occupy fine canaliculi which perforate the bone substance. When the bone-forming cell is finally surrounded by the matrix it is termed an *osteocyte*. Its ultrastructural complexity diminishes although it retains the necessary apparatus for its continuing function in the maintenance of the tissue matrix and the turnover of the minerals of bone.

In addition to these bone-forming cells there are multinucleated giant cells, known as *osteoclasts* or bone destroying cells, which occupy lacunae or hollows on the surface of the resorbing bone trabeculae. The surface of the osteoclast in contact with the underlying bone matrix is marked by an elaborate infolding of the cell membrane, forming a brush border of great complexity, the processes of which are separated by deep clefts in the cytoplasm (Plate 83b). Cytoplasmic vacuoles of various sizes are found, some of which contain apatite crystals apparently ingested by the cell in the course of bone destruction. The underlying bone matrix is rather disorganised, the crystals being more haphazardly scattered than in normal bone, with areas of collagen free of associated crystals. The osteoclast may fulfil its destructive role through the extracellular release of lysosomal enzymes.

Contraction and motility

All cytoplasm has some *contractile* capacity, probably reflecting the wide distribution of actin within cells of every type. In complex animals, contractility has become a function of specialised cells, the *muscle* cells, in which the contractile filament mechanism is elaborately developed. Muscles of different types have their own distinctive specialisations which can be closely correlated with their functional properties. A second system for producing movement has evolved around the microtubular system of the *cilium*. Ciliated protozoa use this system for locomotion, but in higher species the ciliated cells are more typically associated with a protective role, as in the respiratory tract, where it is not the cells that move, but the overlying layer of extracellular mucus. Both of these systems provide elegant examples of intricate cytoplasmic specialisation in association with a highly evolved mechanical function.

10.1 CONTRACTION

Muscle is the source of movement, either for locomotion, through its attachments to the bones of the skeleton, or for internal motility and propulsion, through its presence in the vascular system and hollow organs. The three main groups of muscular tissues are *skeletal* or striated muscle, attached principally to the bones of the skeleton; *visceral* or smooth muscle, found principally in the viscera or internal organs; and *cardiac* muscle, with striations similar to skeletal muscle, but with special features exclusive to the heart.

The cellular unit in *skeletal muscle* is the individual muscle cell, often referred to as a muscle fibre on account of its elongated shape. The term fibre, however, is more correctly reserved for *extracellular* structures with a fibrillar pattern, such as collagen. A single skeletal muscle cell may be up to 100 µm thick and may be centimetres long, extending at times from end to end of the anatomical muscle. There are numerous nuclei which lie close to the cell surface. The main substance of the cell consists of closely packed parallel *myofibrils*, showing a regular pattern of cross-striation with a spacing of 2 to 3 µm. This type of muscle is directly controlled by the cranial or spinal motor nerves and depends on intact innervation for its function and, indeed, its survival.

The unit of *visceral muscle* is the individual spindle-shaped muscle cell, which has a single central nucleus and a wide size range from 10 to 100 µm in length and up to 10 µm in thickness. The precise dimensions depend on the location and function of the cell. Since visceral muscle shows no cross-striated pattern, it is often described as smooth muscle. This type of muscle is innervated by autonomic nerves and in general is not under voluntary control. The cells have an inherent rhythmic activity which does not depend on the integrity of external innervation. Smooth muscle can also be stimulated by hormones and by various biologically active polypeptides.

Cardiac muscle has a branching structure. The single central nuclei of neighbouring cells are separated by partitions known as *intercalated discs*. The cells, however, are cross-striated with a pattern similar to that of skeletal muscle. Cardiac muscle has a strong inherent rhythmic activity, modified indirectly by autonomic nerves through a pacemaker and conducting system, but independent of them for its basic function.

10.1.1   Skeletal muscle

Confusion can sometimes arise from the use of certain histological terms which persist from an earlier era and are still sometimes applied to parts of the skeletal muscle cell. For example, the small amount of cytoplasm surrounding and separating the myofibrils is traditionally known as the sarcoplasm, while the surface region of the muscle cell, containing the nuclei, may be loosely referred to as the sarcolemma. These terms were justified when the resolution limits of histology did not permit more accurate nomenclature. The modern electron microscopist may prefer, however, not to use such terms, since the skeletal muscle cell is just a cell like any other, with cytoplasm, nuclei and cell membrane.

The cell membrane forms an unbroken partition between cytoplasm and tissue fluid. At points along the cell surface (Plate 87a) there are small flask-shaped invaginations of the membrane which indicate *micropinocytosis*. Immediately external to the membrane there is a distinct *external lamina*, reinforced in places by a delicate lacework of collagen fibrils. This lamina is closely applied to each individual cell, surrounding it and separating it from its neighbours. The surface layer of cytoplasm which incorporates the nuclei, along with the cell membrane and the surrounding external lamina, combine to form the image interpreted, at the resolution level of light microscopy, as the sarcolemma (Plates 84, 87a).

The contractile subunits of the muscle cell are the *myofibrils* (Figs. 21, 22), just visible by histology. The individual myofibril, about 1 μm thick, extends the length of the cell. The cross-striated pattern of the myofibril can be seen by electron microscopy to be due to the presence of a highly organised system of parallel thin and thick *myofilaments* (Plates 85, 86, 87), consisting of the proteins actin and myosin respectively. The alignment of these filaments to form regular regions of overlap and interdigitation is responsible for the distinctive repeating pattern of cross-striation. Their movement with respect to each other forms the basis of muscle contraction.

A complex nomenclature exists to identify the components of the *sarcomere*, the repeating unit of cross-striation. Two main bands, light and dark,

are seen with polarised light, the *isotropic*, or I band, and the *anisotropic*, or A band. The region of the myofibril in which only thin filaments are found corresponds to the I band and the region in which thick filaments are present is the A band. The myofibril is thus composed of alternating I bands and A bands. The I band is bisected by the dense Z line which links together the thin filaments. The sarcomere extends from one Z line to the next and consists, therefore, of two half I bands with a complete A band between them. In the A band of the relaxed myofibril there is a central portion, the H zone, which contains only thick filaments. Thickenings at the mid-point of each thick filament align to form the M line which bisects the A band. At each end of the A band there is overlap between the interdigitating thick and thin filaments. At intervals there are minute lateral projections from the thick filaments, which appear as *cross-bridges* linking overlapping thick and thin filaments. These are best seen after special extraction procedures.

When a transverse section of muscle is examined (Plate 84b), the filaments of the myofibril are viewed 'end on' rather than 'side on', appearing in cross-section as dense spots (Plate 87b). The thin filaments are around 7 nm in diameter, the thick filaments about 12 nm. Thus if the plane of section passes through the I band, where only thin filaments are present, a pattern of 7 nm spots is seen. If the section cuts the central part of the A band where only thick filaments are found, cross-sections 12 nm in diameter are seen. If the section passes instead through the region of overlapping thick and thin filaments, a hexagonal pattern can be seen in which each thick filament is surrounded by six thin filaments.

By various methods it has now been shown that the thick filaments represent aggregates of the muscle protein *myosin*, while the thin filaments are composed largely of *actin*. The cross-bridges between filaments represent the enlarged ends of the myosin molecules. When calcium is present, myosin can split ATP, releasing the energy stored within its high energy phosphate linkages. This energy is partly converted to mechanical work by the overlapping filament mechanism of the myofibril. A mechanical force is exerted at the cross-bridges between the thick and thin filaments through a form of *ratchet mechanism*, causing the

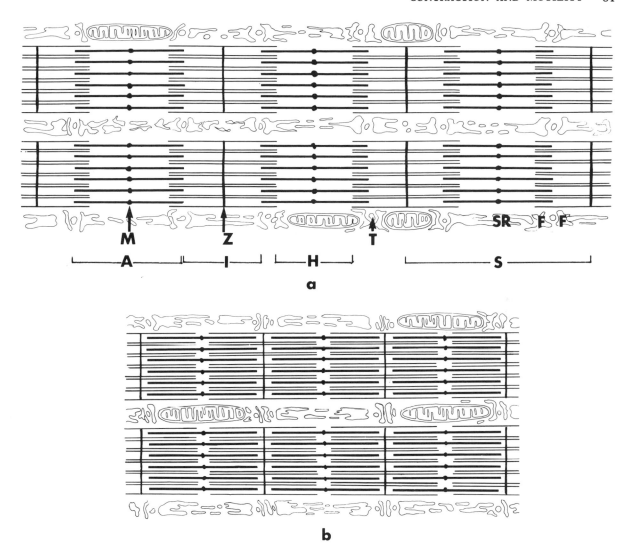

Fig. 21 *Striated muscle, relaxed and contracted*

Diagram a shows two parallel myofibrils of relaxed striated muscle, separated by the cytoplasmic components of the sarcoplasm. The sarcomere, S, and its various subdivisions are identified. The M line and the Z line are indicated by arrows. The smooth endoplasmic reticulum, or sarcoplasmic reticulum, SR, has dilated end pieces known as foot processes, F, which flank the T tubules, T. The triad of T tubule and surrounding foot processes lies at the junction between the A band and the I band.

Diagram b shows the same muscle in the contracted state. Careful examination will show that the thick and thin myofilaments are exactly the same length in both diagrams, although the contracted muscle is one third shorter than the relaxed. This change is produced by a sliding mechanism, altering the relationships of the thick and thin filaments to each other without any change in their actual lengths. The A bands remain the same length, but the I bands and the H zones have become much narrower in the contracted muscle. In extreme contraction, the H zone may in fact disappear completely as the thin filaments meet and even overlap themselves.

thin filaments to slide with relation to the thick and thus leading to shortening of each sarcomere.

During contraction, therefore, the thick and thin filaments themselves do not shorten, but the sliding of the thin filaments between the thick leads to an increase in their area of overlap and a corresponding decrease in the width of the I bands (Fig. 21). The A bands, determined by the length of the thick filaments, remain of constant width. Since the thin filaments slide closer to the M line, the H zone narrows and finally disappears in contraction. When the sarcomere is fully contracted there may be a region of double overlap in the centre of the A band, where thin filaments from either side of the sarcomere overlap each other as well as overlapping the thick A band filaments. Cross-sections of the area of double overlap show twelve thin filaments surrounding each thick filament.

The myofibrils are surrounded by the muscle cell cytoplasm, or *sarcoplasm*, which contains several formed components. The mitochondria vary in size, number and configuration in different muscles according to their energy needs, being most complex in muscles on which severe functional demands are placed, such as the rapidly-contracting insect flight muscles (Plate 85b). The mitochondria tend to lie at corresponding points in each sarcomere, thus reinforcing the repeating pattern already present in the myofibrils. The juxtaposition of mitochondria and myofibrils is not fortuitous, since it represents the most efficient relationship between the source and the consumer of the energy supplied by mitochondrial *oxidative phosphorylation*. Dense 30 nm glycogen particles may often be seen in the sarcoplasm (Plates 33a, 53b, 85a), distinguished by their size from the ribosomes, which measure only 15 nm. A small Golgi apparatus and a few ribosomes are often seen at the poles of the nuclei.

The prominent agranular endoplasmic reticulum of the striated muscle cell is known as the *sarcoplasmic reticulum*. It has a complex but regular segmental arrangement which again corresponds to the repeating sarcomere pattern of the myofibrils (Fig. 22). The cisternae of the sarcoplasmic reticulum lie parallel to the long axis of the cell, winding between and forming a lace-like sleeve of intercommunicating cavities around the myofibrils. In the contraction of the myofibril, *calcium* plays an important part. The sarcoplasmic reticulum binds calcium within its cisternae, releasing it when the cell is depolarised. The presence of calcium in the cytoplasm around the myofibrils exposes the actin binding sites, allowing myosin to split ATP and thus triggering off the sliding mechanism. The reticulum then once again takes up the calcium, allowing relaxation. In some instances, mitochondria share in this function of calcium regulation.

Closely integrated with this predominantly lon-

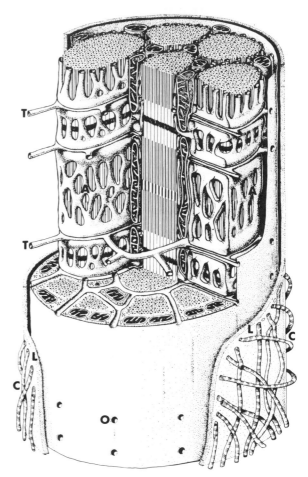

Fig. 22 *Sarcoplasmic reticulum and its relationships*
This diagram shows the relationships of the myofibrils and sarcoplasmic components of a typical striated muscle cell. The smooth or sarcoplasmic reticulum of muscle cells surrounds the individual myofibrils with an anastomosing network of cisternae, A. This system forms a repeating relationship with the transverse tubules, T, located at the junctions between A bands and I bands. The T tubules open onto the surface of the muscle cell, O. This diagram also shows the presence of a surrounding external lamina, L, with reinforcing collagen fibrils, C.

Diagram by courtesy of R. V. Krstić

gitudinal system of sarcoplasmic reticulum are the *transverse tubules* or *T-tubules*. These lie at a constant position in each sarcomere, normally either at the Z line or at the A-I junction. They run at right angles to the sarcoplasmic reticulum (Plates 85a, 86), crossing the muscle cell parallel to the cross striations. As already mentioned, the longitudinal tubules of sarcoplasmic reticulum are also arranged in a segmental pattern. These end in expanded *foot processes*, which flank the T-tubules in each sarcomere. The repeating system, *the triad*, consisting of the T-tubule with its two closely applied sarcoplasmic reticulum foot processes, lies at a constant position in each sarcomere, usually around the Z line (Plate 86). There is no open communication between the T-system and the sarcoplasmic reticulum, although the components of the triad often appear linked by fine dense strands.

The T-tubule, however, reaches and joins the surface membrane of the muscle cell (Plate 85b), so that its narrow lumen communicates directly with the extracellular space. The T tubules provide the pathway for communication between the surface membrane and the deep interior of the muscle cell. Each sarcomere, even in centrally placed myofibrils 50 μm from the cell membrane, is kept in direct touch with the functional state of the cell surface through the T-tubule, so that all sarcomeres can contract in unison when *depolarisation* of the cell membrane occurs.

Muscle function involves two separate biological phenomena. There is a rapidly spreading electrical surface phenomenon of membrane depolarisation, involving ion fluxes. This is termed *excitation.* Following this, there are the consequent chemical and mechanical responses within the cell, which constitute muscle *contraction.* The excitation wave of membrane depolarisation and repolarisation is initiated by the release of acetyl choline at the motor end plate and spreads along the cell within a few milliseconds. Muscle contraction is a much slower process which takes up to 100 milliseconds.

Although electrical stimulation of the membrane surface of the intact muscle cell causes contraction, electrical stimulation of the isolated myofibril is not effective. This indicates that the excitation and contraction processes are separate, although they are coupled in the intact cell. The T-tubules and the sarcoplasmic reticulum are responsible for this process of *excitation-contraction coupling.* If only a tiny portion of the cell surface membrane is depolarised, using a hollow glass micro-electrode, the excitation produced does not spread further. No contraction occurs unless the electrode is placed on the surface of the cell at the A-I junction, the point where the T tubule normally approaches the exterior. When this highly localised surface depo-

larisation affects the T-tubule it causes the related sarcomeres to contract. This reflects the segmental function of the T-tubule in relation to its immediately adjacent sarcomeres.

Skeletal muscle is designed for rapid, powerful but relatively short-lived mechanical action, with a quick cycle time, a clearly defined rest length and a narrow range of movement. This type of muscle depends totally on its *nerve supply* for normal activity (Section 11.1.3). If this is interrupted, the muscle cells will undergo progressive wasting, known as atrophy.

10.1.2 Visceral muscle

The biological role of visceral or smooth muscle is so different from that of skeletal muscle that their structural differences should not come as a surprise (Fig. 23). Visceral muscle is less forceful, slower to contract and slower to relax than skeletal muscle. It has no clearly defined rest length and it can adopt different ranges of movement depending on physiological needs. Unlike skeletal musle, visceral muscle can be isolated from its external nerve supply and still retain an essentially normal functional capability.

The smooth muscle cell membrane is invested on its outer surface by a delicate external lamina, along with related collagen fibrils (Plate 88). This forms a barrier between each cell and its neighbours, except at certain limited areas where their membranes come into close contact. These contact areas, commonly termed *nexuses*, are now recognised as *gap junctions* (Section 2.2.2). The number of close contacts between smooth muscle cells varies in different sites. They provide areas of low electrical resistance, allowing the passage of excitation from cell to cell and thus accounting for the characteristic spread of activity in visceral muscle. At other points on the surfaces of smooth muscle cells there are numerous membrane invaginations or caveolae. The significance of these vesicles is not clear, although it has always been assumed that they represent micropinocytotic activity. Extracellular tracers, however, do not seem to be taken into the cell by this route in significant amounts.

At the poles of the nucleus there are cones of cytoplasm free from filaments, which contain the small Golgi apparatus, the ribosomes and occa-

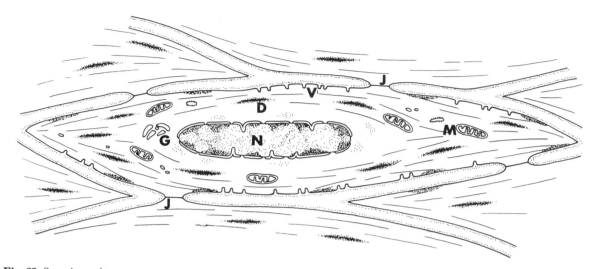

Fig. 23 *Smooth muscle*

The smooth muscle cell is shown with its central nucleus, N, and a filament mass which is not subdivided into regular sarcomeres. Scattered dense bodies, D, alternate with myofilaments, which are also inserted into the cell surface at similarly dense attachment plaques. Occasional mitochondria, M, Golgi membranes, G, and subsurface micropinocytic vesicles, V, are present. The cells are linked in places by gap junctions, J, which allow electrical coupling between cells. External to each cell there is a diffuse external lamina.

sional lysosomes. Mitochondria in visceral muscle lie at random in the cytoplasm and do not show features of high metabolic potential. The well-organised sarcoplasmic reticulum and T-system of skeletal muscle are not seen in visceral muscle, but smooth surfaced tubular cisternal profiles which are scattered through the cell, often in a sub-surface location, probably serve a similar function in relation to calcium ion storage and release.

Most of the smooth muscle cell cytoplasm is occupied by *myofilaments*, but there are no myofibrils and no periodic sarcomere pattern. Both of the muscle proteins, actin and myosin, are present but the proportion of actin to myosin is 30 to 1 instead of 2 to 1 as in skeletal muscle. Until recently, however, it was not even possible to identify with confidence structurally distinct myofilaments corresponding to the thick myosin and thin actin filaments of striated muscle. Conventional specimen preparation techniques for electron microscopy show the cytoplasm to be filled with closely packed longitudinal filaments apparently of variable diameter. Minor variations in filament organisation are related to the physical state of the muscle cell at fixation. Throughout this mass of filaments there are scattered cigar-shaped *dense bodies*, roughly 70 nm in diameter and of uncertain length. The thin filaments are attached

at various points to dense zones on the inner aspect of the cell membrane, representing the points at which the contractile apparatus exerts its pull. These *attachment zones* are the functional equivalent of the Z lines of the sarcomere.

On the basis of this conventional electron microscopic picture it was understandably difficult to correlate the poorly-defined smooth muscle filaments with the highly ordered apparatus of striated muscle. Improved preparation techniques, however, show filaments in smooth muscle which do correspond to actin and myosin. Thin actin filaments form small regular arrays. Thick myosin filaments are much less numerous, consistent with the 30 to 1 actin to myosin ratio. They are stacked in a relatively regular 60 to 80 nm lattice; each is a little over 2 μm long.

The identification of structurally distinct components corresponding to actin and myosin brings smooth muscle into line with the other forms of muscle tissue. Current views of smooth muscle function rely on a *sliding mechanism* involving actin and myosin molecules, activated by calcium, which is similar in its broad essentials to the mechanism now accepted for striated muscle. The system in smooth muscle, however, is capable of generating contractile force over a tenfold range of starting length, a flexibility which is essential for visceral function, but quite inappropriate for the role played by skeletal muscle.

The innervation of smooth muscle is very complicated, but in general, visceral function is controlled by local autonomic nerve plexuses linked to the major autonomic nerve outflows. The small unmyelinated nerves which wind between the muscle cells have multiple synaptic specialisations containing transmitter substances, but nowhere is there the intimate direct innervation which is typical of skeletal muscle. These local nerve plexuses can be disconnected from external autonomic control without prejudicing the basic function of visceral muscle.

10.1.3 Cardiac muscle

Cardiac muscle, the third main histological and functional type, is the muscle of the *heart*. It provides motive force for the circulation by its rhythmic beat. The muscle of the heart may rest for only a fraction of a second at a time and must be capable of rapid, powerful, yet sustained contraction. It must have flexible functional reserves to meet suddenly increased demands during exercise, and the action of every part must be co-ordinated so that the effort is directed to the movement of blood and not wasted by inefficient or disorganised contraction.

Cardiac muscle is distinct from both skeletal and visceral muscle, but combines in its histological structure certain features of both. The cardiac muscle cell is *striated*, displaying the same sarcomere pattern as is seen in skeletal muscle. The nuclei, however, are *centrally* placed. In contrast to the multinucleate arrangement seen in skeletal muscle, the nucleated segments of cardiac muscle are separated by transverse partitions termed *interca-*

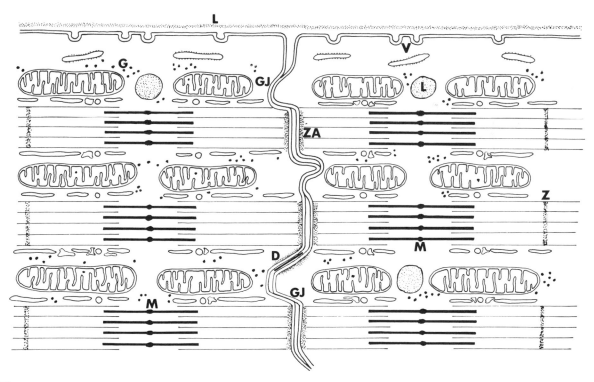

Fig. 24 *Intercalated disc of cardiac muscle*
In its essentials, the pattern of muscle striation in the heart is identical to that of skeletal muscle. The various landmarks of the sarcomere are identified as in Figure 21. The external lamina, L, surface micropinocytotic vesicles, V, cytoplasmic glycogen, G, and lipid, L, are identified. The particular feature shown here, the intercalated disc, is an example of a specialised cell boundary. Notice that the disc replaces the Z line of the sarcomere. Three zones of specialisation are seen, the gap junction, GJ, the fascia or zonula adhaerens area, ZA, into which the actin filaments are inserted, and the desmosome, D, with its intermediated dense band. The specialisations of the disc promote adhesion and communication between the cellular units of cardiac muscle.

lated discs, which mark the boundaries between individual cells (Plates 89, 90).

The myofibrils and myofilaments of cardiac muscle closely resemble those of skeletal muscle. As in skeletal muscle, the cross-striation is due to a repeating pattern of thick and thin filaments arranged in register and overlapping to form the familiar bands. The powerful and rapid contraction and the relative constancy of rest length for which the striated pattern provides are well suited to the physiological role of the heart.

The organisation of the components of the sarcoplasm is not significantly different from that seen in skeletal muscle. The T system and the longitudinal components of the sarcoplasmic reticulum have a similar arrangement, although in cardiac muscle they may show slightly less regularity. The heart is constantly active and uses much energy during prolonged exercise. The *mitochondria* which provide the necessary energy are correspondingly large and numerous and their cristae closely packed, pointing to the importance of oxidative phosphorylation in the economy of the cell. Glycogen and lipid droplets, both fuel material, are often found in the sarcoplasm close to the mitochondria. The few examples of skeletal muscle which show a comparable development of the mitochondrial mass have unusual mechanical functions, calling also for sustained high energy output. They include the highly specialised insect flight muscles.

The *intercalated disc* is the most specific ultrastructural feature of cardiac muscle. Electron microscopy has shown that each disc marks the point of contact between two entirely separate cardiac muscle cells, each with its own nucleus. Cardiac muscle is thus composed of numerous separate cellular units joined together closely at the intercalated discs (Fig. 24).

The cell membranes at the intercalated disc form different specialised areas of contact. There are quite extensive zones where the membranes of adjacent cells come close together (Plate 90), forming an area recognised as a gap junction. This is traditionally termed the *fascia occludens*, by analogy with the zonula occludens of the epithelial junctional complex, but this terminology is misleading and inaccurate, since it is now recognised that the zonula occludens and the gap junction are structurally distinct. As with the *nexus* of smooth muscle, these areas are thought to allow the passage of ions between cells. This provides *electrical coupling*, permitting the rapid spread of surface excitation throughout the muscle mass. At other points on the disc there are specialisations similar to the *macula adhaerens*, or *desmosome*. Presumably these have a mechanical role. The third component of the epithelial junctional complex, the zonula adhaerens, has its equivalent in the *fascia adhaerens* of the intercalated disc. The thin I band filaments of the myofibril are inserted into the cytoplasmic sides of the membranes at these points, causing the characteristic density of the disc under the microscope. These portions of the intercalated disc anchor the contractile apparatus, since the disc replaces the Z line of the sarcomere. The zig-zag form of the intercalated disc is characteristic, the gap junctions and desmosomes tending to lie parallel or oblique to the myofibrils, the adhaerens areas transverse.

The intercalated discs thus have two separate functions. They provide a *mechanical link* between cells, preventing their separation and acting as anchorage points for the contracting myofibrils of each cell. In addition they allow *communication* between cells, making possible the rapid transmission of surface excitation which is essential to the co-ordinated action of the heart-beat.

Cardiac muscle has no direct innervation, except at the specialised region known as the sino-atrial node, where autonomic nerves control the rate of the cardiac pacemaker. Elsewhere, autonomic nerves are not seen between cardiac muscle cells, which can continue to beat rhythmically even in complete isolation.

10.2   MOTILITY

10.2.1   Cilia

While complex animals rely on muscles for movement, a more delicate apparatus exists at the subcellular level. Many protozoa, such as paramoecium, depend for motility upon the repeated whiplash activity of numerous thin projections or *cilia* which can be seen under the light microscope. The combined action of the cilia carries the cell through the surrounding fluid.

In higher forms of life, cilia are present at surfaces where a moving carpet effect is of particular value (Plates 34c, 91, 93). The epithelial cells of the upper respiratory tract are ciliated, providing the surface of the trachea, bronchi and nasal passages with a continuous carpet of cilia in constant motion. Mucus secreted by goblet cells forms a protective surface layer which traps inhaled particles of dust and bacteria. The action of cilia, beating some twenty times a second, then sweeps the mucus back to the larynx where it can be removed by coughing. The mucus layer is swept along at a rate of over a centimetre per minute.

Each cilium is an elongated cytoplasmic extension covered by the apical surface membrane. Measuring from 5 to 10 µm in length and about 0.25 µm in diameter, they can just be individually resolved by light microscopy. The cilium has a central core of parallel subunits which extend, unbranching, from its base to its tip. This core is termed the *axial filament complex*, although the subunits are actually tubular in structure, identical to the microtubules found elsewhere in the cytoplasm. The term *axoneme* is also used.

The detailed arrangement of the tubules of the axial filament complex is clearly seen in transverse section, which allows an 'end on' view of the subunits. Two tubules, each 24 nm in diameter, lie slightly separated in the centre of the complex (Plate 94). Nine peripheral pairs of tubules are evenly spaced around the central pair. One member of each peripheral pair, subunit A, has a denser core than the other, subunit B. Two short arms extend from subunit A of one pair towards subunit B of the adjacent pair (Fig. 25). These arms represent *dynein* molecules. When the cilium is viewed from base to tip, the dynein arms point in a clockwise direction.

The components of the axoneme originate from the *basal body*, a cylindrical structure which anchors the cilium in the apex of the cell and may control and co-ordinate ciliary activity. The basal bodies arise during development from repeated division of the *centrioles*, explaining the obvious fine structural similarities between centrioles and cilia (Section 4.10).

The cilia of a single cell or of an entire region beat together with co-ordinated function, the direction of beat being defined by the orientation

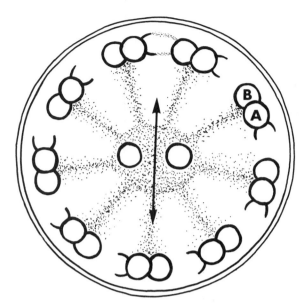

Fig. 25 *Cilium in transverse section*
Each of the nine peripheral subunits of the axoneme consists of two tubular components as shown. The two small arms extending from subunit A are the dynein components which are essential for motility. The direction of ciliary beat is indicated by the arrows lying perpendicular to the plane of the two central tubules.

of the *central pair* of the axial filament complex. When a number of cilia are cut in cross-section in a particular specimen, the orientation of their central pairs is approximately parallel, and the direction of beat lies at right angles to this line.

Ciliary motility is known to depend on the interaction of the dynein molecules with the tubulin components of the axial filament complex. The tubules themselves have no independent contractile capacity. It is presumed that dynein and tubulin interact through some kind of *ratchet mechanism*, leading to controlled bending and straightening of the cilium. The active phase of the beat is a forward stroke during which the cilium remains almost straight, whereas on the recovery stroke it curves on itself. The ATP-ase function of dynein, unlike that of myosin, seems to be independent of calcium ions but to require magnesium ions. Moreover, ciliary beating can continue even after the experimental removal of the surrounding cell membrane, ruling out an action potential phenomenon related to the cell surface.

A rare defect of cilia occurs in a condition known

as *Kartagener's syndrome*, characterised by the presence of non-motile cilia which are apparently structurally normal. Closer study, however, reveals that the dynein arms are missing from the cilia, explaining the failure of motility.

Cilia and microvilli are quite different specialisations, which may at times co-exist on the same cell. Microvilli are relatively static specialisations which increase surface area, while cilia are motile. Microvilli, too small to be individually resolved by light microscopy, form a striated border if closely packed. Cilia are large enough to be resolved as individual structures. While microvilli have a simple core of parallel actin filaments, the highly structured microtubular axonemes of cilia are quite distinctive on electron microscopy. Some confusion is caused by the term *stereocilia*, used to describe the apical processes of the epididymal epithelium. These are no more than unusually elongated *microvilli*, with no motile function. It is important to distinguish clearly between these different specialisations of the cell surface.

10.2.2   Flagella and spermatozoa

The flagellum is a motile cell specialisation closely related to the cilium. It is distinguished by its length, which can be up to 150 μm, and by its spiral action, which contrasts with the simple reciprocating whiplash movement of the cilium. A solitary flagellum provides propulsion for the motile spermatozoon. In cross-section the cilium and flagellum are apparently identical.

The function of the *spermatozoon* is the carriage to the ovum, over considerable distances, of the genetic information contained in the male germ cell nucleus. The sperm has two essential parts concerned with these functions, the nucleus which carries the genetic information and the propulsion unit which pushes it to its destination. Since streamlining of structure is as important here as in racing cars, it is not surprising to find specialisations in the sperm cell designed to minimise any unnecessary drag. During its development from precursor cells in the seminiferous tubule (Plates 26a, 95c) to its mature streamlined form (Plate 95a), the spermatozoon sheds almost all of its cytoplasm, retaining only the nucleus, the acrosome and those components which are essential to

motility. This is in contrast to the sedentary ovum, which has a considerable accumulation of cytoplasmic material for distribution to the cells of the early embryo.

The *head* of the spermatozoon contains the nucleus, in which the chromatin is in a highly condensed, inactive form. The sperm head presumably represents the smallest volume in which this amount of genetic information can be effectively stored. The restriction of nuclear size is clearly in the interests of efficiency, since it reduces drag during movement.

During the development of the sperm in the seminiferous tubule, a collection of material produced by the Golgi apparatus accumulates in a vesicle known as the *acrosome*, which becomes applied like a cap to the pole of the nucleus (Plates 26a, 95b, 95c). The acrosome appears to be a specialised lysosome, concerned with the penetration of the zona pellucida of the ovum at fertilisation and possibly with the initiation of subsequent division of the zygote.

The remaining structural components of the sperm cell are concerned with motility. The axial filament complex of the *sperm tail* originates in one of the centrioles, which forms a basal body at the rear pole of the nucleus. The cross-sectional pattern of this complex is the same as in the cilium (Plate 96a). The axial filament complex, or axoneme, runs through the *mid-piece*, the only significant cytoplasmic area of sperm. Here there are found mitochondria, which are tightly wound around the axial filament complex, providing a readily available supply of ATP for the essential function of movement. The substrates for oxidation are taken up from the surrounding environment.

In the mammalian sperm cell, nine additional broad *dense structures* reinforce the nine tubular doublets of the tail. These additional components, straight and unbranching in a longitudinal section (Plate 96b), terminate at different levels along the tail of the sperm. It is believed that they have no contractile function, but might be extra structural reinforcements for the axial complex. Further along towards the tail, the mitochondrial sheath is replaced and the cell perhaps strengthened by a system of *fibrous ribs* linked longitudinally in the plane of the central filaments. It must be admitted that many of these structural details are poorly understood and their functional significance is unknown.

10.2.3 Bacterial flagella

As described in the preceding sections of this chapter, the engines that produce movement in living things fall into two main categories. The first includes the cilia and flagella, with their microtubular architecture and *sliding-bending* mechanism. The second is the actin-myosin filament system, which produces movement by a pure *sliding mechanism*, typified by skeletal muscle. In both of these systems, from protozoa to mammals, the principle involved is essentially that of a *reciprocating engine*.

It is interesting to note that the bacterial flagellum operates on a different principle. The flagellum itself, structurally extremely simple, is a thin inert filament with no inherent motile or contractile properties. This is driven in a circular motion, like a propellor, by a rotary mechanism located just beneath the bacterial cell surface. This is the only *rotary engine* recognised in living organisms. This difference in principle emphasises the structural disparity between this simple flagellum and the highly ordered cilium and sperm tail.

Communication

Nervous tissue is specialised for the reception, storage, processing and communication of information. The basic unit is the neurone, or nerve cell, together with its specialised supporting cells. The nervous tissues of the body are divided into two broad groups, the central and the peripheral nervous system. The central nervous system consists of the brain and spinal cord, the headquarters of neural activity. The peripheral nervous system is made up of the nerves and associated ganglia which convey messages to and from the central nervous system. Communication between the cellular units of nerve tissue is usually accomplished by activity at specialised areas of contact known as synapses. This chapter will look briefly at the principal structural specialisations of nervous tissue and will then examine one of the important forms of sensory input, the system of photo-reception.

11.1 NERVOUS TISSUE

11.1.1 The neurone

The neurone has a compact cell body, which contains the nucleus and typical cytoplasmic organelles, but has elongated processes, the axon and dendrites, which extend for variable distances to make functional contacts with other neurones. The axon of a nerve cell, although often less than a micron in diameter, may be more than a metre in length.

The specialisations of neuronal cytoplasm are essentially those of a *protein-secreting* cell (Plates 97, 98a). Groups of cisternae of the granular reticulum are found in patches in the cell along with moderate numbers of attached and free ribosomes. These patches form the characteristic basophilic areas of the nerve cell cytoplasm long recognised by light microscopy as the *Nissl bodies.* The Golgi apparatus, also first described in the nerve cell, is of moderate complexity and is distributed in a network around the nucleus without obvious polarisation. The usual fine structural pattern of cross-sectioned Golgi membranes can be seen with the electron microscope. Lysosomes and mitochondria are present in moderate numbers. The significance of the protein-secreting apparatus of the nerve cell is explained by the observation that large neurones replace 30 per cent of their protein every 24 hours. Its purpose is the maintenance of the constant distal flow of axoplasm and the manufacture of transmitter substances.

The cytoplasm within the axon contains relatively few formed structures apart from occasional mitochondria and longitudinally aligned tortuous cisternae of smooth endoplasmic reticulum. There is, however, a complex longitudinally orientated system of *filaments* and *microtubules* (Plates 100, 101). The microtubules are similar to those seen in cytoplasm elsewhere, but are often called neurotubules, on account of their site. Neurofilaments of about 7 nm in diameter vary in their incidence and relative proportions; in addition, 10 nm diameter intermediate filaments are found. In the dendrite, most of the main cytoplasmic organelles may appear, along with tubules and filaments (Plate 98b).

These longitudinally orientated structures in the nerve cell processes are the fine structural basis for the light microscopic neurofibrils, which can be demonstrated with silver impregnations. Their importance is still uncertain, but it is supposed that they are related in some way to the mainte-

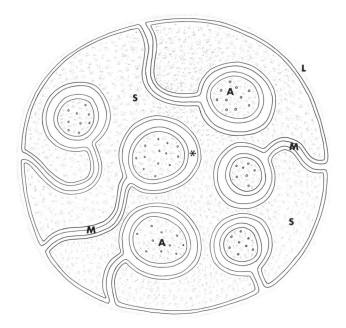

Fig. 26 *Unmyelinated nerve*
This represents a cross-section through a single Schwann cell unit, surrounded by its own external or basal lamina, L. The Schwann cell cytoplasm, S, is invaginated by six unmyelinated nerves, A, each of which is suspended within a mesaxon, M, formed by the infolding of the Schwann cell surface membrane. Each axon is thus surrounded by a narrow space, *, which is continuous with the extracellular space.

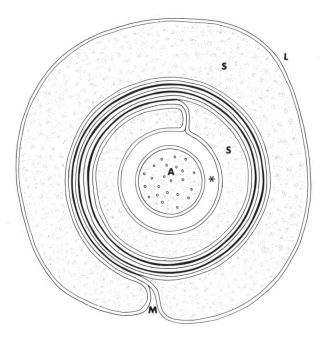

Fig. 27 *Myelinated nerve*
In this cross-section of a single Schwann cell unit, a solitary axon, A, is suspended within the Schwann cell cytoplasm, S, by the mesaxon, M. The membrane layers of the mesaxon have fused to form the characteristic periodic structure of myelin. Although the space between the axon and the Schwann cell has been exaggerated, *, its presence emphasises the separate identity of the two cells involved. There is a surrounding external or basal lamina, as in the unmyelinated nerve.

nance of tensile strength and the transportation of metabolites and, possibly, of formed cytoplasmic components. Various rates of *transport* along axons have been observed, ranging from simple axoplasmic flow, at 2 mm per day, to metabolic transfers at up to several hundred millimetres per day. The orientated structures of the axon might provide structural channels for metabolic transfer or might provide contractile or ratchet mechanisms which could supply the motive force. In this respect it is of interest that fast axonal transport can be abolished by colchicine, which depolymerises microtubules. Possibly the interconnecting cisternae of *smooth reticulum* may be involved in some aspects of metabolic transfer along axons. This interesting area of nerve cell biology is under active study at the present time.

The nerve cell has a high resting membrane potential which forms the basis of its *excitability* and which enables it to conduct impulses. Following focal depolarisation, it quickly recovers its resting potential, but the wave of spreading depolarisation triggered off by this initial localised excitation is propagated across the entire surface of the cell. The rate at which this wave of surface activity, the *nerve impulse*, passes along the axon depends mainly on the size and character of the axon. In axons of small diameter, the rate of conduction of the nerve impulse may be from one to two metres per second but in axons of large diameter it may be over 100 metres per second. In this way information, expressed as the activity of a single cellular unit in the nervous system, is conveyed without loss over long distances.

11.1.2   Nerve axons

A mixed peripheral nerve contains the axons of cells carrying impulses with motor, sensory and autonomic functions. The cell bodies lie in the spinal cord, the dorsal root ganglia and the sympathetic chain. The *axon*, commonly but imprecisely referred to as the nerve fibre, is surrounded by a cellular sheath composed of a chain of satellite *Schwann cells* lying end to end. Two groups of axons are recognised by light microscopy, myelinated and unmyelinated. The *myelinated* axons are distinguished by the presence of a fatty sheath which is absent from the *unmyelinated* axons. At the ultrastructural level, these two types of nerve axon have distinctively different relationships with their associated Schwann cells.

The *unmyelinated* axon (Plate 100a) has a simple relationship to the Schwann cell which is best seen in electron micrographs showing a transverse section of a nerve. The axon lies in a tunnel within the Schwann cell (Figs. 26, 28). It appears to reach this position by invaginating the surface of the Schwann cell, carrying with it a *mesaxon* of Schwann cell surface membrane. Although the axon is surrounded by the Schwann cell, the surface membrane of each cell is intact, so that there is no cytoplasmic continuity between them. In a typical unmyelinated nerve, several axons are carried in this way within each Schwann cell tube, each suspended by its own mesaxon composed of two layers of Schwann cell membrane. Since the Schwann cell is much shorter than the axon it surrounds, each axon is enclosed by a succession of satellite Schwann cells along its length. The axon remains shielded by these cells to its finest branches, being naked only when approaching the nerve terminal. The Schwann cell tube itself is surrounded by a typical *external* or *basal lamina*, which separates it from the surrounding stroma.

The *myelin sheath* is recognised in electron microscopy by its marked affinity for osmium. Myelin consists of a closely packed spiral of regular lipoprotein lamellae (Figs. 27, 28). Peripheral nerve myelin displays, in section, a major periodic spacing of 12 nm and an intermediate minor spacing (Plate 101). Evidence from other techniques such as X-ray diffraction, however, indicates that the periodicity of myelin in vivo may be nearer 18 nm. During the development of a myelinated nerve, the axon concerned, initially unmyelinated, is the only occupant of its Schwann cell sheath (Plate 100b). The myelin is formed by growth in length of the mesaxon, the surplus membrane material becoming wrapped round and round the axon. The resulting close-packed double layers of Schwann cell surface membrane fuse tightly together to produce the periodic myelin pattern. As the myelin sheath gains thickness, the cytoplasmic components of the Schwann cell are pushed peripherally to form a thin rim on the

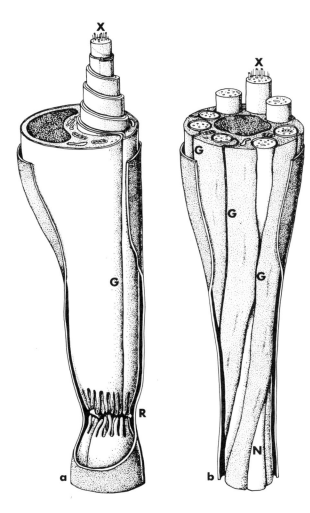

Fig. 28 *Myelinated and unmyelinated nerve fibres*
These three-dimensional representations show the relationships between the nerve axons and the surrounding Schwann cell. In diagram a, a single axon is shown with the surrounding spiral of myelin cut away to demonstrate its formation from the Schwann cell surface. The tubules and filaments running longitudinally within the axon are shown projecting from its cut end, X. The longitudinal groove, G, marks the origin of the mesaxon at the Schwann cell surface. Where one Schwann cell stops and the next begins is the area of specialisation known as the node of Ranvier, R. At this point the myelin sheath terminates and recommences, leaving a small segment of the axon exposed. Diagram b shows the corresponding features of the unmyelinated nerve. Multiple axons are shown, similarly cross-sectioned, X. In this case, there are multiple longitudinal grooves, G, indicating the surface origins of the mesaxons. On occasion, a nerve axon, N, is not completely buried within the Schwann cell, but lies exposed at its surface. In both diagrams, the entire Schwann cell tube is shown enveloped by its surrounding external or basal lamina, which has been cut away to reveal the underlying detail.

Diagram by courtesy of R. V. Krstić

outside of the myelin, where the Schwann cell nucleus can also be found. Myelination is apparently an active cellular process, involving a dynamic relationship between the sheath cells and the axon. The fully formed myelin sheath is still part of the living satellite cells, structurally distinct from the axon which it encloses. The pathological process of *demyelination* initially affects the Schwann cells rather than the axon, which remains intact although it may eventually suffer secondary damage.

Since each Schwann cell encloses only part of the length of the axon, the myelin sheath is the product of many Schwann cells, lying end to end. The nodes of Ranvier are the junctions between adjacent segments of myelin, marking the limits of successive Schwann cells (Figs. 28, 29). At the node, the axon is virtually unshielded apart from the basal lamina of the Schwann cell. The axon thus comes into direct contact with the tissue fluid, while at other points it is insulated by the myelin lamellae.

In terms of function, the myelin sheath is a vital specialisation. Myelinated axons, normally of large diameter, conduct the nerve impulse much more rapidly than unmyelinated ones, since the nerve impulse jumps from node to node at the points where the axon is exposed. In general, the Schwann cells provide an essentially unbroken sheath for all nerve cell processes in the peripheral nervous system. Apart from their insulating function, it is likely that they have an important metabolic role in relation to the neurones; presumably the distance between the axon and its cell body makes necessary this additional auxiliary life support system.

11.1.3 The synapse

The point of functional contact between two nerve cells or between a nerve and some other excitable cell type is called the *synapse*. The function of the synapse is to transfer neuronal activity from cell to cell, making possible the data-processing functions of the nervous system. Each neurone has synaptic contacts with a great many others, forming a complex network with countless possible functional circuits. These contacts, in addition, may be either excitatory or inhibitory in their effect.

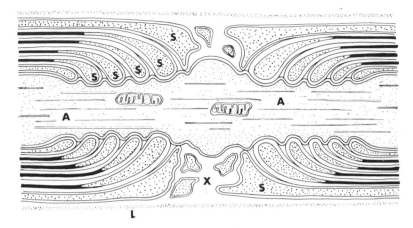

Fig. 29 *Node of Ranvier*
The myelinated nerve axon, A, is continuous through the node. The myelin sheath, however, terminates on either side, since the node represents the point at which two Schwann cells meet. A narrow gap, X, exists into which small tongues of Schwann cell cytoplasm project. Here the axon is bathed in extracellular fluid, whereas elsewhere it is protected by the Schwann cell and its myelin sheath. The areas of Schwann cell cytoplasm, S, have a distinctive pattern produced by the termination of successive layers of the myelin sheath as the node is approached. Note that the continuous external or basal lamina, L, surrounds the entire structure.

The synapse can be regarded as a special type of contact specialisation (Plates 98b, 102). Various morphologically and functionally distinct types of synapse are now recognised, but they share a number of common structural features (Fig. 30). Since the nerve impulse passes normally in only one direction along a multi-neuronal pathway, the contacting parts of adjacent neurones which comprise the synapse can usually be identified either as *pre-synaptic* or *post-synaptic* components. The narrow intercellular space is termed the *synaptic cleft* or *gap*. There is no cytoplasmic continuity between the cells at this point.

The pre-synaptic element usually forms a small expansion in which there is often found a cluster of mitochondria, suggesting localised metabolic activity. Close to the pre-synaptic membrane lies a group of small membrane-bound *synaptic vesicles* each about 40 nm in diameter. In the motor end-plate, a simple synapse between nerve and skeletal muscle (Fig. 31, Plate 103), these apparently empty vesicles contain acetylcholine, although elsewhere in the nervous system various other transmitters may occur in similar vesicles. In some synapses, dense-cored vesicles occur (Plate 102b). These may contain a catecholamine or a polypeptide transmitter substance. The inter-cellular space or synaptic cleft measures 15 to 30 nm in width. In the commonest types of synapse there is generally

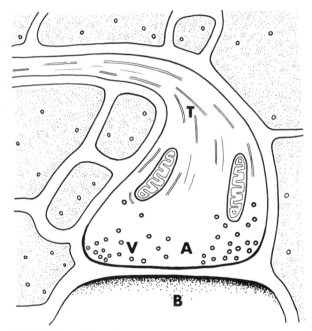

Fig. 30 *Synaptic specialisations*
This diagram shows the pre-synaptic component of a synapse, A, separated by a narrow gap from the post-synaptic component, B. The pre-synaptic terminal contains numerous small synaptic vesicles, V, in which transmitter substance is stored. Both pre- and post-synaptic membranes show localised specialisation. Longitudinally sectioned neurotubules, T, are shown in the nerve axon approaching the terminal. When a nerve impulse reaches the terminal, the synaptic vesicles discharge their contents into the intercellular space, leading to depolarisation of the post-synaptic membrane. In this way the nerve impulse is passed from cell to cell.

an accumulation of dense material on the cytoplasmic side of both the post-synaptic and the pre-synaptic membrane. Type I synapses, usually *excitatory*, have a more marked post-synaptic thickening and a wider gap than type II synapses, which are generally *inhibitory* in function. A further distinction lies in the morphology of the synaptic vesicles, usually round in excitatory synapses and flattened in inhibitory ones.

The synaptic vesicles contain transmitter substance which is released into the synaptic cleft when a nerve impulse reaches the pre-synaptic terminal. The transmitter produces localised depolarisation of the post-synaptic membrane which may be sufficient to trigger off a nerve impulse. After its release, the transmitter is either rapidly *destroyed* by the action of an enzyme, such as acetyl cholinesterase, or is *taken up* again by the nerve terminal, as may happen with noradrenalin.

The neuro-muscular junction, or motor end plate, has similar characteristics. Acetylcholine is stored within synaptic vesicles in the terminal branches of the motor nerve (Plate 103). The surface of the muscle cell in contact with the nerve terminals is extensively infolded, maximising the area exposed to the released transmitter substance (Fig. 31). Cholinesterase is located at this elaborate interface.

The ability of the synapse to transmit the nerve impulse in only one direction is due to the structural and functional distinction between the pre-synaptic component, which releases the stored transmitter, and the post-synaptic component, which responds to this transmitter. The characteristic slight *delay* in transmission of the nerve impulse across the synapse is accounted for by the time taken for the discharge of the synaptic vesicles, the diffusion of transmitter across the intercellular cleft and the depolarisation of the post-synaptic terminal. Drugs which *block* synaptic transmission are now important in medicine. In general, such drugs take advantage of the fact that the extracellular portion of the nerve pathway is more accessible to the pharmacologist than the

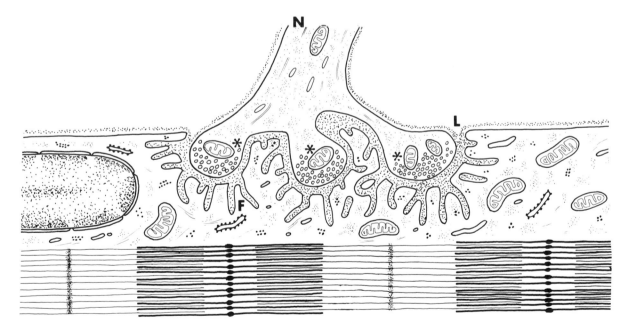

Fig. 31 *Motor end-plate*
This diagram shows the specialised junction between a motor nerve, N, and its muscle cell. The nerve terminals, *, are embedded in grooves on the muscle cell surface. The gap between the muscle and the nerve is filled by material in continuity with the external lamina of the muscle cell and the nerve cell. Complicated infolding of the muscle cell membrane, F, is characteristic of this site. Each active region of the nerve terminal, *, contains numerous synaptic vesicles and small groups of mitochondria. When a nerve impulse reaches the nerve terminal the synaptic vesicles are discharged, leading to depolarisation of the muscle cell membrane at this complex infolded region. The transmitter is then rapidly inactivated by cholinesterase located at this interface.

complex intracellular mechanisms involved in the propagation of the nerve impulse.

This pharmacological approach is invalidated in another class of synapse. This relies not on *chemical* transmission across a synaptic gap, but on *electrical transmission*, through ionic flow across a gap junction. These electrical synapses are rare in mammals, but occur more commonly in lower species. The electrical synapse has no delay in transmission comparable to that of the chemical synapse and is not accessible to extracellular pharmacological blocking.

11.1.4　Neuroglia

The neuroglial cells are the *supporting* cells of the central nervous system, analogous to the Schwann cells of the peripheral nerve. Three main cell groups are described, the oligodendrocytes, the microglia and the astrocytes. The *oligodendrocytes* are most numerous in white matter. They are involved in the formation of *myelin* sheaths in the central nervous system and are thus the direct Schwann cell equivalent. The oligodendrocytes have a patchy arrangement of nuclear chromatin and the cytoplasm contains free ribosomes, a moderate amount of granular endoplasmic reticulum, some lysosomes and occasional multivesicular bodies. A single oligodendrocyte may have several processes contacting different axons in the brain, forming for each a separate myelin sheath. In the central nervous system the periodicity of the myelin is approximately 10 per cent smaller than in peripheral nerves, in keeping with the observation that the two types of myelin are immunologically distinct.

The *microglial* cells are generally believed to be the phagocytic cells of the central nervous system and are said to be of mesodermal origin, in contrast with the other cell types of the brain which originate from the ectodermal neural tube. Microglial cells are identified as small cells with dense cytoplasm, but it has been suggested that these may simply be variants of oligodendrocytes. Since many of the phagocytic cells seen in brain damage have now been shown to originate from the blood monocytes or from blood vessels rather than from within the glial cell population, there is uncertainty regarding the nature of the microglia.

The usual connective tissue elements seen elsewhere, such as fibroblasts and collagen fibres, are excluded from brain tissue. The *astrocytes* instead are the main supporting cell type in the central nervous system. They have a round pale nucleus with an even chromatin pattern. The astrocyte cytoplasm is pale (Plates 97, 98b, 99), with scattered ribosomes and little granular endoplasmic reticulum. The Golgi apparatus is of moderate size and there are scattered glycogen particles and bundles of fine filaments, 6 to 10 nm in diameter. Astrocytes have relatively little cytoplasm around the nucleus but have an extensive stellate mass of peripheral processes from which the name of the cell is derived. The processes of different astrocytes make contact with one another at areas which may display junctional specialisations suggesting the possibility of communication between these cells. The astrocyte cytoplasm forms an interconnecting *cellular network* of great complexity and delicacy, with numerous thin lamellar extensions which penetrate and often surround groups of neuronal processes. Small astrocyte processes often flank areas of synaptic contact between neurones and may at times form *complete sheaths* around synaptic complexes.

The absence of connective tissue and the close packing of cell processes in the brain limit the *extracellular space* to clefts of little more than 20 nm in width. Although the total *volume* of this extracellular space is small, there is an enormous intercellular contact *surface* area, which may well have important implications with respect to brain functions. Astrocyte surfaces form a large part of this physiologically important interface.

The astrocytes have a distinctive relationship with the thick-walled, non-fenestrated cerebral *capillaries* (Plate 99). Astrocyte *processes* are applied to the external surface of the capillary basal lamina, forming a virtually complete *glial sheath*. The gaps between the components of the perivascular glial sheath account for less than 1 per cent of the total capillary surface area. While the phenomenon of the *blood-brain barrier* can now be explained in part by special properties of the capillary, this wall of astrocyte cytoplasm between vessels and neurones must also influence metabolic exchange between blood and brain.

Relative to its cytoplasmic volume, the astrocyte

has perhaps the greatest *surface area* of any cell in the body. Its fine lamellar processes control up to 60 per cent of the network of intercellular gaps in the grey matter and the close relationship which it bears to synapses suggests a possible metabolic involvement in neuronal function. Astrocytes may *insulate* the nerve cells and *isolate* synapses from one another. In addition they may *regulate* the extent and content of the narrow intercellular diffusion space in the brain, through selective uptake and volume changes. When their strategic position in nervous tissue is considered, it is difficult to believe that the astrocytes are limited to a purely mechanical supporting role.

In the study of the central nervous system the techniques of electron microscopy have both advantages and disadvantages. The improved resolution emphasises the complexity of brain tissue, while the essentially two-dimensional nature of thin sections limits the display of three-dimensional organisation. New patterns, however, are beginning to emerge, which will in time lead to a fuller understanding of the functioning of the brain.

11.2 PHOTORECEPTORS

Light is so familiar a component of the environment that its importance tends to be overlooked. Nevertheless, the whole of life as it has evolved on the earth depends in different ways upon the energy of light. In the *chloroplasts* of green plants the reactions of *photosynthesis* take place, forming the essential first link in the chain of life on which the more complex animals depend for their entire nutrition. Animals use light to provide *information* to assist their movement. The evolution of *photoreceptors*, the eyes, has made it possible for the nervous system to become more fully informed about the environment. Although the chloroplast and the eye use the energy of light for totally different purposes, the components concerned with the trapping and transformation of light energy show ultrastructural specialisations of a surprisingly similar nature.

The *chloroplasts* are discrete membrane-limited structures lying within the cytoplasm of plant cells. They contain the green pigment *chlorophyll*. The internal structure of the chloroplast varies greatly in different species of green plant, but their essential common feature is the presence of parallel *membrane lamellae* associated structurally with the chlorophyll molecules. In the higher plants the lamellae of the chloroplast form a number of discrete specialised packages termed *grana*, in which the chlorophyll is localised (Plate 104a). The intervening *stroma* contains fewer membranes. These run between the grana and link them together. Between these components homogeneous pale starch granules may be found. Chloroplasts, like mitochondria, appear to have a certain degree of independence within the cell, since they contain their own genetic material in the form of DNA.

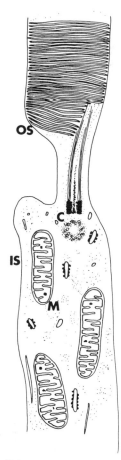

Fig. 32 *Retinal rod photoreceptor*
The outer segment, OS, and the inner segment, IS, of the photoreceptor are connected by a narrow portion which is, essentially, a cilium. The outer segment is in fact a derivative of the cilium. This originates in a centriole, C, lying at the end of the inner segment. The closely packed membrane lamellae which fill the outer segment represent the location of the visual pigment, where light is trapped and transduced.

Photosynthesis can take place only in the presence of chlorophyll, associated with the membrane lamellae of the chloroplast, suggesting that the enzymes concerned may be spatially organised on the complex lamellar template. The reactions of photosynthesis use the radiant energy of light to promote the combination of the simple molecules of water and carbon dioxide, resulting in the synthesis of carbohydrate and the release of oxygen. Without this reaction, life as we know it would cease to exist.

The *photoreceptor cells* of the eyes of animals far apart in the evolutionary scale show essential similarities of fine structural specialisation. In all of them, the light-trapping cells contain numerous membrane lamellae or tubules showing a high degree of spatial organisation, the membranes being associated closely with molecules of the photosensitive pigment rhodopsin. The similarity in principle to the chloroplast is clear.

In the vertebrate eye (Fig. 32) the photoreceptor cells are the rods and cones, each of which has an inner and an outer segment joined by a narrow bridge. The *inner segment* of the photoreceptor cell contains its main cytoplasmic components, including significant numbers of mitochondria. This part of the cell forms a synapse with the first neural link in the visual pathway. The *bridge* between the inner and outer segment is structurally similar to a cilium, although lacking the central doublet (Plate 104c). It is no doubt involved in the functional communication between the two parts of the photoreceptor. In the *outer segment*, a highly evolved derivative of the cilium, closely packed parallel membrane lamellae stacked like a pile of coins fill all the available space (Plate 104b). Integrated with these lamellae are the rhodopsin molecules, which consist of the pigment retinene, closely related to vitamin A, in association with the protein component opsin. It is here that the radiant energy of light is trapped through the reversible dissociation of rhodopsin and transduced or converted into cellular activity. This leads to the initiation and propagation of a sensory nerve impulse. The efficiency of the light-trapping mechanism embodied in the retinal photoreceptor is so great that in some cases stimuli amounting to the simultaneous reception of only a few photons are sufficient to generate a visual sensation.

Unsolved problems

12.1 TRANSMISSION MICROSCOPY

In an introductory text such as this, where the emphasis must lie on the simple presentation of basic material, it is all too easy to give the impression that there are few problems in ultrastructure that are not already 'cut and dried'! This is not so, despite the advances already made. There are many ultrastructural findings which cannot be explained in functional terms and there are many important aspects of cell function for which no related structural specialisation can be identified. It is these unsolved problems that provide the electron microscopist with the stimulus for further research.

The *cell organelles* still present many mysteries. For example, the complexities of the *nuclear pore* apparatus must relate in some way to the regulation of nucleo-cytoplasmic exchanges. The nature of this regulatory function and the significance of the pore structure in relation to its function remain obscure. Likewise, the *Golgi apparatus* is now so familiar that we often overlook our ignorance of its basic functions. Its special role in secretory cells is understandable enough, even with our limited grasp of the enzymic processes involved, but the presence of the organelle in almost every type of cell implies some role which is more basic to cell functions than the packaging of secretion. The *annulate lamellae* provide a third example of a highly organised structural entity, the function of which is not understood. Careful study of other aspects of cell structure will reveal similar areas of ignorance on many such matters.

When we consider *specialised cells*, we can again identify many unsolved problems. For example, the astonishing structural variation in the mor-

phology of *spermatozoa* from species to species illustrates morphological specialisation which baffles rationalisation. Presumably the many differences in size, shape, mitochondrial numbers and arrangement, acrosome morphology and other details have some evolutionary significance and can in some way be correlated with function, but so far such explanations have not been worked out.

There are two interesting cell types from the intestine which illustrate this point. The first of these is the *Paneth cell*, once generally supposed to secrete digestive enzymes, for which its ultrastructure is certainly appropriate. Consistently, however, doubts have been cast on this suggestion and no digestive enzyme could be positively identified in the distinctive granules. Recent evidence now suggests that the protein secretion of this cell is lysozyme, a natural antibacterial agent also found in tears. The Paneth cell, in some species at least, also seems capable of phagocytosis of microorganisms in the intestinal crypt. Could this cell play some part in maintaining the environmental hygiene of the crypt? The role of the Paneth cell cannot be explained by ultrastructural analysis alone, but electron microscopic data are contributing to the better understanding of the problem.

The second mystery cell from the intestine is known as the *tuft cell*. This cell type is found also in the mucosal surfaces of the stomach, colon, airways, biliary and urinary tract. It bears a prominent apical tuft of microvilli longer than those of adjacent cells. The filamentous cores are particularly well defined, extending deep into the apical cytoplasm. The cell apex shows evidence of micropinocytosis. There is no apparent secretory product. As yet, no firm evidence exists as to the function of these distinctive cells, whose identity as a class was only formally recognised within the last ten years. One attractive suggestion proposes a sensory or receptor function with respect to the luminal contents. What it might be sensing and what its response might be remain completely unknown.

One can point to many other examples of differentiated cell types with distinctive ultrastructural characteristics for which there remains no clear functional correlation. What, for example, do the characteristic *endothelial granules* contain? What is the significance of the intimate association between the epithelial cells of the *thymus* and their neighbouring lymphocytes? In skin, electron microscopy shows occasional cells, the *Merkel cells*, with all of the ultrastructural characteristics of APUD cells, such as the typical small dense endocrine granules. What role could such an endocrine cell have in the skin? The epidermal *Langerhans cell* is now believed to be related to the macrophage and to play some part in the immune responses of the skin. This cell has distinctive granules (Plate 105) whose function is totally unknown, but whose appearance is sufficiently characteristic to be used as a diagnostic marker for certain types of disease.

It is perhaps less surprising that certain distinctive biological functions should have no clearly recognisable structural correlates. The present inability to relate *chromatin morphology* to cell function and to the details of heredity suggests only that we are using the wrong technique. That problem is perhaps better solved by the techniques of the molecular geneticist than by those of the electron microscopist. Another example is provided by those cells which generate rhythmic activity, acting, in a sense, as *biological clocks*. No structural explanation for such behaviour can be brought forward.

12.2   SCANNING MICROSCOPY

So far, all of the unsolved problems mentioned are those raised by transmission electron microscopy of thin sectioned material, but there are parallel problems to be found in surface studies of cells and tissues by scanning electron microscopy. These relate mainly to unexplained structural surface features, although equally there are aspects of cell function which cannot yet be correlated with surface data.

Unexplained structural surface features include variations in normal mucosal patterns, seen for example in *rectal mucosa*. Variations in the patterns of furrows and crypts, all thought of as normal, may in time prove to have functional significance related perhaps to diurnal rhythms, ageing processes, or developing disease. It is only by continuing research that such correlations can be made. The scanning electron microscopic view of *connective tissue fibres* and their relationships may in time be seen to have more significance than is at present recognised. There are many questions still to be asked about the significance of the surface topography of cells and tissues.

This small selection of unsolved problems may serve to underline the *limitations* of electron microscopy. In general, however, the fine structure of cells provides a reliable and practical guide to their identity and function. The cell biologist and the anatomist can work together to relate these findings to the molecular physiology and biochemistry of the cell, while the pathologist can put this understanding to practical use in the more precise diagnosis of disease, as described in the next chapter.

APPLICATIONS AND INSTRUMENTATION

This final section of the text presents the authors' personal view of some practical applications of electron microscopy in medical diagnosis and research. It concludes with a brief outline of the technology and apparatus involved and a simple guide for the beginner to the interpretation of the electron micrograph. This is not a detailed practical handbook for the would-be specialist in electron microscopy, but an informal guide to the technique for beginners and for interested students who might wish to understand how such pictures are produced.

The electron microscope in the study of disease

The study of *disease* in a scientific laboratory is known as *pathology*. The pathologist may specialise in different directions. Medical microbiologists and virologists are concerned with the infecting organisms which can cause illness; haematologists deal with disease of the blood; chemical pathologists deal with the biochemical changes of disease; clinical immunologists study the immune mechanisms in relation to disease; morbid anatomists and histopathologists investigate the structural changes of disease. These broad subdivisions of pathology are flexible enough to accommodate even wider interests in medical science. For example, the pathologist may be involved in histochemistry, or in radioimmunoassay, or perhaps may have an interest in the emerging field of chronobiology, which takes account of time as a relevant dimension of living systems.

Whatever his speciality, the hospital pathologist has two main functions. The first is to provide a *service for patients* under medical care, through the diagnosis of disease or the assessment of the progress of treatment. The chemical pathologist, haematologist, microbiologist and immunologist examine various kinds of specimen including blood, urine, faeces or sputum. The histopathologist receives tissues removed at operation and identifies disease processes by histological examination. These laboratory services form the essential foundation on which modern patient care is built. The second function of the hospital pathologist is to advance the *understanding of disease*. Progress in diagnosis and treatment depends on the success of research, not only in the ward but also in the scientific laboratory.

However varied may be the duties and interests of the different specialists within the disciplines of pathology, the electron microscope can be of use to them all. Its value in research has already been touched upon in the earlier chapters of this book, but its usefulness is not confined to research. There are now many *practical uses* for electron microscopy is the diagnosis of disease.

13.1 ELECTRON MICROSCOPY IN THE CLINICAL LABORATORY

Perhaps the earliest contribution was in the field of *virology*. There is virtually no virus large enough to be clearly resolved by the light microscope. At first the existence of these organisms had to be inferred from their behaviour, rather than demonstrated in structural terms. Viruses caused disease, passed through filters, damaged cells in culture and induced immune responses, but could not actually be seen. X-ray diffraction studies showed that the main structure of a virus consisted of regularly stacked protein subunits formed into patterns often with either helical or cubic symmetry. Ordered stacking such as this allows the complex structure of the virus particle to be built from simple subunits, just as complex architectural details can be achieved by the organised stacking of simple, standard bricks. In this way the limited coding ability of the viral nucleic acid is put to best structural use.

The development of the techniques of *negative staining* (Section 14.6.1) made possible the study with the electron microscope of the structure of virus particles, allowing virus identification and structural classification. Many viruses (Plates 116d, 116e), were found to have either a rod-shaped or helical form, corresponding to helical symmetry, or an icosahedral form, corresponding to cubic symmetry. Whole families of viruses have now been identified by a combination of morphological and immunological testing.

This academic groundwork led quickly to the use of the electron microscope in the diagnosis of *viral disease.* One clinical situation in which this technique was once used lay in the differential diagnosis of lesions resembling smallpox. Here, the importance of rapidly establishing an accurate diagnosis was so great that the delays of tissue culture or immunological virus typing were unacceptable. Direct examination by electron microscopy of fluid from a blister can allow the virologist to identify the infection within minutes as either smallpox, chickenpox or herpes, but since smallpox is now officially extinct, this skill is no longer of immediate practical importance. Less dramatic, but also of clinical use, is the identification of the pustular virus infection known as *orf* (Plate 106c), which affects sheep handlers and for which immunological testing is ineffectual. In encephalitis, or inflammation of the brain, the recognition of virus particles in thin tissue sections may be of value. In *herpes simplex encephalitis,* the virus particles are discrete round structures containing a central nucleoid. Similar viruses (Plate 106a) are seen in *cytomegalic disease,* a common infection in childhood. In *subacute sclerosing panencephalitis,* the finding of the twisted threads of measles virus helps to reinforce the clinical and laboratory diagnosis. Current interest centres on new families of viruses recently found to be associated with *gastroenteritis* in infants and adults. Electron microscopy has been essential in recognising cases such as these, by examination of samples of faeces.

For the immunologist, the bacteriologist and the biochemist, electron microscopy is still a research tool rather than a diagnostic aid. Negative staining has permitted the visualisation of individual macromolecules such as the *immunoglobulins,* providing pictorial confirmation of their molecular shape, as predicted by other studies. The haematologist is now increasingly interested in electron microscopy, particularly in the study of the *leukaemias.* The ultrastructural details of the leukaemic cells can sometimes help to the identification of the type of leukaemia, which may in turn help to determine the most effective treatment and to predict the outcome more precisely. Of particular interest to the haematologist is the technique of scanning electron microscopy (Section 14.3.2), which can be used to study the changes in *red cell shape* in various abnormalities of the peripheral blood.

13.2   ELECTRON MICROSCOPY IN HISTOPATHOLOGY

Until now, diagnostic service *histopathology* has relied almost exclusively on the light microscope, for valid historical reasons. A century ago the foundations of histopathology were laid through the recognition of specific patterns of disease under the light microscope. Not surprisingly, this is the tool which has since dominated both the research and the diagnostic functions of the histopathologist. Today, however, less than thirty years after the first effective use of the electron microscope in tissue studies, its potential in diagnosis is becoming apparent. This is particularly the case in the field of *oncology,* the study of *tumours.*

13.2.1   Tumour diagnosis

There are two main ways in which ultrastructure can serve in the histopathological diagnosis of tumours. Its first and most obvious role is in the *assessment of the histogenesis,* or tissue of origin of a malignant tumour. In histological terms, one can often recognise, in the pattern of the cells or in their individual features of differentiation, some histological echo of the parent cell type. Hence a cancer arising in squamous epithelium will often show obvious histological keratinisation: a glandular mucosa on the other hand may give rise to an adenocarcinoma, a cancer with a well-marked glandular pattern.

In the same way, many tumours betray their differentiation patterns and their origins in their fine structure. The *squamous carcinoma* retains many of the ultrastructural features of the normal *keratinocyte,* such as prominent desmosomes and cytoplasmic tonofilaments. The *adenocarcinoma* retains the ultrastructural surface specialisations typical of a *gland* lumen, while exocrine mucus secretory activity may often be seen. The malignant *melanoma* retains the *premelanosomes* which characterise its parent cell, the normal melanocyte, even though there is no pigment to be seen under the light microscope (Plate 107a, 107b).

Many endocrine tumours retain the structural and functional characteristics of their cell of origin. Among these are the *islet cell tumours* of the pancreas. Those originating from the beta cell may secrete insulin in excess, leading to hypoglycaemic attacks, while others may produce gastrin, which leads to gastric hypersecretion and severe duodenal ulceration. The structural differences between the parent cells are sometimes repeated in those different tumours. In the well-differentiated *insulinoma,* for example, the distinctive crystalline granules typical of the normal beta cell may be seen (Plates 108d, 42). Carcinoid tumours of the intestine, phaeochromocytomas of the adrenal medulla and parathyroid tumours may mimic the normal endocrine gland cells in fine structure. In certain cases of *lung cancer* the cells, despite their bronchial origin, are able to produce inappropriate hormones such as ACTH. In these cases, however, the cells contain small dense endocrine secretion granules resembling those seen in the pituitary gland and other APUD cell types. The origin of these small cell cancers of the lung is thought to be the endocrine cell population of the respiratory mucosa, which resembles the intestinal endocrine cell population in the gut.

In a small but important group of cases, histological examination fails to provide a basis for tumour classification, usually because the tumour cells have deviated too far from the characteristic differentiation pattern of their parent cell. Electron microscopy may help in such problems. For example, some of the endocrine tumours mentioned above may be unrecognisable by histological examination, but may retain their distinctive ultrastructural characteristics (Plate 108a, 108b). A very undifferentiated squamous carcinoma, histologically quite atypical, may still retain desmosomes and tonofilaments in a characteristic ultrastructural pattern (Plate 107c). It is sometimes difficult to distinguish histologically between a poorly differentiated lymphoma, a tumour of lymphoid cells, and a poorly differentiated carcinoma: in this case, the persistence of desmosomes between the tumour cells suggests an epithelial

origin. The rhabdomyosarcoma, a malignant tumour which differentiates towards skeletal muscle, is often very poorly characterised in histological terms, but may sometimes still be identified on electron microscopy by the presence of small interdigitating bundles of distinctive thick and thin myofilaments.

The malignant *spindle-cell tumour* may also resist accurate histological diagnosis. This histological pattern is quite common in various different types of tumour, which can as a result be impossible to distinguish from one another by light microscopy alone. Electron microscopy, however, may still show the distinctive ultrastructural features of a fibroblast, a smooth muscle cell, a squamous cell, an endocrine cell, a melanocyte (Plates 107a, 107b), a Schwann cell or a meningeal cell (Plate 108e). In cases like this a more precise identification can often be made from ultrastructural findings that is possible with histology alone.

The second potential application of electron microscopy to tumour diagnosis involves the systematic study of common tumours, with the aim of uncovering previously unsuspected fine structural differences. In this way, groups of tumours at present looked on as homogenous may be further subdivided and *reclassified* on the basis of ultrastructural criteria. Reclassification of this kind might give a new basis for predicting behaviour or for suggesting particular forms of treatment. Ultrastructural studies might help to explain those cases in which the clinical outcome following histological diagnosis turns out to be unexpectedly favourable or unfavourable.

Thyroid cancer provides a histological analogy. The three main groups of thyroid cancer traditionally identified by light microscopy were follicular carcinoma, papillary carcinoma and anaplastic or undifferentiated carcinoma. The outlook for the third group was considered to be bad. In time, however, a distinctive sub-group was recognised by certain specific histological features, such as the presence in the stroma of an amorphous eosinophilic substance known as amyloid. These cases, identified now as a new entity, *medullary carcinoma,* were found to have unusually good survival figures. An advance in diagnosis was thus made by relating particular histological features to a distinctive pattern of clinical behaviour.

In the future, similar advances may well be made by correlating newly defined ultrastructural features with clinical data. Curiously enough, the same example would serve equally well to illustrate the principle. In medullary carcinoma of the thyroid, the tumour cells, although histologically nondescript, have the ultrastructural characteristics consistent with their origin from the thyrocalcitonin-secreting *C cells* of the thyroid (Plate 108c). These features include the presence of the typical small dense endocrine granules seen in the C cell, but absent from thyroid follicular cells. Ultrastructural examination is now recognised as a reliable method of distinguishing between medullary and anaplastic thyroid carcinoma.

In summary, the value of ultrastructural pathology in the scientific study of cancer is well established. The electron microscopist can help, at times, to reach a more precise diagnosis in problem cases when histopathological methods have been exhausted. As our understanding of malignancy improves, the choice of treatment will widen through the introduction of new anti-cancer drugs. It will then become even more important to define the origins, differentiation patterns and behaviour of malignant tumours. It may well be in this direction that the main clinical applications of electron microscopy will lie in the future.

13.2.2   Ultrastructural features of other diseases

There are other well-recognised applications of electron microscopy in pathology. In the field of *renal disease,* electron microscopy has become virtually routine in all renal biopsies (Plates 109, 110). The histological interpretation of glomerular disease has always been restricted by the limited resolution of light microscopy, but now our knowledge of glomerular ultrastructure allows a new approach, based on the recognition of ultrastructural abnormalities in each component of the filtration interface. We can recognise abnormalities of *podocyte foot processes,* of the *basal lamina,* of the *endothelial cells* and of the *mesangium. Dense deposits* in and around the filtration basement membrane may have distinctive features that can help in reaching a diagnosis. Correlations can be made between these findings and the identification by

immunofluorescence of immunoglobulins, immune complexes, complement and fibrin. Abnormal relationships of *mesangial cells* may be recognised in some types of glomerulonephritis. Electron microscopy can help also in assessing the extent of ultrastructural glomerular damage in the so-called *minimal lesion,* characterised by proteinuria without any light microscopic glomerular abnormality (Plate 109a).

Other applications of electron microscopy in histopathology include the identification of the abnormal glycoprotein, *amyloid* (Plate 112b). This substance is stained by certain coloured dyes, but such techniques are not wholly dependable. Electron microscopy allows the reliable identification of a distinctive fine fibrillar pattern in amyloid. In the rare diseases of skeletal muscle, various distinctive ultrastructural features can help the pathologist to classify the lesions. In the study of the various inborn errors of metabolism, particularly the *storage diseases,* electron microscopy often provides an elegant demonstration of the nature of the cellular abnormality. Stored metabolites, such as glycogen or complex lipids, present very distinctive appearances on electron microscopy (Plate 111, 112a). *Whipple's* disease is a rare abnormality involving the intestinal mucosa, in which large foamy macrophages accumulate. Electron microscopic examination has shown that numerous small bacteria are present in and around these cells, but that the macrophages seem unable to deal with them effectively (Plate 112c).

Abnormalities of the highly specialised *intestinal* epithelial cell may cause impairment of absorption of foodstuffs. If a single enzyme such as lactase is absent from the surface membrane of the microvilli there is a failure to digest and absorb lactose from the diet. This can lead to diarrhoea and serious malnutrition in infancy, although the absence of the single enzyme may produce little significant ultrastructural change in the epithelial cells. In contrast, certain individuals are sensitive to dietary proteins, most commonly to the wheat protein, *gluten.* This damages the intestinal epithelium, causing severe histological and fine structural abnormalities such as flattening of the villous pattern, with disorganisation and shortening of the microvilli (Plate 113b). This abnormality, often called *idiopathic steatorrhoea* in adults and *coeliac disease* when it occurs in children, is characterised by a malabsorption syndrome, in which all of the digestive and absorptive activities of the intestine are impaired. Mucosal surface abnormalities such as these are particularly well studied by scanning electron microscopy, which is becoming increasingly used in the investigation of human disease (Plate 114).

13.2.3 Limitations of ultrastructural diagnosis

Light and electron microscopy can solve different kinds of diagnostic problem. For the histopathologist, the *strength* of electron microscopy lies in its ability to reveal structural information beyond the reach of light microscopy. Its *weakness* lies in the unavoidable limitations placed on the area examined and on the sampling of a lesion.

The diagnostic pathologist relies heavily on the lowest magnifications of the *light microscope* for the assessment of the overall pattern of tissue architecture. This is particularly important, for example, in some diseases of pancreas, breast and lymph nodes, where abnormalities of individual cells seen at higher power may wrongly suggest a sinister character in a benign disease. Wide areas of tissue can be quickly examined to pick out localised abnormalities. For example, the outlook in many kinds of cancer depends on the extent of secondary spread, or metastasis. The use of low magnification allows the pathologist to examine quickly many different sections, in order to pick out small deposits of tumour cells in lymph nodes.

In situations such as these *electron microscopy* is worse than useless. Tissue architecture cannot be assessed from an electron microscope grid. Focal abnormalities will be unlikely to appear in the small sample taken for electron microscopy — five tiny blocks may represent the entire ultrastructural sample of a 5 kg tumour mass. Limitations on time available may demand that a case be assessed on the basis of micrographs from a couple of dozen representative cells. It is these limitations which will ensure that the discipline of histopathology remains rooted in histological technique, whatever additional information may be contributed by ultrastructural examination. Finally, even in cases which are apparently suitable for electron microscopy, there will remain unresolved problems. There are, for example, various tumours which give no ultrastructural evidence of any specific differentiation pattern. Such cases show that every technique has a limiting *threshold* beyond which it cannot contribute further information.

Other cellular techniques are being developed. Immunocytochemical methods allow the precise identification of substances such as hormones or specific marker proteins, which may in time prove even more useful in the functional characterisation of tumours. Such techniques, of course, now have their own ultrastructural counterparts. When used appropriately, electron microscopy seems likely to will continue to provide a powerful tool for the investigation and diagnosis of human disease.

Techniques and applications

This chapter is not intended to serve as a laboratory manual for the practical electron microscopist, since there are various detailed texts on technical procedures which will better serve this purpose. It is intended, rather, to provide a background of understanding of the methods employed by the electron microscopist, for the benefit of those who may not have the opportunity or the inclination to spend time personally acquiring these technical skills. The principles of construction of the electron microscope are summarised and the various modern modes of operation of electron column devices are described. In addition, the basic principles of specimen preparation are outlined to allow the general reader to appreciate the special needs and problems of the electron microscopist.

14.1 LIGHT AND ELECTRON MICROSCOPY

The purpose of a microscope is to produce a *magnified image* of a specimen in order to obtain more information about its structure. Any conventional microscope has three essential parts, a source of *illumination,* a *lens* system to provide magnification and some means for *observing* the final magnified image. The *light microscope* uses a beam of visible light, generally produced by an electric lamp, to illuminate the specimen. This light is concentrated by a glass condenser lens which focuses it on a small area of the specimen by bending its rays. The details contained in the illuminated specimen are then magnified by the objective lens and the image formed by this system is observed through the eyepiece lens. The rays of light which form the magnified image fall directly on the light-sensitive retinal receptors of the

microscopist and are registered as a visual image. The specimens most commonly used in biology are 5–10 µm thick sections of fixed tissues or cells, mounted on glass slides, stained with dyes and examined with transmitted light.

Since a microscope is designed to allow the observation of details not visible to the unaided eye, the most important measure of the efficiency of its performance is the fineness of detail which it permits the observer to distinguish. This aspect of the performance of a microscope is called its *resolving power.* The resolution obtained in a microscopic image is determined in figures as the minimum distance between two points which can just be distinguished by the use of the microscope. From theoretical considerations based on the laws of optics it can be shown that the resolving power of any microscope depends essentially on two variables, the *wavelength of the light* which is used to illuminate the specimen and the *numerical aperture* of the objective lens, the main magnifying lens in the optical system of the microscope. The glass lenses used in the light microscope can now be produced to exacting standards and lens design does not effectively limit its resolving power. Only the first of these factors, the wavelength of visible light, remains to limit performance.

Visible light, the only form of electromagnetic radiation to which our eyes respond directly, has a clearly defined range of wavelength which extends from 450 nm at the blue end of the spectrum to 700 nm at the red end. The theoretical limit of resolving power of a microscope is half the wavelength of the light used to illuminate the specimen, so that when visible light is used no details finer than around 0.2 µm can be clearly distinguished, however perfect the design of the microscope may be. The limitation lies purely in the physical nature of the light.

It is important to realise that this limitation to resolution which has been described has nothing to do with *magnifying power*. A lens system can be designed to produce any required degree of magnification, simply by making a much larger final image. No matter how large the final image is made, however, no more meaningful information can ever be obtained from it than the *resolving power* of the system will allow. When the limit of resolution is reached, any finer details of the image remain indistinct and further optical magnification will do no more than magnify the blur. Resolving power rather than magnification is the most important measure of the performance of a microscope.

Light rays are not the only form of electromagnetic radiation which can be used to form an image of a specimen, although they are the only form to which the *human retina* can directly respond. For example, a *beam of electrons* can be used instead of a beam of light. In an electron microscope, the effective wavelength of an electron beam can be as small as 0.0025 nm. The second variable factor, which in practice determines electron microscope resolution, the design of the *lens system,* has been so technically demanding that only within the last few years has it been possible for commercially available instruments to achieve a resolution better than 0.25 nm.

The use of *electrons* presents a number of fundamental practical problems which have now been solved in various ways. Electrons are small negatively charged particles which are very readily absorbed and scattered by any form of matter. It is possible to produce and sustain a beam of electrons only in a *high vacuum,* since even air alone will scatter electrons. The electron microscope must therefore be a closed system in which a vacuum of at least 10^{-4} Torr (10^{-4} mmHg) must be maintained. In modern instruments, 10^{-6} Torr is usual, and 10^{-11} Torr can be obtained for special purposes (Section 14.3.3). The beam is produced within the vacuum chamber by passing a high voltage electric current through a tungsten filament and shielding with metal apertures all but a small pencil of the electrons produced.

With this form of radiation glass lenses are valueless, since they scatter all the electrons, but the negative charge of the electron makes it susceptible to *magnetic fields.* The electron lens consists of coils of wire wound on a hollow metal cylinder designed in such a way that an axially symmetrical magnetic field is produced in the centre of the lens by an electric current passing through the coil. The carefully structured magnetic field produced by an electron lens forces the electrons to spiral around a central axis. The electron beam, passing through this field as it travels along the microscope column, is deflected by a variable amount depending on the current flowing through the lens coil. Thus a variable resistance with a rotary switch control allows the beam to be focussed and makes image formation possible.

By comparison with optical lenses, the *electron lens* is very imperfect—the electronic equivalent of a pinhole camera. The resolving power of a microscope using such a lens is limited in practice to about 50 times the wavelength of the electron beam. Thus in a transmission electron microscope with an accelerating voltage of 100 kV, we might hope to achieve a resolution approaching 0.2 nm. Recent advances in electron lens design have permitted the leading manufacturers to guarantee a resolution better than 0.25 nm in their more expensive instruments, although mechanical, thermal and electronic stability then begin to present new problems of design.

For the biologist, however, such figures are nearly meaningless. *Tissue sections* are already so imperfect that they themselves are now the principal barrier to the attainment of anything better than 15 nm resolution. Significant improvements in specimen preparation techniques will be required before the biologist can take full advantage of even the currently available levels of instrument resolution.

14.2 BEAM–SPECIMEN INTERACTIONS

So far, we have described how an electron beam can be produced and manipulated within a column of lenses. We must now consider what happens when the beam strikes a specimen, such as a thin section, placed in the microscope stage. This is shown diagrammatically in Figure 33.

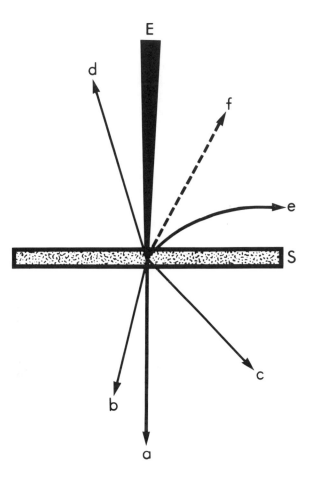

Fig. 33 *Beam-specimen interactions*
This diagram shows a typical specimen, S, in the path of the beam of electrons, E, in the electron microscope column. When the beam strikes the section, there are various possible consequences. Most of the beam, a, passes straight through the specimen without deviation to reach the viewing screen or the camera below. Some electrons, b, lose some of their energy as they pass through the specimen and are deflected a little way from the axis of the beam. These are termed inelastically scattered electrons. Other electrons, c, undergo elastic scattering through impact with atoms of the specimen. These electrons do not lose energy but deviate widely from the original path of the beam. Many such electrons are in fact backscattered, d, instead of being transmitted. In some cases, a high energy electron of the beam is absorbed by an atom of the specimen and a lower energy electron, e, is emitted. This secondary electron signal is particularly useful in forming images in the scanning electron microscope. Finally, when an electron of the beam dislodges an electron from one of the atomic shells of an atom in the specimen, the empty space in the atom's structure is filled by an electron from an outer shell. When this happens, the atom emits an X-ray of characteristic energy and wavelength, f, which can be detected by appropriate apparatus.

Most of the electrons of the beam *pass through* the specimen without hindrance, their properties remaining unchanged. Others collide with various atoms in the specimen, losing a proportion of their energy. Elections whose properties have been slightly altered in this way are known as *inelastically scattered* electrons. When carefully analysed, the lost energy can give information on the chemical nature of the specimen. Still other electrons are widely dispersed, following collisions which do not result in energy loss. These are the *elastically scattered* electrons. Some of these high energy electrons of the incident beam are *backscattered*, being picked up above the specimen surface. Some electrons of the beam provoke emission of *secondary electrons* at lower energy from the specimen. This emission is of particular importance in the technique of scanning electron microscopy. Finally, when electrons from the beam displace electrons from the inner shells of atoms, *characteristic X-rays* are emitted which can be used for elemental analysis of the specimen.

Every aspect of this complex interaction between beam and specimen provides *information* of some kind. Different instruments use different parts of this information to provide their output. The conventional electron microscope uses the transmitted electrons to produce a magnified visual image, the contrast of which is provided by the scattering of electrons mentioned above. The scanning microscope can be adapted to use widely differing signals to produce meaningful data on the specimen under examination. X-rays can be analysed to provide information about elemental composition, while their distribution can be mapped, providing a pictorial representation of elemental distribution. Just as light microscopy now goes far beyond simple visual image formation, so electron microscopy now gives the morphologist more than just structural information.

14.3 ELECTRON COLUMN SYSTEMS

14.3.1 The conventional transmission electron microscope

In the conventional electron microscope the electron lenses fall into two groups, the *illuminating* system and the *image-forming* lenses. The illuminating system consists of one or two condenser lenses which focus the electron beam on the specimen to provide a spot of concentrated illumination. A magnified image of the illuminated specimen is then formed by the second group of lenses, of which there are usually at least three. The most important of these is the *objective* lens. The virtual image produced by the objective lens is then magnified by the intermediate and projector lenses. By varying the current passing through these different lenses it is possible to focus the image and alter the magnification. A more elaborate microscope incorporates one or more additional lenses in this image-forming system in order to give greater flexibility, particularly for specialised techniques such as electron diffraction, used for studying periodic materials such as crystals. In a modern microscope, the combinations of lenses can be microprocessor controlled for extra flexibility and convenience.

The electron image produced in this conventional way is an expression of the *differential scattering* of electrons by areas of different density in the specimen through which the beam is transmitted. The variations in density are perhaps better described as areas of atomic number contrast, since the electron scattering power of elements is proportional to their atomic number. Since there are now other kinds of electron microscope available which form images in different ways, this familiar mode is best specified as *conventional transmission* electron microscopy (CTEM) although the abbreviation TEM is acceptable for most purposes. It should be noted again that the image is formed in a CTEM by the use of electron lenses, which are subject to the various defects of lens systems of any kind, such as spherical and chromatic aberration.

Before the information in the electron image can be assessed, it must be converted into a visible form, since the eye of the microscopist cannot respond directly to the electron beam. By projecting the electron image on to a *fluorescent screen* the energy of the electrons is transformed into visible light through excitation of the chemical coating of the screen. Areas of the section which scatter electrons are often described as *electron dense*. Thus an electron–dense area in the original specimen will appear as a dark area in the magnified image

on the viewing screen. An area with little dense material allows more electrons to pass through, so that they reach the screen and produce a bright patch of light. In this way, the patterns of electron scattering present in the specimen are translated into patterns of light and darkness which constitute the image on the screen. The screen can then be viewed directly by the observer with the unaided eye, or with a binocular light microscope, which further magnifies the details of the screen.

A *permanent record* of the image can finally be made on a photographic plate or film, which responds to electrons as well as to light. For this purpose, a camera is placed under the screen of the microscope. When the screen is moved out of the path of the beam, the electron image falls instead on a photographic film. After the film has been exposed, it can be removed from the microscope and developed and fixed in the usual way. It then becomes a negative which can be printed in the form of a black and white photograph, the *electron micrograph*. In practical terms, the micrographs are the main permanent record of material studied by the electron microscope whereas the convenient light microscope slide, quickly re-examined and easily marked and shown by projection, makes photographic recording of light microscopic material slightly less important for record purposes.

Since the maintenance of a high vacuum in all parts of the electron microscope is essential for its operation, a *vacuum pumping system* is an integral part of all machines. A simple rotary pump is used to clear most of the air from the column of the microscope, while high vacuum pumping is completed by *diffusion* pumps or *ion* pumps. To maintain a high vacuum, it is essential that the microscope should be free from air leaks from the atmosphere. Every section of the microscope and every moving control must be sealed to prevent leaks, usually by the use of rubber or plastic rings between adjoining metal surfaces. Another technical feature of microscope construction is a *water cooling system* for the pumps and the lenses, since in both of these places it is important to prevent overheating. In some cases a circulating pump, refrigeration plant and filter system forming a closed circuit are included in the design of the microscope. In modern machines, miniaturisation of electronic systems has helped to reduce the size of complex control circuits, while automation makes routine maintenance and operation more simple in many respects.

14.3.1   The scanning electron microscope

The best known type of scanning electron microscope (SEM) provides a method for the examination of the surface structure of thick specimens. This mode of operation is more correctly known as *surface scanning* electron microscopy (SSEM). The typical surface scanning electron micrograph has a distinctive three dimensional appearance. Although the earliest instruments of this type were devised in the mid 1930s, considerable electronic refinement was required before a commercially viable scanning microscope first appeared in the 1960s.

The illuminating system of the scanning electron microscope is similar to that of the conventional transmission electron microscope. The electron beam, however, is compressed by the action of one or more condenser lenses into a much narrower pencil of electrons, sometimes termed an *electron probe.* The strength of the final condenser lens is adjusted in order to focus the pencil of electrons on to the surface of the specimen. The electron probe is then electromagnetically *deflected* to and fro by *scanning coils,* the spot of electrons being made to sweep line by line across a rectangular area of the specimen surface. The specimen, generally inclined at an angle to the scanning beam, is usually situated at the foot of the microscope column. It can be of any thickness, since the beam is not required to pass through it in order to form the magnified image, as is the case in the transmission electron microscope. In fact, the scanning microscope usually has no image-forming lenses beneath the specimen, since image formation is achieved by television electronics.

As the beam strikes the surface of the specimen, high energy *backscattered* and low energy *secondary* electrons are given off in variable numbers, depending on the surface contours of the specimen. A collecting system usually diverts the backscattered electrons and picks up the secondary electron signal which results from this interaction. The varying signal given off from moment to moment is *amplified* and then used to vary the intensity of the display on a *cathode ray tube,* scanned synchronously with the illuminating beam. In this way a picture is built up on the screen which corresponds, point by point, to the surface contours of the area of the specimen which is being scanned by the beam. Since the final display of the scanning microscope is in the form of an image on a cathode ray tube, the permanent record of the image is made by photographing the screen with a conventional camera.

Magnification in the scanning microscope can be expressed as the ratio of the scan length on the *screen,* which remains constant, to the scan length on the *specimen surface,* which is variable. To increase magnification, the area of the specimen scanned is reduced, thereby increasing the *ratio* of the scan lengths. For example, if the beam scans a 1 mm square area of the specimen and the resultant image is displayed on a 10 cm square screen, the magnification achieved is $100 \times$. If the area scanned is reduced to 0.1 mm square, the magnification is increased to $1000 \times$. As with every other microscopical technique, the effectiveness of increased magnification depends on the available *resolution* of the system, generally around 5 to 10 nm in most modern production instruments. Resolution in this type of microscope is limited by the *diameter of the spot* formed by the condenser lens system, and by the *ratio of signal to noise* at the detector. These factors are related to the size and the brightness of the electron source. In time, the conventional tungsten filament may be replaced by the much brighter *field emission gun,* allowing great improvements in the resolution available in the scanning electron microscope.

It will be seen from this brief description that the term 'scanning' in the scanning electron microscope refers to the combination of the technology of electron microscopy with that of television electronics. Modern developments in electron microscopy are concentrated on exploiting this combined approach to image formation.

It is customary to compare the conventional transmission electron micrograph with the image of a slide formed by the familiar laboratory light microscope. In the case of scanning microscopy, a corresponding comparison can be made. The image formed by the surface scanning electron microscope owes its striking three dimensional effect to a combination of depth of focus and high resolving power. The best analogy in light microscopy is the three-dimensional surface image formed by the binocular dissecting microscope. In practical terms, scanning microscopy is widely used in the study of hard tissue, natural surfaces such as the gut, cut surfaces of soft tissues, and isolated cells of various types. Numerous applications in anatomy, botany, zoology, geology, materials science and forensic medicine are now recognised, while applications in the diagnosis of disease may emerge in the future.

14.3.3 The scanning transmission electron microscope

If the orthodox surface scanning electron microscope, as described above, is provided with a suitable holder for *thin specimens* and if an electron detector is placed *below* the specimen, the transmitted electron signal can then be used to form an image of the section. The image, produced without image-forming lenses, is in all essentials identical to the image of the same section which one would obtain from a conventional transmission electron microscope. The term applied to this mode of operation is *scanning transmission* electron microscopy (STEM). Today this option can be provided, as suggested above, by adaptation of an existing surface scanning microscope; by modification of an existing conventional transmission microscope; or by a custom-built scanning transmission electron microscope, of which there are now some commercially available models. There are several advantages to be gained from this technique by comparison with conventional transmission electron microscopy, which relies on an image-foming lens system.

The image produced by a conventional transmission electron microscope is formed by unscattered and inelastically scattered electrons, the contrast being largely attributable to those elastically scattered electrons which fail to reach the fluorescent screen. The electrons which form the image are mostly those which have passed right through the specimen without interacting. Moreover, some of the information generated by the interaction of the beam and the specimen has been lost, since all that this system can record about the scattered electrons is the fact of their absence from the image.

The *first advantage* of the scanning transmission microscope lies in its ability to handle these differently scattered electrons. It is possible to design a detector system which collects independently each type of electron passing through the specimen. An image can be formed either with the unscattered electrons alone, or with the scattered electrons alone, or with combinations of the two signals either by addition or subtraction. Much more of the information produced by the interaction of the beam and the specimen is now available than could be gathered with the conventional transmission microscope.

The *second advantage* of the scanning transmission system is its freedom from the defect of chromatic aberration. This defect is familiar in light microscopy as coloured fringes produced by uncorrected light optics. While passing through the specimen, some electrons are slowed down as a result of their interactions. These slower electrons are brought to focus at a point different from those which have not lost energy. This results in a blurring of detail, which becomes progressively more severe with increasing specimen thickness. In the scanning transmission microscope there are no image-forming lenses in which this aberration can arise. In theory, therefore, the scanning transmission electron microscope can provide clear images from much thicker specimens.

A *final advantage* of the scanning transmission electron microscope lies in the manipulation of contrast by electronic means. This makes it possible to produce worthwhile images from unstained biological material which would be virtually invisible under the operating conditions of the conventional transmission electron microscope.

As indicated above, it is possible to adapt a conventional transmission electron microscope to produce a scanning transmission image. The limiting factor in such conversions is beam intensity. Under low-resolution operating conditions, with quite large spot sizes, the meaningful electron *signal* is large, while the random electronic fluctuation which is always present in any system, known as *noise*, is trivial by comparison. In other words, the signal-to-noise ratio is satisfactory and the picture is sharp and clear. However, at higher magnifications, as the spot size is reduced to give increased resolution, the meaningful signal is also reduced and the constant electronic noise becomes relatively more important. The signal-to-noise ratio becomes progressively less satisfactory, resulting in a snow-storm effect, with reduced picture quality and poor resolution.

The only way to overcome this limit to performance is to produce a brighter source of illumination. The new custom-built scanning transmission electron microscopes avoid this problem by the use of a very bright and very small electron source formed by a *field-emission electron gun*. A negative field applied to a sharp tungsten point under very high vacuum conditions supplies an intense steady stream of virtually mono-energetic electrons. It is possible with a field emission gun (FEG) to produce effective spot sizes smaller than 0.5 nm without serious loss of signal intensity. This makes possible very *high resolution* STEM operation closely approaching if not matching that of the modern conventional transmission electron microscope.

There is, however, one major problem in the operation of an instrument of this type. The field-emission gun can only perform effectively with a vacuum of 5 to 6 orders of magnitude better than that of the conventional microscope. This has called for the development of pumping systems capable of reaching operating pressures of 10^{-11} Torr (10^{-11} mmHg). The most sophisticated FEGSTEM instruments are very expensive. The ultra high resolution scanning transmission electron microscope may become the standard instrument of the future, but years of development lie ahead before it can replace conventional technology. An intermediate resolution STEM instrument, on the other hand, might soon prove competitive with the smallest models of CTEM.

14.3.4   The high voltage electron microscope

When the term *high voltage* is applied to electron microscopy, it implies accelerating voltages far in excess of the 60 kV normally used in the examination of thin sections. In most conventional instruments a range of from 20 kV to 100 kV is available for different purposes, whereas high-voltage elec-

tron microscopes can be operated at from 3 to 10 million volts. The technique is now proving of interest to biologists.

If the accelerating voltage used to produce the electron beam is increased, two effects are noticed. Since the electrons have higher energy, they are more likely to pass straight through the specimen than to interact with it and be scattered. Increased penetrating power is thus a major advantage of high-voltage microscopy. However, since contrast in the image is largely due to differential electron scattering, this increased penetration is accompanied by a marked loss of contrast. The specimen, on the other hand, is less likely to suffer damage from bombardment by the beam than at lower voltages, since fewer destructive interactions occur between the incident electron beam and the delicate specimen. The recent interest in high-voltage microscopy stems from this combination of *increased penetration* and *reduced specimen damage.*

High-voltage electron microscopy presents a number of special technical problems. High-voltage microscopes are usually installed in specially designed separate buildings, rather than fitting in to a more conventional setting. The microscope requires much more *space* than a conventional instrument, to allow for suitable insulation of the high-tension supply. There is, in addition, a *radiation* hazard. When an electron beam of such high energy strikes any object, X-rays of considerable range and penetrating power are emitted. The various apertures within the column, the specimen itself and the viewing screen of the microscope are all sources of dangerous radiation which must be blocked by thick shielding. For example, the lead glass through which the screen is examined is up to 20 cm in thickness in the 1 MV microscope and the segments of the column itself are so massive, due to lead shielding, that a fork-lift truck is employed to dismantle the microscope. Despite such precautions, it is still necessary at times to operate the microscope by remote control from a shielded console distant from the column.

It is possible with the high voltage microscope to achieve acceptable levels of resolution from sections up to several microns in thickness, as opposed to the conventional 50 nm sections more familiar to the biologist. For example, at one million volts, a section 1 µm in thickness will permit a resolution of around 3 nm. In sections of this thickness, *superimposition* effects present a major problem, but stereoscopic techniques allow this to be overcome. For this purpose a micrograph is made of a particular field. The section is then *tilted* through a small angle relative to the beam and a second exposure is made, on a fresh film, of the same field. The two micrographs can then be viewed as a *stereopair,* to give a three-dimensional presentation of the details within the full section thickness. This can be of value in the interpretation of fine structure within the cell, since it eliminates the need for the reconstruction of three dimensional details from serial thin sections. In particular, details of the cytoskeleton have been studied by this technique, using specimens which consist of monolayers of whole cells, fixed and dried without embedding in plastic.

In theory, at least, the high voltage microscope could allow the study of living *hydrated cells* under the electron beam, impossible in the conventional electron microscope on account of poor specimen penetration and marked beam damage. A prerequisite for such studies is the provision of a suitable chamber in which the cells can be exposed to the beam while being maintained in a moist environment. The cells are sandwiched in a thin moist layer between two thin films, thus isolating them from the column vacuum. The high voltage beam can penetrate the films, the atmosphere maintained between them and the full thickness of the cell, revealing its internal structure to some extent. So far, however, little new information has been obtained with such techniques. High voltage electron microscopy is at present used more by the materials scientist than the biologist; the technique, however, is now recognised as having important biological applications and will be increasingly used for biological research.

14.4 ROUTINE SPECIMEN PREPARATION

14.4.1 Special requirements

The use of electrons in microscopy imposes particular demands not only on microscope design, but also on *specimen preparation.* Since electrons are scattered and absorbed so readily, the sections of tissues used for the light microscope are too thick for electron microscopy. Sections about 50 nm in thickness are now routinely obtained, through the use of plastic *embedding* media which are less volatile and give more support to the tissue than wax and with the help of modern *ultramicrotomes* which are more delicate in their construction and operation. Equally important are refinements in *fixation* technique. While the standard histological fixatives preserve the structure of cells well enough for routine light microscopy, the preservation is not satisfactory when more critical examination of the material is possible at higher resolution. The details of cell structure which are made visible by the electron microscope can be preserved successfully only by careful control of the pH, osmotic pressure and ionic strength of the fixative.

14.4.2 Fixation

It has been found that the most useful general fixative combination for electron microscopy is a solution of *glutaraldehyde* followed by postfixation in *osmium tetroxide,* often known as osmic acid. The deposition of osmium in the tissues is an important factor in the enhancement of contrast. Its high atomic weight ensures that any structures in which osmium is deposited during fixation appear dense in the electron image. Such structures are described as osmiophilic. The fixative must contain dissolved salts and sometimes sucrose to bring the *osmotic pressure* and ionic composition close to the level found in normal tissues, in order to avoid tissue damage during fixation. Sucrose allows the osmotic pressure to be increased without affecting ionic composition. The structure of the cell is damaged by uncontrolled acidity during fixation, but this can be prevented by suitably *buffering* the fixative to pH 7.4. The commonest fixatives use either phosphate, veronal acetate, S-collidine or cacodylate buffer systems. Buffered solutions of formaldehyde can be used for special purposes, such as for fixing tissues prior to certain types of electron histochemical procedures, and for routine fixation in histopathology, allowing subsequent EM studies to be carried out in cases of interest.

To be effective, fixation must begin as soon as possible after the circulation of blood to the tissue has ceased. When specimens are being obtained from a laboratory animal, the tissue can sometimes be fixed by dripping the fluid directly on to the organ which has been exposed at operation. Alternatively the tissue can be removed fresh and chopped immediately into small pieces in a drop of fixative to ensure rapid and complete penetration of the fixative into the centre of each small piece. It is commonly accepted that blocks of tissue for electron microscopy should always be smaller than a 1 mm cube, especially when using an osmium-containing fixative, which penetrates tissues slowly and with difficulty. The use of *perfusion fixation* is more satisfactory, particularly for small experimental animals in which accurate dissection of tissues can be difficult and time-consuming. In perfusion fixation, the fixative, which is passed under pressure through the blood vessels of the anaesthetised animal, replaces the circulating blood and comes into immediate and intimate contact with the tissues. The material can be dissected at leisure once it has been fixed in this way. It has been customary to use chilled solutions to avoid, as far as possible, unwanted chemical extraction of components of the cells, but there is now evidence that certain structures such as microtubules are preserved more satisfactorily by fixation at room temperature. It is important that tissues should be handled at all times with the maximum possible care, avoiding crushing and using only the sharpest of razor blades for dissection.

14.4.3   Embedding

After fixation, normally complete within one to three hours, the pieces of tissue are washed in buffer solution and then prepared for *embedding*. Wax embedding methods used for histology are not suitable for electron microscopy since wax does not support tissue sufficiently for sectioning and would evaporate in the high vacuum of the microscope when exposed to the heating effect of the electron beam. Plastic materials have now been developed which do not have the same defects. Since these materials, like wax, are not usually miscible with water, prior *dehydration* of the tissue is essential. This is achieved by passing the material through increasing concentrations of ethanol, finishing with absolute ethanol. A *clearing* agent

such as propylene oxide or xylol, miscible both with alcohol and with the embedding medium, is usually required to provide a transition to the final stage of the processing.

The *plastic* takes the form of a liquid *monomer* solution, which is sufficiently fluid to penetrate readily through the tissues. The tissue is placed, along with its identification code number, in a small gelatin capsule or rubber mould filled with liquid embedding medium. The process of *polymerisation*, or curing in the case of the modern resins, is then induced by heating in an oven to temperatures up to 100°C. The polymerised form of the embedding medium is hard and durable. The tissue is surrounded and infiltrated by this supporting material, forming a convenient block which is ready for sectioning (Fig. 34).

In the past, the process of polymerisation could take up to three days, but current methods permit adequate hardening of the block within an hour at 100°C if circumstances demand. With some sacrifice of quality, tissues removed at biopsy from a patient in hospital can be processed and examined within two hours, making electron microscopy available, if required for diagnosis, in a way not previously possible. A 24-hour schedule is more usual and is roughly equivalent to the time taken for routine paraffin processing.

The first effective embedding media were acrylic resins such as *methacrylate,* familiar as perspex. This was too volatile and unstable in the electron beam and caused excessive tissue damage during the process of polymerisation. The methacrylates were replaced by the *epoxy resins,* such as Araldite and Epon, and their more recent successors Spurr and Maraglas, or by a polyester material such as Vestopal. For special purposes, such as lipid preservation and some histochemical work, modern *water-soluble* media such as Durcupan and ureaformaldehyde may be required.

14.4.4   Section cutting

Electron microscopy demands sections of about 50 nm in thickness from these hard blocks, since the electron beam is less able to penetrate thicker specimens, chromatic aberration results in loss of resolution and there is confusion from overlapping images. Ordinary microtome knives are not suitable

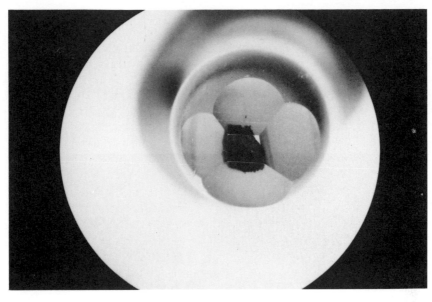

Fig. 34 *Resin block prepared for sectioning*
This shows an end-on view of a block of tissue embedded in plastic. The small irregular black area is the osmium-stained tissue within its surrounding plastic embedding medium. The surrounding resin has been trimmed away on the four sides of the tissue, exposing a blockface which is almost rectangular in shape. Sections will now be cut from this exposed face of the tissue.

for cutting such thin sections since it is impossible to polish a metal edge finely enough to obtain good sections. *Glass* or *diamond knives* are used instead. Glass knives are made by breaking triangular shapes from plate glass strips, either manually, using glass cutters and pliers, or mechanically. There are *knife-making* instruments available which make it possible to produce consistently satisfactory glass knives with little trouble. The sharp edge of the freshly broken angle of a glass knife is extremely fragile and is often badly damaged after only a few sections have been cut. Diamond knives, though expensive, are more durable and can be used for serial sectioning and for cutting dense tissues such as bone.

Once the knife has been made, it is provided with a trough which allows a fluid reservoir to lie behind the cutting edge so that sections, once cut, will float on the surface in a ribbon (Fig. 35). The trough is made of metal or tape and is attached firmly to the shoulder of the knife. The fluid which fills the bath is a mixture of acetone and water, or simply water alone.

Before sections are cut from the block of tissue, the gelatine capsule is first removed by soaking in water. The block is then mounted in the microtome

chuck and the tissue is exposed for sectioning (Fig. 34) by trimming off the tip of the block and forming a small pyramid with the tissue at its apex. At this point, a 2 μm thick section can be taken and mounted on a glass slide for light microscopy. This is stained with methylene blue to allow localisation of areas of interest for thin sectioning. The block face is then trimmed down to include only those structures which are of importance for electron microscopy. The scale of operation in all of these procedures is much more delicate than is the case with conventional histological techniques. A binocular microscope is an essential part of any *ultramicrotome*.

The microtomes used for thin sectioning must be designed to strict tolerances. The machines may be either manually operated or semi-automatic. The knife with its bath in place is clamped in the microtome and the block is fixed in a movable chuck. The block must advance towards the knife at each successive stroke by only 50 nm in order to allow a section of this thickness to be cut. This carefully calibrated advance can be produced in two ways. A mechanical advance involves a carefully machined micrometer screw and lever system. A thermal advance uses a calibrated electrical

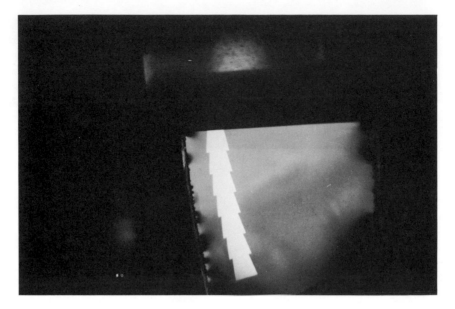

Fig. 35 *Ribbon of ultrathin sections*
This photograph shows a ribbon of sections lying on the surface of a water bath behind the edge of a glass knife. The sections are seen by reflection of incident light. This ribbon will now be picked up on a 3 mm diameter copper grid and will be stained for examination under the electron microscope.

Photograph by courtesy of LKB

current to heat the metal arm of the microtome, producing controlled expansion. Not surprisingly the ultramicrotome is sensitive to draughts, vibrations and sudden temperature changes. It is best, therefore, to have a separate quiet room for section cutting, as free as possible from such disturbances. Considerable skill is involved in the production of consistently satisfactory results.

As the sections are cut they float from the knife edge (Fig. 35) in a continuous *ribbon* on the surface of the bath. When viewed in reflected light, their interference colours serve as an approximate guide to section thickness. Satisfactory sections appear grey or silver, while slightly thicker sections, still useful for some purposes, are gold in colour. Compression introduced during sectioning is removed by exposing the sections to chloroform or xylol vapour as they lie on the surface of the trough.

Tissues for electron microscopic study are mounted for examination by spreading the sections over the surface of a finely perforated *copper grid,* about 3 mm in diameter. The sections are placed on the grid by laying the grid on the ribbon as it floats on the surface of the bath. The sections

adhere to the grid by surface forces. In this way, small areas of the section are stretched between the metal cross bars of the grid which give sufficient support to the section without the need for an intervening layer, equivalent to the glass slide in light microscopy. As a result, scattering of electrons by unnecessary layers is eliminated since the electron beam need only pass through the section.

14.4.5   Contrast enhancement

If sections prepared as described above are examined in the conventional transmission electron microscope without further treatment, the inherent *contrast* in the tissue is so low that little detail can be distinguished. In light microscopy, contrast can be enhanced by the use of coloured dyes which are bound by chemical reactions to different parts of the section, producing areas distinguished by their different colours. Colour, however, is not a property of the electron image. Contrast in the electron microscope relies instead on the *differential scattering* of electrons, which depends in turn on the presence of staining material of high atomic weight. Fixation of the tissue with osmium solutions adds some contrast, but in general, contrast must be further increased by exposing the section to a solution of a *heavy metal salt*, such as lead hydroxide, acetate or citrate, uranyl acetate or phosphotungstic acid. The heavy metal is taken up preferentially by membranes, granules and other distinctive structures, causing them to scatter more electrons. Such components appear dense in the final image. It is usually necessary to post-stain in this way to gain sufficient contrast for the adequate investigation of fine structure. A procedure in common use involves staining initially in a saturated solution of uranyl acetate in 50 per cent ethanol, followed by an aqueous solution of lead citrate. The alcohol is said to facilitate the entry of the heavy metal into the substance of the thin section. Staining 'en bloc' with uranyl acetate added to the alcohol used for dehydration can also contribute useful contrast enhancement.

Contrast also depends on the *operating conditions* of the microscope. If a *small aperture* is placed in the column at the objective lens, the contrast is increased, while the same effect is obtained by operating the microscope at a *lower voltage* than the usual 60 to 80 kV. Unfortunately the use of lower accelerating voltages reduces resolution, increases the heating effect of the beam and damages the specimen more quickly. It has been calculated that the dose of irradiation received by the area of a section under examination for the time taken to produce a single micrograph is equivalent to that encountered at a distance of 30 metres from a 10 megaton nuclear explosion. Not surprisingly, it is now recognised that *beam damage* causes significant degradation of fine structure and introduces further limits to the effective resolving power of the electron microscope in thin section work.

14.4.6 Support films

In most cases, the thin sections for transmission microscopy are mounted on grids without any additional support, but when particulate specimens such as viruses are to be examined by negative staining techniques, a *support film* must be provided to carry the particles. Similar films may also be used to give extra support to the more delicate water-soluble embedding media and to allow uninterrupted ribbons of serial sections to be examined on special wide-mesh or single-hole grids. The ideal support film is mechanically strong, is of uniform thickness and has a low electron scattering capacity. Thin plastic films can be made from formvar (polyvinyl formal) and from parlodion (nitrocellulose). These are reinforced by a layer of carbon, produced by vacuum evaporation of carbon from a carbon arc. Pure carbon films can also be made. The thickness of a suitable carbon support film is up to 10 nm, while plastic films are usually in the range of 10 to 40 nm.

14.5 SPECIMEN PREPARATION FOR SURFACE SCANNING ELECTRON MICROSCOPY

Hard specimens need little attention, but most other tissues must be fixed and dehydrated prior to examination. Glutaraldehyde, formaldehyde and osmium tetroxide are in common use. The tissue may then be dehydrated using alcohol or acetone, followed by simple *air drying*. This, however, produces considerable shrinkage artefacts and other distortions due to surface tension forces. *Freeze-drying* avoids most of these disadvantages, and *critical-point drying* is now widely accepted as a satisfactory procedure for avoiding such defects.

If a liquid in equilibrium with its vapour is heated in a closed pressure vessel, there is a *critical temperature* specific for each substance, above which the vapour cannot be liquefied irrespective of the pressure applied. One can also increase the *pressure* within such a vessel to the *critical value* at which there is no difference in density between the liquid and vapour phases. When both temperature and pressure are simultaneously adjusted to these critical values, the liquid is said to be at its *critical point*. The most popular substance for this purpose is carbon dioxide, although nitrous oxide or fluorocarbon fluids may equally well be used.

The tissue is first dehydrated in acetone and is then placed in an open boat containing acetone, within a pressure vessel. Liquid carbon dioxide is then passed under pressure into the chamber, where it mixes with and displaces the acetone. The vessel is now warmed gradually to the critical temperature of CO_2, 36.5°C, while the pressure rises to the critical value. At this point the vapour phase and the liquid phase co-exist, providing conditions of zero surface tension. The chamber is then slowly opened, allowing the CO_2 to vent as gas. This leaves the tissue completely dry, without ever having been exposed to the severe surface tension forces involved in any air drying procedure. The dried tissue is finally fixed to the specimen carrier with *electrically conducting* paint or glue.

The next important procedure is the provision of an electrically conducting *surface coating.* Biological samples are poor electrical conductors, and can accumulate substantial static charges under bombardment by the beam of electrons. This *charging* of the specimen can seriously impair image quality and resolution. The specimen carrier with the tissue attached to it is placed in a vacuum chamber and a thin layer of metal is evaporated onto its surface. Gold, palladium and aluminium are commonly used metals. The aim is to provide the specimen with a thin evenly distributed surface metal layer. Low-vacuum *sputtering techniques* are now popular for this purpose. After coating, the tissue is ready for examination. After examination in the scanning electron microscope, specimens can be re-processed for light microscopy and transmission electron microscopy, making possible detailed correlation of the different images.

14.6   SPECIAL TECHNIQUES FOR ELECTRON MICROSCOPY

14.6.1   Negative staining and shadowing

Small particles such as viruses are best examined on a thin carbon or plastic support film. Their intrinsic contrast, however, is so low that special staining procedures are required (Figs. 36, 37).

Negative staining involves suspending the virus in a solution of a heavy metal salt, such as uranyl acetate or phosphotungstic acid. A drop of the suspension is placed on a grid covered by a support film and the surplus is blotted off. The stain forms a thin dense layer over the film, outlining the virus particles and penetrating their surface recesses. This results in a finely detailed negative contrast image of the virus (Plates 116d, 116e) which appears pale against a dark background. Individual macromolecules can be visualised by this technique.

Particulate specimens can also be studied by heavy metal *shadowing techniques* (Plates 116b, 116c). The grid is placed in a vacuum chamber and metal is evaporated from an electrically heated filament. Through appropriate masking, the metal falls at a predetermined angle on to the specimen surface. The particles of the specimen gather a deposit of opaque metal at one side and cast a

Fig. 36 *Negative staining*
This represents two virus particles lying on a thin film stretched across two bars of an electron microscope grid. A thin dense layer of negative stain is spread across the surface of the specimen. This outlines the surface detail allowing a negative contrast image to be formed by transmission electron microscopy.

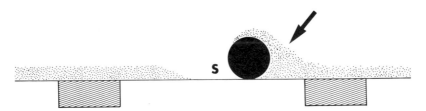

Fig. 37 *Shadow casting*
This represents a round structure lying on a thin film stretched across two bars of an electron microscope grid. A layer of heavy metal has been deposited at an angle indicated by the arrow. When viewed in the transmission electron microscope the area without heavy metal, S, appears as a reverse contrast shadow cast by the object. If the angle of shadowing is known, the length of the shadow provides a measure of the height of the specimen.

'shadow' on the other side, just as snowdrifts form around boulders. Given the angle of shadowing, the size and shape of the particles can be calculated from measurements of the length of the shadow.

14.6.2   Replication, freeze fracturing and freeze etching

Suitable surfaces may be studied by transmission electron microscopy through the use of *replicas*. A cast of the surface is made using a thin carbon or formvar layer. This replica is then stripped from the surface and shadowed with heavy metal, thus enhancing the contrast provided by the surface detail of the replica. Frozen surfaces can be replicated (Plate 1). Of particular interest are the surfaces exposed within tissues and cells by the process of mechanical *fracturing* of the frozen tissue. There is a tendency for the tissue to fracture along natural boundaries, such as membranes, thus exposing their outer or inner surfaces for examination. Further information may be gained by allowing some of the ice to sublime away under vacuum from a freeze-fractured surface before making the replica. This process of *freeze-etching* exposes otherwise hidden details of cellular and subcellular structure. These are of particular interest, since the tissues have not been distorted by the conventional fixation or dehydration procedures used for most other microscopical techniques and are thus closer to their natural state (Plate 1).

14.6.3   Tracer techniques

The thin sectioning technique used for transmission electron microscopy is the standard method of specimen preparation nowadays for most purposes in biological laboratories. However, in recent years ultrastructural techniques have been developed which have extended the range of the electron microscope beyond the limits of pure morphology. Most of these techniques have evolved from existing microscopic procedures modified to meet the special demands of electron microscopy. An early example is the use of silver and osmium impregnation methods in conjunction with electron microscopy for the demonstration of the Golgi apparatus. The position of the electron-dense

heavy metal deposit allows accurate identification by electron microscopy of the cytoplasmic structures which represent the Golgi image in the light microscope.

By using electron-dense colloidal particles as markers, a number of interesting studies of cell behaviour can be undertaken (Plate 115a). The processes of phagocytosis by different cells can be studied by following the progress of fine particles of colloidal thorium or gold injected into the circulation. The individual colloidal particles, too small to be visible directly by light microscopy, are immediately obvious in the electron micrograph, on account of their density. The particles can be seen adhering to the surfaces of macrophages and their segregation and isolation within cytoplasmic lysosomes can be followed. The extent to which intestinal cells in experimental animals ingest particulate matter from the gut at different stages in their development from newborn to adult has been studied with similar marker methods. The molecule of ferritin is particularly valuable in this context. Although it is a biological macromolecule, it still has sufficient intrinsic density on account of its high iron content to be directly visible by electron microscopy. The ferritin molecule has been used to demonstrate continuity in frog skeletal muscle between the T tubules and the extracellular space. The living muscle is placed in saline containing ferritin for a period of time and is then processed for electron microscopy. Control muscle cells placed in saline alone have no ferritin molecules within the T system, while the accumulation of ferritin within the T tubules of muscle exposed to the experimental solution indicates that free passage exists in this case between the extracellular fluid and the lumen of the tubule. Solutions of lanthanum salts can be used in a similar way.

A more recent tracer technique uses the enzyme *peroxidase* which is readily obtainable in a pure form from various sources, such as horse-radish. The position of the peroxidase molecules can be determined in tissues by incubation with a suitable reaction medium, leading to the formation of an electron-dense precipitate at the site of enzyme action. In this way studies can be made of the properties of a diffusion space or of the rate of transport across a surface.

Antibodies labelled with fluorescent dyes have been used for years in light microscopy to localise specific antigens within tissue sections. Labels suitable for electron microscopy have now been developed. Early workers used ferritin as a marker, but the most adaptable of these techniques has proved to be the *peroxidase-labelled antibody* procedure. Horse-radish peroxidase is conjugated with a specific antibody, which is then allowed to bind to its appropriate antigen in the tissue. The bound conjugate can then be demonstrated by providing the substrate for the peroxidase reaction and examining the tissue to identify the location of the reaction product. This technique seems likely to provide the key to a wide range of problems involving the subcellular localisation of antigens. An adaptation of the peroxidase labelling procedure involves the use of lectins, molecules which bind specifically to particular carbohydrates. Peroxidase-conjugated lectins such as concanavalin A are currently popular tools for investigation of cell surface receptors.

14.6.4 Autoradiography

Another technique adapted from light microscopy is *autoradiography* (Plate 116a). Localisation of radioactive material in a section of tissue is made possible by spreading a thin layer of photographic emulsion over the section (Fig. 38). The emulsion in contact with an area containing radioactive material becomes exposed to the action of the radiation, causing the formation of a latent image localised to the position of the emitting substance. When the emulsion is then developed and fixed as in conventional photography the areas exposed to radioactivity appear as dense clusters of silver grains overlying the histological section. The position of the grains in relation to the underlying histological details indicates the site of the radioactive tracer. The number of grains in a cluster gives a rough indication of the strength of the source of radioactivity and the amount of radioactive material present in a given area.

By autoradiography, using the DNA precursor thymidine labelled with the radioactive isotope of hydrogen called tritium, it is possible to identify the cells which are engaged in the synthesis of DNA at a particular time, thus providing the basis of research into the kinetics of cell division and cell population renewal. The uptake of radioactive sulphur and its conjugation with mucus has been shown, by autoradiography of the goblet cell, to take place in the Golgi apparatus. Labelled carbohydrate molecules have also been shown to become incorporated in newly synthesised mucus within the Golgi apparatus, whereas radioactive amino acids are incorporated into the protein component of mucus within the endoplasmic reticulum at the base of the goblet cell. With careful planning of experiments, many dynamic cellular processes can become accessible to study through autoradiographic techniques.

14.6.5 Cytochemistry

Light microscopic histochemistry and cytochemistry are based on reactions in tissues which can be studied by the use of coloured dyes. In some cases, such as the test for mucin or the PAS reaction, the dye concerned reacts directly with specific chemical groupings in the material to be demonstrated. The coloured dye thus becomes bound to the areas of the specimen containing the material of interest, which can then be localised by light microscopy. In other cases, enzymes present in the cell can be

Fig. 38 *Autoradiography*
This represents a specimen, S, stretched between the bars of an electron microscope grid. The specimen has been covered by a thin layer of photographic emulsion, E. Two radioactive sources in the specimen emit radiation at all angles as shown by the arrows. This radiation causes exposure of silver grains in the emulsion within an area around the radioactive source. After development, the specimen is examined on the transmission electron microscope. The position of silver grains in relation to tissue details indicates the localisation of radioactivity within the cell.

demonstrated indirectly by the reactions of dyes with the products of their specific enzymatic action. Enzyme histochemistry has become an important branch of light microscopy, since it allows an insight into cell function at the molecular level. The function of a cell is related to the enzymes it contains and abnormalities may be reflected in diminution or absence of enzymes. There have been many efforts to adapt these cytochemical reactions for use in electron microscopy, in order to localise more precisely the position of the enzymes in the cell in relation to the known components of fine structure.

The nature of electron microscopy imposes restrictions on cytochemical procedures. Suitable markers must be found for electron microscopic localisation. The colourful dyes on which much of light microscopic histochemistry is based are not suitable as stains for electron microscopy. They are organic molecules of low density, visible by their colour on light microscopy, but invisible under electron microscopy on account of their low electron scattering power. A coloured image cannot be produced by electrons. Nevertheless, certain organic dyes such as ruthenium red can be used to stain the cell coat for electron microscopy (Plate 115b, 115c). Adequate contrast is made available to mark the stained area by the affinity of the bound dye for the osmium which is used for post-fixation. The PAS reaction can also be demonstrated by electron microscopy, by the use of a silver proteinate compound instead of the coloured Schiff reagent (Plate 33b).

The most suitable enzyme reactions for electron cytochemistry are those which can be adapted to produce precipitates of heavy metal salts at the reaction site. Among the many reactions which have been demonstrated successfully by electron microscopy are the phosphatase group, including acid and alkaline phosphatase techniques, utilising lead salts (Plate 26b). The use of the acid phosphatase method has made possible the ultra-structural localisation of the enzyme in sections of tissue, allowing the identification of lysosomes. Alkaline phosphatase has been localised to the surface of the microvilli in the small intestinal epithelium. It seems likely that techniques will eventually be developed to demonstrate many enzymes of biological interest within the cell.

14.6.6 Microanalysis

When the electron beam hits the specimen, electrons from the beam dislodge inner-shell electrons from the atoms present in the specimen. The vacancies produced must then be filled by electron transitions from outer shells. These transitions are accompanied by the emission of X-rays of specific *wavelengths* and characteristic *energies,* which provide the basis for the identification of the elements involved. The X-rays can be collected and analysed according to their wavelength (*wavelength dispersive analysis*) or according to their energy (*energy dispersive analysis*), using apparatus attached to the electron microscope column. Under suitable conditions, as little as 10^{-19} g of an element can be identified in a thin section. The position of material present in adequate concentration can be localised to within 10 nm.

An example of this technique is provided by Figures 39a and b. In this case, a patient was found to have a malignant liver tumour, part of which was removed for microscopic examination. The surrounding liver was found to contain deposits of dense granular material of unknown identity. These granules were easily seen on electron microscopy and were submitted for microanalysis. The X-ray spectrum (Fig. 39a) of the control area shows the presence of various peaks, mostly due to elements in the specimen holder and its surroundings. These contaminating peaks are unavoidable in the system used. An adjacent area containing the granular material, however, showed additional peaks (Fig. 39b). These are of the energies characteristic of the element thorium. On detailed questioning it was then discovered that the patient had been wounded in the war and had been X-rayed using a thorium-containing substance called thorotrast to outline a wound sinus. The thorium had been taken up and stored by the liver, where its radioactivity may have contributed to the development of the liver cancer thirty years later.

Many technical problems remain to be solved. In particular, valid results in many instances demand the use of frozen ultrathin sections, examined in the microscope in the frozen hydrated condition, to minimise elemental mobility. The techniques for *cryo-ultramicrotomy* are still in their infancy, although rapid progress has recently been

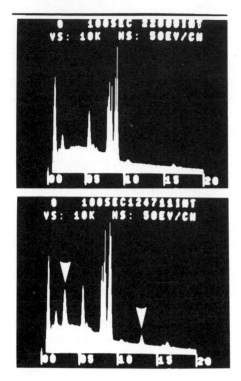

Fig. 39a, b *Energy dispersive X-ray analysis spectra*
Figure 39a is the control, Figure 39b the test. These spectra were taken from tissue from the liver, the test being an area containing granular material of unknown nature and the control being a nearby area of similar tissue without the deposits of granular material. The horizontal axis of the spectrum defines the energy of the X-rays, on a scale from 0–20 KeV. The vertical axis displays the number of X-rays of a given energy. The various peaks in the control correspond to the elements present in the specimen, the grid and the microscope pole-pieces. These peaks are also seen in the test. The new peaks, arrowed in the test, correspond to the specific energies of the characteristic X-rays of the element thorium. The granular material was later confirmed to be thorotrast, a preparation of thorium dioxide used on this patient in the past as a contrast medium for X-ray examinations.

made. Many important applications of X-ray microanalysis are now emerging in *forensic medicine*, allied to the use of surface scanning microscopy. For thin section work, however, the best results in future may demand the new field emission gun scanning transmission electron microscope (FEG-STEM), allowing the use of narrower, brighter beams of electrons. In microscopes such as this, a further analytical technique will become available. This is *electron energy loss analysis*, or spectroscopy, (EELS) which makes use of information ignored by the conventional transmission electron microscope. It is still in the early stages of development.

14.7   CONCLUSION

In the last ten years the versatility of the electron column has increased. Various special instruments with new and distinctive characteristics, such as scanning and analytical systems, have been perfected through the rapid advance of electronics technology. In this way, the limited horizons of the ultrastructural biologist of the 1950s and 1960s have been transformed. Our understanding of cell structure will remain rooted in basic thin section techniques, but the modern biologist must now be prepared to adapt to a new and much wider range of technology, capable of answering questions which could not have been asked 20 years ago.

15

Critical examination of the electron micrograph

The three essentials in the interpretation of an electron micrograph are an adequate basis of *theoretical knowledge* of cell structure, a systematic technique for *reading* the electron micrograph and an awareness of *artefact*. Theoretical knowledge may come from the earlier chapters of this book. The awareness of artefact and the skills of reading the micrograph are best learned by experience, but the present chapter contains some simple guidelines which may assist the beginner to acquire that awareness and these skills.

15.1 READING AND INTERPRETING THE ELECTRON MICROGRAPH

15.1.1 Identification of specimen and technique

It is important to read whatever *caption* is provided. The information conveyed by a good caption will include details of the species from which the tissue was taken, the organ involved, the cell type and the experimental or clinical data which are relevant to the understanding of the micrograph. In addition it will provide *technical information*, including the fixative, embedding medium and staining technique used, as well as the magnification of the print. An informed assessment demands the fullest use of this information. For example, minor degrees of imperfect preservation can be accepted in a study of immersion fixed human biopsy material, but would perhaps be considered unacceptable in an experimental situation in which perfusion fixation could have been employed. Certain histochemical procedures can only be carried out on fresh tissue, prior to fixation, while others will only work effectively with fixatives not primarily intended or suitable for ultrastructural preservation. In situations such as these the investigator must strike a balance between the theoretical ideal of tissue preservation and the imperfect, but attainable results which are determined by the practical constraints of the investigation. In general, a technically imperfect but useful result is more satisfactory than nothing at all.

The caption will also define the operating conditions and the particular technique of microscopy used to produce the picture. Today the term 'electron micrograph' no longer automatically means an image produced by a conventional transmission microscope from a thin section, as was the case with most tissue work in the past. The increasing diversity of electron column techniques and of specimen preparation has made it essential that the caption should be carefully studied for *technical data* on the instrument used. Scanning transmission microscopy can show details of specimens which might previously have been undemonstrable by conventional methods; high voltage microscopy calls for the evaluation of stereopairs to make the most of the details available from thick sections; surface scanning microscopy has introduced a three-dimensional effect which could previously only be hinted at by replica techniques and shadowing, perhaps accompanied by freeze-fracturing procedures.

The caption, in other words, is invaluable. If it falls short of expectations, the information outlined above must be sought elsewhere, such as in the section concerning materials and methods, which will be found at the start of most scientific papers.

15.1.2   Magnification and relative scale

Knowledge of the magnification is essential for interpretation. In most cases this will be provided, but an experienced observer can usually make a fairly accurate guess by picking out different features of the tissue and relating them to familiar images from personal experience. In practical terms, the exact figure is often unimportant and the observer simply classifies the micrograph as falling into a low, medium or high magnification range. Certain relatively standard components can be used as built-in scale lines, just as in histology. The nucleus of a cell, a nearby red corpuscle or lymphocyte, or a clearly identifiable unit of cellular organisation such as a secretory acinus will indicate the lower range of magnification. The higher range is indicated by the clear identification of trilaminar membrane structures, ribosomes, collagen periodicity, parts of the formed cellular organelles such as mitochondria and cell contact specialisations. The beginner should try to gain experience of different tissue types over a wide range of magnifications in order to build up a concept of relative scale at the subcellular level. Numerical magnification figures are of little practical use without some familiar standards against which to make comparisons.

The magnification may be quoted in two ways, both used in this book. The *numerical value* may be given in the caption and a *scale line* may be provided on the micrograph. These two methods are interchangeable, since the length of any scale line is directly determined by the magnification of the micrograph. The most commonly used scale line is the micron or micrometre line, usually accompanied by the abbreviation 1 µm, or often simply µm. The various features of the micrograph can be compared with this scale to determine their dimensions.

The 1 µm scale line can be easily converted into a magnification factor. Its length should be measured in millimetres. Taking Plate 3 as an example, this will give a measurement of around 26 mm. This means that any part of this specimen which is one micrometre across will measure 26 mm on the micrograph. Simple proportion indicates that the magnification factor is expressed by the ratio of 26 mm to 1 µm. Since 26 mm are equal to 26 000 µm, it follows that the magnification of the print can be expressed as 26 000 ×. Hence, as a general rule, the magnification factor of a micrograph may be expressed by measuring the length of a 1 µm scale line in mm, and multiplying that figure by 1000. Of course, if the scale line is stated to represent 0.5 µm or 0.1 µm, the appropriate extra multiplying factor of 2 or 10 must be included in the calculation.

Conversely, when magnification is stated only in numbers, this can be equally readily converted in a 1 µm scale line. This is done by dividing the numerical value of the magnification by 1000, and stating this figure in millimetres. Thus a magnification of 10 000 will call for a scale line 10 mm in length.

There are smaller units of magnification which sometimes appear on micrographs in the highest range of magnification. The nanometre (nm) is now the standard unit. There are 1000 nm in each micron (µm). The Ångstrom Unit is still sometimes used. There are 10 Ångstrom Units in each nanometre (nm), and 10 000 Å in each micron (µm). In a very high magnification micrograph it might be convenient to indicate magnification by a scale line of 10 nm. At a magnification of one million times, such a scale line would measure 10 mm in length.

No magnification figure is ever exact. In normal use, a variation of at least ±5 per cent can be expected from the official scale reading of an electron microscope, while further uncertainties are introduced by the enlargement of the negative and by actual physical shrinkage of both negative and positive materials during photographic processing. If absolute standardisation is required, special care must be taken in all of these matters and the microscope must be calibrated before and after use with a standard specimen, such as a replica of a diffraction grating or a section of a crystal with a spacing of known periodicity. Even with such precautions, a potential error of two or three per cent will always remain. Far greater uncertainties than this are introduced into thin sections by compression artefacts, which distort the dimensions of the tissue components, as well as by shrinkage of tissues during fixation and dehydration. Electron micrographs of fixed tissues are an *unreliable* source of accurate dimensional data.

15.1.3 Making the most of the micrograph

The *whole area* of the micrograph must always be examined, despite a natural tendency to concentrate attention on the central features. The information recorded in the micrograph is only a tiny fraction of that contained in the tissue and none of it should be ignored. A systematic scan of the margins of the print as well as the centre should always be made. Intentional or unintentional bias by the microscopist can seriously diminish the significance of any collection of micrographs. Striking features, which make pleasing micrographs or are easy to focus upon, may become over-represented at the expense of less obvious, but perhaps more typical areas of the tissue. Errors in interpretation due to *selection* or bias on the part of an inexperienced observer are a particular hazard of electron microscopy, where high magnification implies a strictly limited coverage of a tissue sample. For this reason, it is essential always to ensure that any detailed ultrastructural study rests on a basis of *wide tissue sampling* and adequate numbers of representative low magnification survey pictures. In experimental and clinical work, the importance of adequate controls cannot be overstated.

15.1.4 Preliminary tissue identification

There are various clues to tissue identity which one can pick up from an unknown micrograph. The distinctive anatomical features of epithelia of various kinds and of nerve, muscle and connective tissue allow a tentative identification of different structures even before the details of cytoplasmic organisation are considered.

In epithelia the regular, orderly and close-packed appearance of the cells gives an indication of their nature. Extensive contact surfaces, limited intercellular spaces, the absence of an intercellular matrix and the frequent occurrence of contact specialisations are pointers to an epithelial identity. A limiting basal lamina related to a layer of polarised cells, with a clearly defined base and apex, is typical, as is the presence of microvilli on the apical free surface.

Areolar, or loose connective tissue, has a distinctive open appearance, with sparse cells, widely separated by intercellular material and with few contact specialisations. The presence of collagen indicates connective tissue. The irregular outlines of the cells may often lead to confusing plane of section effects, with small islands of cytoplasm indicating cells which have been sectioned in a plane which misses the nucleus. Blood vessels, lymphatics and small nerves are common features of loose connective tissue in many sites. Other specialised tissues, such as bone, cartilage, muscle and nerve are identified even at low magnification by their distinctive characteristics readily learned by experience. One should aim to become familiar with the characteristic patterns of widely different tissue types as soon as possible, since this experience will greatly assist the interpretation of electron micrographs.

After the initial assessment of general tissue architecture, the structural features of *individual cells* can be studied, noting surface specialisations of any kind, such as microvilli, cell contacts or membrane invaginations. The observer should examine the various organelles in a systematic fashion, noting the predominant features and thinking equally of what organelles are absent or poorly developed. In this way the many known associations between structure and function can serve as clues to cellular identity.

Much more doubt may arise in the interpretation of high resolution electron micrographs. At this level, one becomes increasingly aware that molecular patterns are distorted by processing, so that the final image, which relies so much on heavy metal stains, bears an uncertain relationship to the molecular organisation of the living state.

15.1.5 Tangential sectioning effects

The plane of section determines the appearance of any feature shown in a conventional thin section. An obscure or unfamiliar image may often be the result of an *oblique* or *tangential* plane of section. When a membrane is cut obliquely it no longer displays its familiar well-defined linear image, but appears *blurred*. When mitochondria are tangentially cut, their internal cristae may be clearly defined, while their limiting membranes are indistinguishable. This effect must not be confused with injury of the mitochondrion, or with rupture of its wall. Similarly, areas of obliquely sectioned cell membrane must not be taken as evidence of cell damage or focal rupture. Such effects are simply a reminder that an electron micrograph prepared from a thin section represents a *single plane* through a three-dimensional structure.

In some cases, the unconventional plane of section can give *additional* structural information not available from the more characteristic view. The most striking example is the appearance, in a tangential section, of pores in the nuclear envelope or in a fenestrated capillary. The plane of section permits a surface view of the pores which are normally only pictured in the more conventional sectioned view. Similarly, when cisternae of the granular endoplasmic reticulum are 'grazed' by the plane of section, patterns of ribosomes on their surfaces can be distinguished which are not apparent in other planes of section. Examples of this phenomenon are found to be numerous if looked for carefully.

15.1.6   Special aspects of surface scanning micrographs

Several special considerations enter the interpretation of the scanning micrograph. It is important to recall that the specimen is usually examined at an angle to the beam, giving a characteristic angled aerial view of the surface contours. This produces a marked *foreshortening* effect, which can distort the relative sizes of different features. Magnification figures express the linear enlargement of the frame width, but cannot be used to make direct measurements from such prints, since the perspective distorts the other dimensions to such an extent. Accurate measurements can only be made from *stereo-pairs* viewed under a measuring stereoscope. As in transmission microscopy, any attempt to make absolute measurements from scanning micrographs must be accompanied by control exposures of an accurately calibrated test specimen, examined under identical operating conditions. Even variations in the thickness or height of specimens will lead to variations in the effective final magnification. As a result, the scanning microscope is more often used for qualitative than for quantitative studies in biology, particularly in view of the considerable *dimensional changes* pro-duced by the drying stage of tissue processing for SEM.

It is important to remember that *staining effects* are virtually absent from surface scanning micro-graphs using secondary electron emission to form the image. The identity of particular structures must always be inferred from their contours, rather than from their staining intensity, in contrast to histology and transmission electron microscopy. Backscattered electron imaging, on the other hand, can be used to provide atomic number contrast, using metal stains such as silver. Whatever the signal being displayed, however, the scanning micrograph demands above all from the observer an acute *three-dimensional* awareness. Constant exposure only to the two dimensions of the conventional transmission micrograph can dull this faculty in the beginner.

15.2   THE RECOGNITION OF ARTEFACT

15.2.1   Desirable and undesirable artefacts

A histological artefact is any appearance or image which is not present in the living state but is introduced by the techniques of tissue preparation or examination. Hence all observations made on fixed and stained tissues are observations of artefact: much of conventional histology is concerned with the interpretation of systematic artefacts such as protein precipitation and dye binding. Despite this, however, the picture of conventional histology is now accepted as a tolerably close approach to some aspects of the organisation of the living tissues. The artefacts which make this possible may be regarded as desirable, since they are consistently reproducible and have proved meaningful in the general context of biology and medicine.

In the same way, the *desirable artefact* of electron microscopy has now become almost universally accepted. This artefact is also consistently *reproducible*, since the essentials of cell structure have now been confirmed in numerous different species and tissues. It is *meaningful* to modern biology and medicine since it can be rationally related to normal cellular function and to disease.

It is only at the highest magnifications that the uncertainties of systematic artefact become a major limitation to interpretation. It is questionable whether a living membrane, in constant physical and biological motion, is as clearly defined a structure as its trilaminar image suggests. At this level there is a kind of biological uncertainty principle; to observe the molecular patterns of life directly, we must first make them visible, but the means which we use often alter or destroy the original structural reality.

Considerations such as these have provided the impetus for the development of the new techniques of ultrathin *frozen sectioning*, *freeze fracturing* and *freeze etching*. By these means one can record certain structural features of cells which have been exposed to no more than a carefully controlled sudden freezing, a procedure that at least some cells and organisms can survive intact and alive. The most encouraging aspect of this pursuit of the artefact-free specimen has been the wide agreement between the structure of such frozen cells and the familiar images of conventionally fixed and embedded tissues. So far, at least, the desirable artefacts provided by conventional techniques have been found to represent a *tolerable approximation* to the living state.

Apart from the systematic artefact upon which conventional microscopy depends, there are other artefacts which may be introduced by faulty technique. Such defects are *undesirable*, since they are random in occurrence, may obscure useful information and may add to the difficulties of interpretation. Technical artefacts are particularly common in electron microscopy; the resolution of the technique is so great that imperfections which would remain unnoticed on light microscopy immediately become obvious. In the course of fixation, embedding, sectioning, staining and microscopy there are many possible defects, introduced by technical errors, which may at times mislead the observer or impair the quality or value of the results. An outline of some of these unwelcome artefacts is given below.

It is clear that the final electron micrograph must be examined with full awareness of all of these possible artefacts, since failure to distinguish them may lead to errors of interpretation. The student will rarely encounter, in set reading, any serious artefacts, since normally only those pictures which are technically adequate are published. On the other hand, it is important that those who may wish to interpret a collection of routine micrographs obtained during a research project should be familiar with possible sources of technical artefact. Confusion and errors of interpretation can be avoided only by an awareness of the *limitations* of the instrument and of the specimen at different levels of resolution.

15.2.2   Fixation and processing artefacts

Soon after the circulation is halted, complex biochemical changes begin to take place which lead to cell death and eventual dissolution. There are structural alterations which accompany these changes. The process of breakdown, called autolysis, is arrested by *fixation*. Although this kills the cell, it preserves its structural integrity by halting biochemical destruction of the cell components. Fixation normally relies on the stabilisation of structural protein and the inactivation of destructive enzymes by chemical means.

In order to forestall the structural damage of autolysis it is best to initiate the fixation of tissues as soon as possible after the interruption of the blood circulation. The optimum results for electron microscopy are obtained by perfusion fixation, which replaces the circulating blood by fixative solution without any form of intervening trauma to the tissue. It is thought that even a short period of anoxia may induce significant alterations in fine structure. Anaesthesia itself may modify particular details of cell structure. It is important to remember, however, that a certain proportion of cells have reached the end of their life span at a given time in any tissue. These 'physiologically' dying cells may show changes of fine structure which are not the result of artefact but are a natural event. In the intestinal epithelium, worn-out cells are shed constantly from the tip of the villus and it is common to encounter structural changes here which must not be confused with fixation damage or with the effects of disease.

Delays in fixation are usually associated with a range of changes in fine structure to which it is difficult to set clear limits. In general, fixation damage is seen as the presence of pale or empty areas within a cell, accompanied by clumping of cytoplasmic structures. There is an unacceptable coarseness of detail which will be obvious only after experience of examining the particular tissue. Other indications of damage are broken membranes, swollen and disorganised mitochondria, vacuolated cytoplasm and distended cisternae of the endoplasmic reticulum. The normal fine granular pattern of the nucleus may often be lost, being replaced by irregular patchy aggregations of chromatin with abnormal empty areas. Minimal damage of this type is of course very difficult to distinguish from normal fine structural variation. It has been suggested that delays in fixation may produce their characteristic structural effects on the cell by making its components more sensitive to damage during subsequent processing, rather than by causing immediate structural disruption.

Delay in fixation is not always enough to account for 'fixation damage'. Deviations from the accepted normal limits of fine structure may be induced by a fixative which is of unsuitable osmotic pressure or pH. Particularly damaging are hypotonic and acidic fixatives. The embedding process, particularly with the methacrylate method now for this reason seldom used, may itself cause damage to fine structure in well-fixed tissue. This *polymerisation damage* seems to result from the rupture of fine structural components by chemical crosslinking during the polymerisation process. Damage to fine structure may also be caused by the use of blunt knives and by crushing caused by clumsy manipulation while the tissue is being cut into pieces of a suitable size. If the blocks of tissue are too large, the fixative in which they are immersed may not penetrate to the centre of the block sufficiently rapidly to arrest the processes of autolysis.

Other processing faults may lead to technical artefact. Inadequate dehydration, often due to moisture contaminating the absolute alcohol, may result in failure of the embedding medium to penetrate the tissue, which is then poorly supported after polymerisation. Errors in the embedding schedule or in the composition of the resin mixture may result in a block which is too hard, too soft or of irregular consistency, making it difficult to produce thin sections of good quality. The sections may break up, making microscopy impossible, or

pits and holes may destroy the details of fine structure.

15.2.3 Sectioning artefacts

The ultrathin sections used for electron microscopy suffer from two common types of defect. A poor knife with a rough cutting edge causes *scoring* of the section, which appears on examination as parallel lines running in the direction of cutting. The edge of the glass knife is so delicate that any touch will damage it, so that knives must be handled with extreme care to avoid this type of artefact. The second common artefact in thin sections is '*chatter*'. This is a regular transverse banding or rippling of the section, parallel to the edge of the knife, often so fine that it is visible only on examination with the electron microscope. This artefact may make the sections worthless for examination at low or medium magnification. Its cause may be found in slackness in the mounting of the knife or the block, or in external vibrations affecting the ultramicrotome. The periodic variation of section thickness which leads to this banded appearance may be aggravated by faults in the consistency of the block.

Compression of thin sections is inevitable during cutting. Secretion granules, which are normally round, appear oval or flattened as a result of compression and other parts of the tissue are equally, if less obviously, affected. Compression may therefore cause serious inconsistencies in measurements of cell components. Compression artefact is an important limiting factor in accurate calculations of dimensions of membranes and other structures. Compression artefact can be corrected in part by the use of xylol vapour to expand the sections as they lie on the surface of the trough after cutting. *Contamination* of the sections by dirt after cutting is a troublesome source of artefact. Dirt from the trough or a finely particulate or microcrystalline precipitation from the staining solution may obscure details of structure. Staining contamination is often so widely distributed over the section that no free area remains suitable for recording. Contamination of sections not only obscures details but affects image stability at high magnification, causing drift and leading to secondary contamination of the microscope column, with loss of resolution.

15.2.4   Artefacts in transmission microscopy

The exposure of a section to the electron beam causes considerable heating due to the absorption of energy from the beam. If the beam is focussed too strongly on the section by the condenser lens, the section may stretch and tear. Prolonged exposure to the electron beam burns the surface of the section, causing sublimation of the embedding medium, and blurs the outlines of the tissue. During irradiation by the beam there is a fine *contamination* deposited in a layer on the surface of the section, leading to progressive clouding of the specimen. Much of this originates from back-streaming of oil vapour in the pumping system. In addition, the great intensity of the irradiation to which the specimen is subjected during ordinary examination is itself enough to destroy the finer details of tissue organisation. Thus on prolonged examination of a thin section there is a steady loss of resolution. Many of these artefacts can be reduced by the use of a cooling device, although this may introduce fresh problems: prolonged exposure of sections to the beam in the presence of traces of water vapour can lead to etching and thinning of the specimen if an anticontamination device is used.

Two other common artefacts in electron microscopy, *drift* and *astigmatism*, may lead to loss of definition in the micrograph. Drifting of the image may be caused by movement of the specimen or by dirt in the column which may accumulate static charge and deflect the beam. Uneven heating of the section due to dirt on its surface, or tearing at splits in the section, may also cause movement of the image. If there is image drift, the micrograph will appear blurred, since exposures of two seconds are routinely used for recording the image on the photographic plate. Astigmatism, a defect in the electron optical system of the microscope, causes a fine blurring of the image in one direction. Astigmatism becomes a particular problem at high magnification, when small defects in the optical system of the microscope are made more obvious. Astigmatism is commonly caused by dirt on an aperture or lens, where it can have an effect on the electron beam.

Failure to focus the image correctly before recording it is a common fault in electron microscopy. At the exact point of focus the image formed by the microscope lacks contrast. Contrast can be increased by putting the objective lens slightly out of focus. When this is done, interference fringes form around the components of the image, giving a pleasing but false appearance of sharpness. The final micrograph lacks resolution and clarity of fine detail when taken under such conditions.

15.2.5   Artefacts in scanning microscopy

Since the magnifications used are often relatively low and since sections are not often examined, fixation artefacts are not resolved as they are in conventional transmission electron microscopy and are therefore less important. However, since the specimen is dehydrated and usually examined without the use of a supporting resin, the dehydration process is very important. Freeze-drying and critical point drying are now routinely used, since they avoid the artefacts of air drying from the liquid phase. The unsupported tissue is brittle and any damage done during imperfect dehydration will show as cracks, varying from large gaps in the tissue to fine lines, punctures or collapsed areas, seen at high magnification. Such artefacts are all usually fairly easy to recognise.

Another artefact caused by imperfect preparation is called charging. It is due to electrons accumulating on the surface of the specimen and failing to drain to earth. This is caused either by imperfect or cracked coating or by a bad contact between the specimen and its holder. Charging may appear as bright areas, either large or small, or as lines on the image, or as distortions of the shape of the image. Such appearances should not often be seen by the student in published works, but they present significant problems to the beginner in the production of good micrographs and are therefore worthy of mention.

15.2.6   Photographic artefacts

The final recording of the electron image on photographic plate or film is an important part of electron microscopy. The viewing screen of the microscope, which is examined directly by the observer, generally gives poor resolution so that the photographic film must serve as the main permanent record of the specimen. Various artefacts may result from faulty exposure or from damage to the emulsion, such as scrapes and chips. The final electron micrograph printed from the negative can only be as good as the photographic enlarger on which the print is made. Poor lenses can lose much of the fine definition, which may be present in the negative, yet may never reach the print. Dust or dirt on the glass of the negative carrier may mar the enlargement and may be wrongly blamed on a negative fault instead of being recognised as an easily corrected defect of technique.

ELECTRON MICROGRAPHS

This final section consists of a collection of electron micrographs, which illustrate many of the points discussed in the text. The micrographs have been grouped according to the main chapter headings, although often a single plate may illustrate more than one aspect of biological morphology. Appropriate cross-references, detailed labelling and extensive captions have been employed to reinforce various lessons of interpretation for the novice.

THE CELL

Plate 1 *Cell membrane*

This is a transmission electron micrograph of a carbon replica made of the fractured and etched surface of a block of frozen tissue. The striking image contrast has been obtained by shadowing the surface of the replica with heavy metal. This technique shows some of the molecular detail of the cell membranes in a specimen which has not suffered the solvent extraction artefacts conventional fixation, dehydration, embedding and sectioning. For this reason, these details are thought to be an accurate representation of the living state.

The micrograph shows detail of the intercellular junctions in the stratum spinosum of the epidermis. A gap junction (G) is identified by the presence of a rounded patch of closely-packed membrane particles. These represent the individual macromolecules responsible for the functional properties of the gap junction, in particular its unusually high permeability to ions. One desmosome (D_1) has fractured obliquely. Another (D_2) has fractured transversely, showing a row (R) of aligned particles lying between the two outer cell membrane components of the desmosome. This represents the intercellular dense component of the sectioned desmosome.

Randomly distributed particles (P) can be seen on the P face, the surface of the inner leaflet of the cell membrane, at points where gap junctions and desmosomes are not in evidence. These particles correspond to protein macromolecules traversing the full thickness of the membrane, as proposed by the fluid mosaic model of membrane structure.

The cytoplasm adjacent to the fractured membranes shows the presence of numerous tonofilaments (T). Several dome-shaped structures lying near the cell surface represent the membrane-coating granules (M) of the keratinocyte.

D_1	Desmosome, obliquely fractured
D_2	Desmosome, transversely fractured
G	Gap junction
M	Membrane-coating granules
P	Particles associated with the cell membrane at non-specialised areas
R	Row of aligned particles, at mid-point of desmosome
T	Tonofilaments

Tissue Epidermis of calf snout. Freeze-etched.

Magnification 92 000 ×

Refer to Plates 2, 5, 6, 27b, 73
Sections 2.1, 2.2, 4.8, 8.3, 14.6

Micrograph by courtesy of R. Leloup

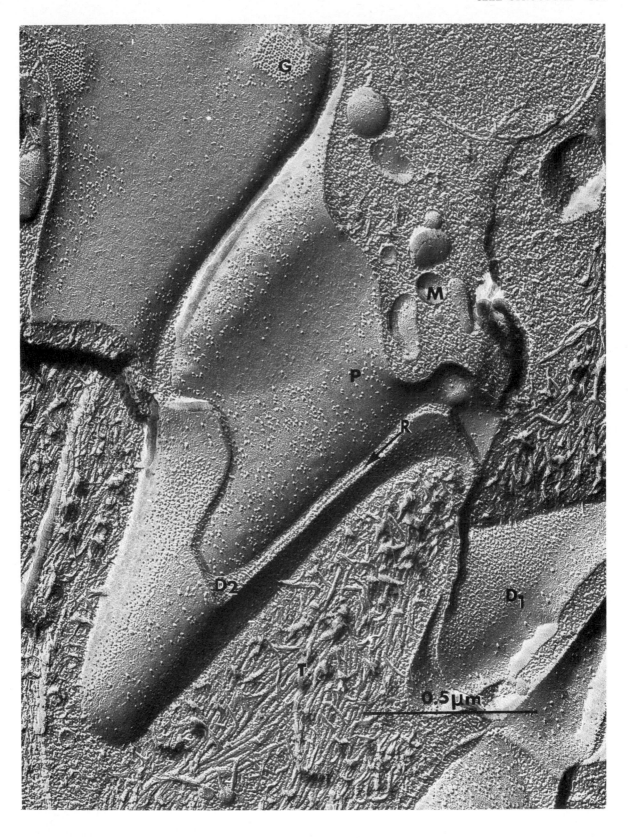

Plates 2a, 2b *Cell surfaces*

Scanning electron microscopy (SEM) is often used to illustrate the surface features of tissues and cells. In these two scanning micrographs, the features of individual cells are clearly seen. In Plate 2a, a group of cancer cells (C) is surrounded by cells of the blood, including red corpuscles (R) and white cells (W). The red corpuscles have become spiky as a result of the processing. The cancer cells form an adherent group, with irregular surface projections. In Plate 2b, the individual surface projections of a cancer cell are seen at higher magnification. This technique emphasises the three-dimensional reality of cells and their membrane specialisations.

C   Group of cancer cells with numerous surface projections
R   Red blood corpuscles, crenated by processing
W   White blood cells

Tissue Cancer cells from human serous effusion. Glutaraldehyde and osmium fixation, uranyl and lead staining.

Magnification   Plate 2a 5000 ×
                  Plate 2b 31 000 ×

Refer to   Plates 1, 34, 47, 69, 72, 114
             Sections 14.3, 14.5

Micrographs courtesy of K. Saleh

Plate 2c *Sectioned tissue surfaces*

Scanning electron microscopy is not always used to look at natural surfaces. This plate shows the surface view of a section through a small intestinal crypt. Clusters of Paneth granules (P) can be seen in the centre of the crypt.

P   Paneth granules

Tissue Human small intestine, fixed in formalin, embedded in paraffin wax. Section cut, then dewaxed. Air dried. Carbon/platinum coated.

Magnification   4500 ×

Refer to   Plates 1, 34, 47, 69, 72, 114
           Sections 2.1, 2.2, 14.3, 14.5

Plate 2d *Epithelial surfaces*

This scanning electron micrograph shows another unusual view of small intestinal mucosa. here the surface epithelial sheet of a villus has partly stripped off from the underlying connective tissue. It has then folded back on itself, allowing a view of several aspects of epithelial morphology. The usual, luminal aspect of the enterocytes can be seen (A) with the polygonal outlines of the cells clearly visible. At the broken edge of the epithelial sheet, the lateral aspect of the enterocytes can be seen (L) showing the cells as columnar in shape, with a polygonal apex and a tapering contour towards the base of the cell.

A further view of this basal aspect of the sheet of cells is seen where it has turned upside down (B). Here the cells are seen as pegs, with irregular circular bases. Particularly interesting are the intercellular clefts seen between the bases of the cells (*). These various features reflect the absorptive role of the cells.

A   Apical aspect of enterocytes
B   Basal aspect of enterocytes
L   Lateral aspect of enterocytes
*   Indicate intercellular clefts between enterocytes

Tissue Human small intestine. Fixed in formalin. Embedded in paraffin wax. Dewaxed and critical point dried. Gold coated.

Magnification   1000 ×

Refer to   Plates 1, 34, 47, 69, 72, 114
           Sections 2.1, 2.2, 6.1, 14.3, 14.5

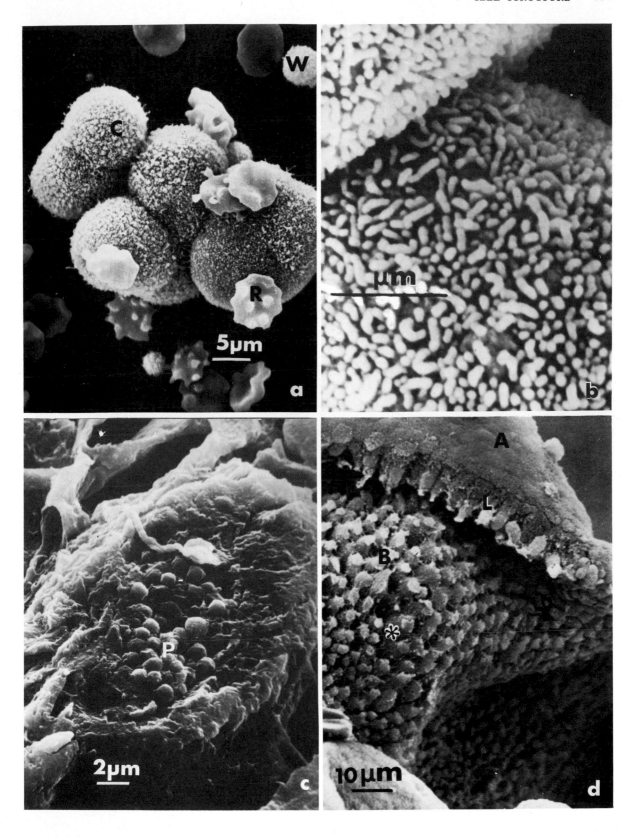

Plate 3 Cell components

The plasma cell seen in this low power micrograph displays many of the main cytoplasmic components which are commonly encountered in the study of cells. The appearances seen are typical of an active protein-secreting cell. The prominent cisternae of the granular endoplasmic reticulum lying in the peripheral cytoplasm are filled with finely granular or flocculent material of appreciable density (GER). This appearance represents stored antibody newly synthesised in the endoplasmic reticulum. The large Golgi apparatus (G), formed from several units of Golgi structures, is typical of this cell type. Lamellae, vacuoles and vesicles are all seen, the vesicles being particularly numerous. The mitochondria are not prominent (M).

The plasma cell lying in connective tissue is separated by a significant space (S), from its neighbours. A small portion of a fibroblast (F) is present and a few collagen fibrils are seen in the upper left hand corner of the micrograph. The rounded contours of the cell and its solitary appearance are evidence that this cell is an independent unit, rather than part of a gland with many related cells in close contact. The appearance of this micrograph is that of loose connective tissue.

In a number of respects this micrograph illustrates the principles and importance of tangential section effect. The surface membrane of the cell at the point marked C has been cut obliquely, losing the sharp well-defined appearance which is seen elsewhere and which is typical of any membrane. This appearance must not be interpreted as rupture of the cell surface. There is still an area of diffuse appreciable density representing the obliquely sectioned membrane. If the membrane were ruptured the damaged ends would perhaps be seen and the escape of cytoplasmic components from the cell might be observed. The nucleus (N), also shows tangential sectioning effect. It is small, and its outline is diffuse while the nuclear envelope is not resolved clearly in its typical form as seen in other micrographs. This suggests that the main bulk of the nucleus lies outside the plane of section, and the present view is the result of a 'grazing' section through the edge of the nucleus. The oblique angle which this imposes on the membranes of the nuclear envelope makes it impossible to resolve its structure clearly and the perinuclear cisterna is not well demonstrated as a result. The oblique cut, however, does permit a semi-surface view of the nuclear envelope to be obtained at its margins. The 'face view' of the nuclear pores, surrounded by their annuli, is obvious at the points arrowed.

The cisternae of the granular endoplasmic reticulum also show areas of tangenital section. In the lower right-hand corner of the micrograph, the endoplasmic reticulum is present, but its cisternae are cut obliquely. In this way a 'surface' view of the ribosomes (R) is obtained as they lie in association with the membranes limiting the cisternae. Rosettes and spirals of ribosomes are numerous, suggesting that the members of each group may be functionally associated with each other. In this way tangential effects can be used to obtain information not otherwise readily available concerning parts of the cell.

C	Cell membrane cut obliquely
F	Fibroblast
G	Golgi apparatus
GER	Granular endoplasmic reticulum
M	Mitochondrion
N	Nucleus
R	Ribosomes
S	Connective tissue space with collagen fibrils
Arrows	Indicate nuclear pores seen in face view

Tissue   Human small intestine. Osmium fixation, lead staining.

Magnification 26 000 ×

Refer to   Plates 15, 68
Sections 5.1, 7.2

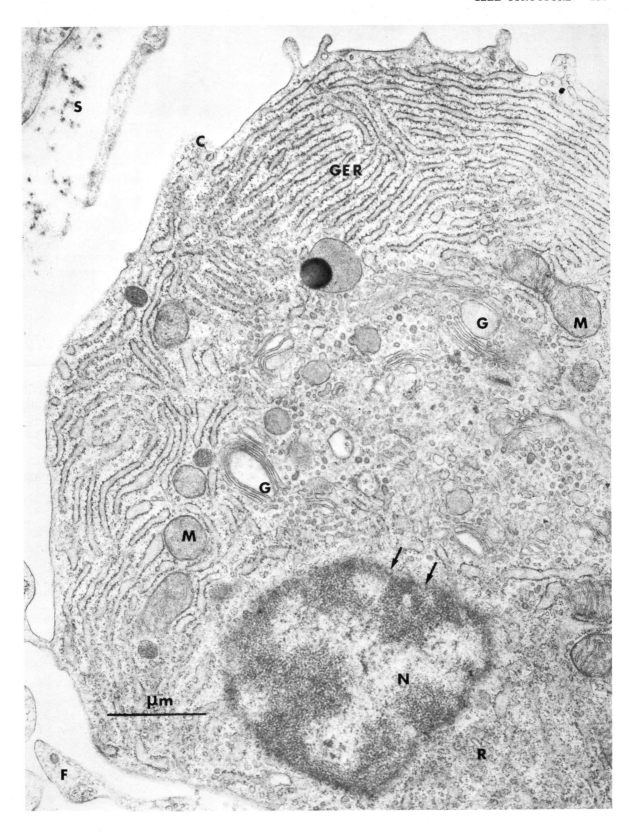

Plate 4a *Trilaminar membranes*

The apical part of this submucosal gland cell from fowl proventriculus shows aspects of the ultrastructure of cell membranes. Two adjacent cells are shown (A, B) and their lateral cell membranes can be seen meeting in a convoluted way at C. Each membrane is trilaminar, although where either membrane turns obliquely to the plane of section (O), the detail of its fine structure becomes blurred. At the apical part of the contact surface between the two cells there are two junctional specialisations. The more apical is a close junction or zonula occludens (ZO) and immediately below this is a zonula adhaerens (ZA). There is an area of increased electron density at the cytoplasmic face of the zonula adhaerens, although this cannot be resolved as filamentous structures in this micrograph. The trilaminar membrane structure can also be seen bounding the stubby microvilli that protrude from the apex of these cells (MV). Cisternae of granular endoplasmic reticulum (G) are studded with ribosomes and a mitochondrion is present (M).

The trilaminar outer and inner mitochondrial membranes are just visible (arrows) although the angle of the section does not allow the mitochondrial cristae to be made out, since they lie parallel to the plane of the page. Notice that the cell surface membrane is markedly thicker than the membrane of the mitochondrion or of the endoplasmic reticulum.

A   Secretory cell of proventriculus
B   Adjacent secretory cell
C   Lateral cell membranes showing trilaminar structure
G   Granular endoplasmic reticulum
M   Mitochondrion
MV   Microvilli
O   Oblique plane of section of lateral cell membranes
ZA   Zonula adhaerens
ZO   Zonula occludens
Arrows   Show trilaminar pattern of mitochondrial membrane

Tissue Fowl proventriculus, submucosal gland cell. Glutaraldehyde and osmium fixation. Uranyl and lead staining.

Magnification   80 000 ×

Refer to   Plates 6, 48, 89, 90
             Sections 2.1, 2.2, 5.2

Plate 4b *Micropinocytotic vesicles*

This micrograph shows the apical border of an intestinal absorptive cell from the gut of a suckling neonatal mouse. Many newborn mammals absorb whole antibody molecules from the milk, thus acquiring passive immunity to infection. Such absorption occurs through the process of micropinocytosis, involving the invagination of the cell apex to form a down pocket between microvilli (V_1). Vesicles (V_2) then pinch off and become isolated in the cell. Their contained antibodies are then transported across the cell and released into the circulation. This selective form of micropinocytosis involves a marked fuzzy thickening of the cytoplasmic surface of the membrane, indicating localised molecular specialisation. This process stops as the infant mouse matures. After infancy, micropinocytosis no longer plays any significant part in intestinal absorption.

V_1 Micropinocytotic vesicle pushing down from the cell apex
V_2 Vesicle completely isolated from the apex of the cell

Tissue Neonatal mouse small intestine, glutaraldehyde and osmium fixation. Uranyl and lead staining.

Magnification   70 000 ×

Refer to   Plates 14, 48, 53
             Sections 2.2, 6.1

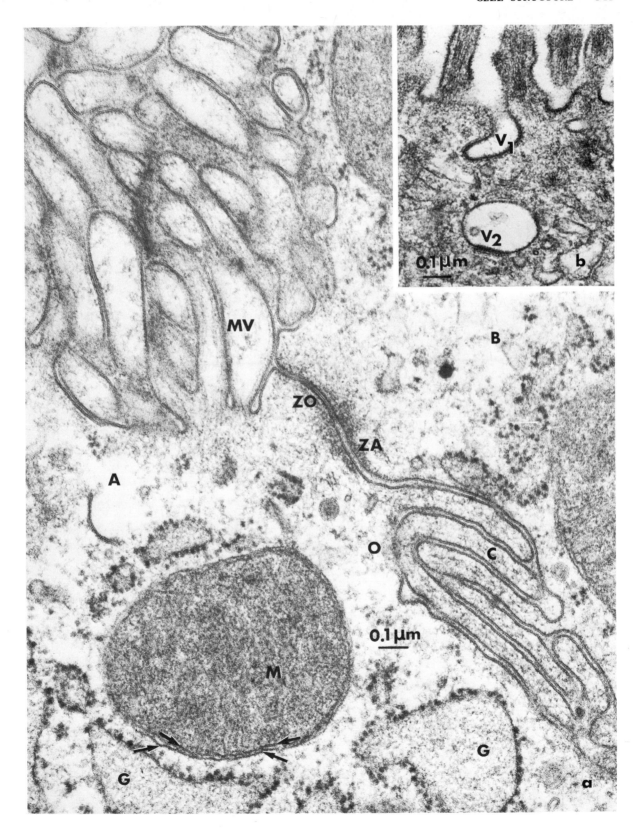

Plate 5 *Cell surface specialisations*

This micrograph of the stratum basale (basal layer) of human skin shows how the interdigitating cytoplasmic processes (P) from adjacent keratinocytes, or epidermal cells, have their contact strengthened by the presence of many desmosomes (D) identified by the associated dense filament bundles. Hemidesmosomes (arrows) can likewise be seen along the membrane of the cell base, strengthening the contact between the membrane and the basal lamina (BL). Keratin tonofilaments (T) can be seen extending inwards from desmosomes and hemidesmosomes, towards the centre of the cell. In one cell, the nucleus has been cut to show the contrasting patterns of euchromatin (E) and heterochromatin (H) along with two nucleoli (N), whereas in an adjacent cell the nucleus has been cut obliquely (X). Notice how the filament bundles in these cells loop around the nucleus and interconnect with desmosomes to form a complex intracytoplasmic skeleton.

Below the epidermal layer, features of connective tissue can be recognised. These include collagen fibrils (C) and an oblique section through a fibrocyte, identified by its irregular outline, and by its granular endoplasmic reticulum within the cytoplasm (G).

BL	Basal lamina
C	Collagen fibrils of dermis
D	Desmosome
E	Euchromatin
G	Granular endoplasmic reticulum within fibrocyte of dermis
H	Heterochromatin
N	Nucleolus
P	Interdigitating processes of keratinocyte
T	Tonofilaments
X	Obliquely sectioned nucleus
Arrows	Show hemidesmosomes at the cell base

Tissue Human skin. Glutaraldehyde and osmium fixation, uranyl and lead staining.

Magnification 17 000 ×

Refer to Plates 1, 6, 73, 74
Sections 2.2, 8.3, 9.1

Micrograph by courtesy of C. Skerrow

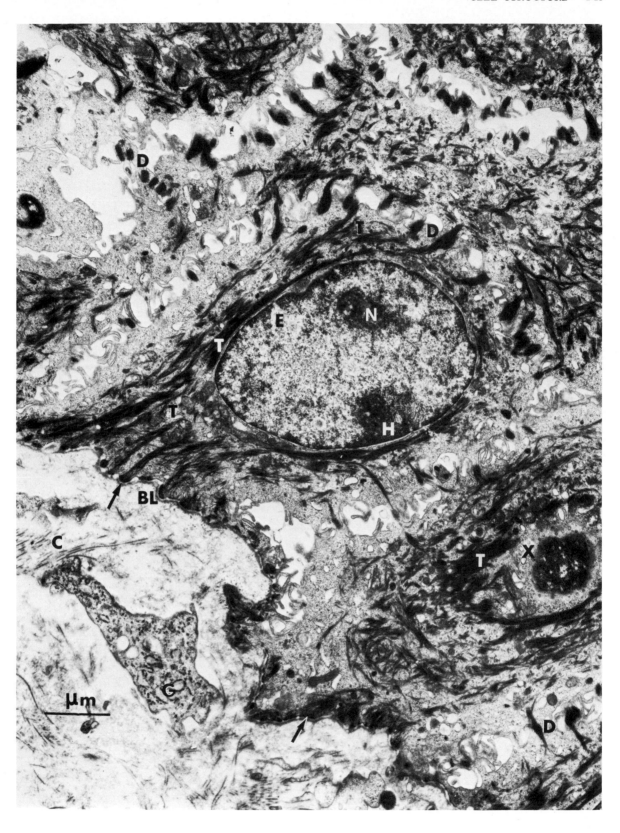

Plate 6a *Cell surface specialisations*

This micrograph shows a longitudinal section of the apical parts of several small intestinal crypt cells. A Paneth cell (P) is identified by its characteristic secretion granules. The adjacent cell is an undifferentiated crypt cell (U), with a few apical microvilli (MV) extending into the lumen (L). A group of mitochondria is identified (M). This cell is attached to its neighbours by prominent junctional complexes (J), associated with well-marked interconnected desmosomes (D). An apical granule (G) serves as an indication that crypt cells have some secretory role, as well as undertaking the function of stem cells.

D	Interconnected desmosomes
G	Apical secretion granule
J	Junctional complex
L	Lumen of intestinal crypt
M	Mitochondrion
MV	Microvilli
P	Paneth cell
U	Undifferentiated crypt cells

Tissue   Human small intestinal crypt. Glutaraldehyde and osmium, fixation, uranyl and lead staining.

Magnification   28 000 ×

Refer to   Plates 4, 5, 14, 47, 48, 90
Sections 2.2, 4.8

Plate 6b *Cell surface specialisations*

This electron micrograph shows part of the cytoplasm of a squamous cell from the epidermis. The elaborate arrangement of desmosomes with associated tonofilaments (T) underlines the important part played by cell adhesion in the biology of the skin. The nearby cytoplasm contains many free ribosomes (R) while only a few are attached to the membranes of the granular endoplasmic reticulum. Notice the influence of plane of section on desmosome morphology. Desmosomes cut at right angles are clearly defined (D$_1$) while those sectioned obliquely are fuzzy in outline (D$_2$).

D$_1$	Desmosome sectioned at right angles
D$_2$	Desmosome sectioned obliquely
R	Ribosomes in the cytoplasm, not attached to membranes
T	Tonofilaments

Tissue   Human epidermis. Glutaraldehyde and osmium fixation.
Lead and uranyl staining.

Magnification   30 000 ×

Refer to   Plates 4, 73, 107c
Sections 2.1, 2.2, 8.3

Micrograph by courtesy of C. Skerrow

Plate 6c *Cell surface specialisations*

This high magnification transmission electron micrograph shows clearly the layered structure of the desmosome. The prominent dense zone on either side (D) is the attachment plaque of the desmosome, into which the keratin tonofilaments (T) are inserted. The pale interspaces of the two trilaminar cell membranes are clearly seen (S), while the median density of the desmosome, the intercellular component, can just be made out (arrows).

D	Desmosome attachment plaque
T	Tonofilaments forming a dense feltwork
S	Pale interspace of the trilaminar membrane pattern
Arrows	Indicate median extracellular component of the desmosome

Tissue   Human oesophagus. Glutaraldehyde and osmium fixed. Uranyl and lead stained.

Magnification   140 000 ×

Refer to   Plates 4, 73, 107c
Sections 2.1, 2.2, 4.8, 8.3

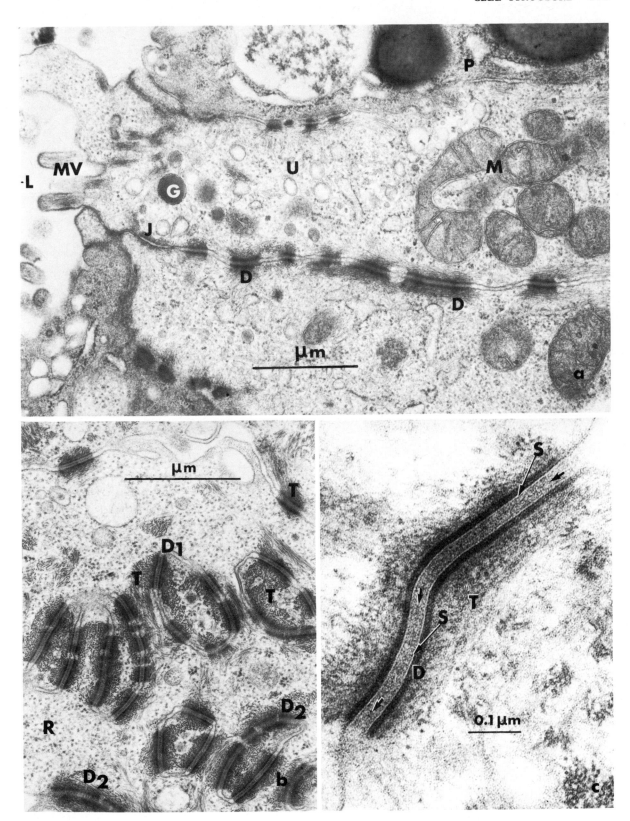

Plate 7a *Cell coat*

This electron micrograph shows the apical parts of two densely stained colonic absorptive cells on either side of the tip of a goblet cell, with mucus granules (G) on the point of discharge into the lumen. The absorptive cells, or enterocytes, have an elaborate striated border consisting of closely packed microvilli extending outwards from the cell surface. External to the tips of these microvilli lies a felt-work of fine filamentous material, (F) forming an external cell coat. Since these extracellular filaments consist of glycoprotein, the coat is often termed the glycocalyx. It is manufactured by the underlying epithelial cells. It should be noted that the glycocalyx is only rarely seen as clearly as this, and that the micrograph has been slightly overexposed to enhance the contrast of the fuzzy coat. Finally, notice that numerous tiny round vesicles form streamers between the microvilli of the striated border. These probably originate from the absorptive cells but their significance is unknown.

F　　Fine filaments composing the glycocalyx
G　　Granules of mucus in the act of discharge into the lumen

Tissue　Guinea pig colon. Glutaraldehyde and osmium fixation. Uranyl and lead staining.

Magnification　13 000 ×

Refer to　Plates 33b, 34, 38, 39b, 49, 113b, 115
　　　　　Sections 2.2, 5.1, 14.6

Plate 7b *Cell coat*

This electron micrograph has been stained by a special procedure to demonstrate carbohydrate-rich material. The carbohydrate chemical groupings have been oxidised by periodic acid and have then been reacted with a silver proteinate compound, which results in an electron dense deposit at the reactive site. Note the reactivity of the filaments (F) of the glycocalyx and the presence of small secretion granules of reactive material close to the cell surface (G). The micrograph shows the apical part of a malignant cell of colonic epithelial origin.

F　　Fine carbohydrate-rich filaments of the glycocalyx.
G　　Small apical carbohydrate-rich secretion granules

Tissue　Human colonic adenocarcinoma. Glutaraldehyde and osmium fixation. Periodic acid silver proteinate stain for carbohydrate.

Magnification　41 000 ×

Refer to　Plates 33b, 49, 113b, 115
　　　　　Sections 2.2, 14.6

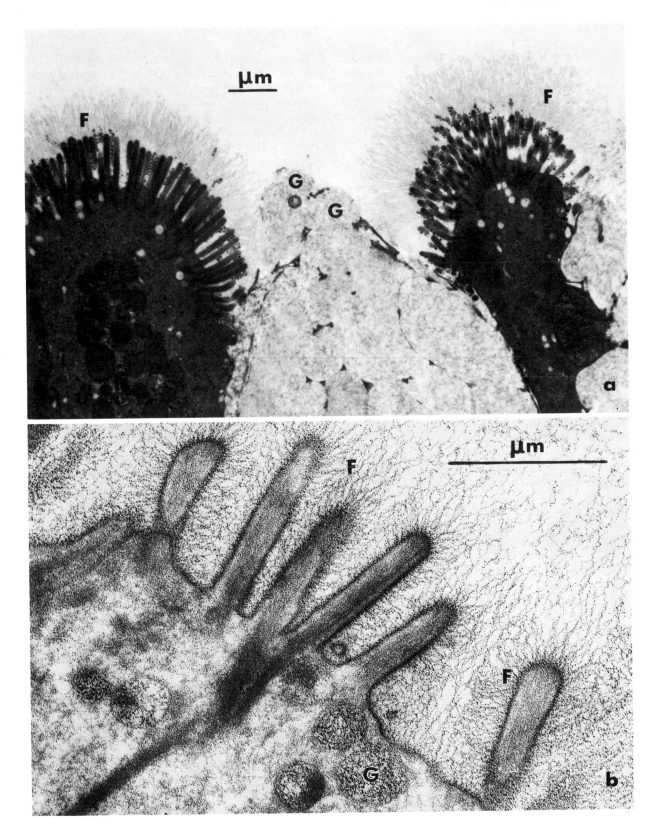

CELLULAR MEMBRANE SYSTEMS

Plate 14 *Intestinal epithelial cells*

This micrograph shows parts of the apical cytoplasm of two intestinal columar absorptive cells, to illustrate the complexity of cellular membrane systems. The cells are actively engaged in fat absorption. Closely packed microvilli form the apical striated border of these cells. The lateral intercellular boundary runs vertically, but the two cell membranes are elaborately interdigitated (I). The junctional complex is seen at the apex of the contact surface (J). Occasional small desmosomes can just be seen (D).

Within the cytoplasm, granular or rough endoplasmic reticulum (R) is present as parallel strands, but in rather small amounts. Much more prominent is the agranular or smooth reticulum (S), which occupies much of the cell apex. The Golgi apparatus (G) is not particularly clearly seen, since it is loaded with absorbed fat, but its groups of cisternae can be made out above the nucleus. Mitochondria (M) with their complex membrane configurations and lysosomes (L) with their single limiting membrane are of typical appearance. The nuclei show clearly the peripheral clumps or blocks of heterochromatin.

The progress of fat absorption can be clearly seen in this micrograph. Notice the absence of micropinocytosis. Fat is not absorbed in bulk, but at molecular level, in a partially hydrolysed state. Resynthesised triglyceride appears in the apical cytoplasm as droplets within the agranular reticulum (arrows). These droplets are carried to the Golgi system, where they accumulate in quite large aggregates (*). Finally, lipid droplets are trans-ported to the lateral surface of the cell and are discharged into the intercellular space (X). From here, these lipid droplets, or chylomicrons, pass to the lamina propria and are carried off by the intestinal lymphatics.

D	Small desmosome
G	Golgi apparatus
I	Interdigitating lateral cell boundaries
J	Junctional complex
L	Lysosome
M	Mitochondrion
R	Granular endoplasmic reticulum
S	Smooth endoplasmic reticulum
X	Lipid droplets or chylomicrons in the inter-cellular space
Arrows	Indicate lipid droplets within smooth endo-plasmic reticulum
*	Indicates lipid aggregated in the Golgi apparatus

Tissue Mouse small intestinal epithelium. Glutar-aldehyde and osmium fixation; uranyl and lead staining.

Magnification   13 000 ×

Refer to   Plates 4, 6, 15, 18, 27a, 36, 38, 47, 48
Sections 2.1, 2.2, 4.2, 4.4, 6.1

Micrograph by courtesy of H. S. Johnston

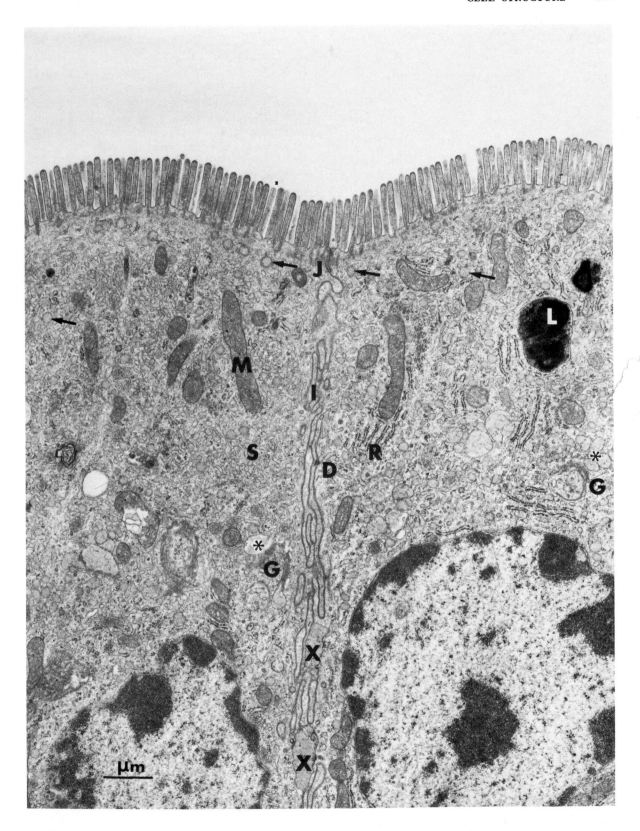

Plate 15 *Golgi apparatus*

This low-magnification view of a single plasma cell allows the entire layout of its complex membrane systems to be appreciated in relation to the nucleus. As the perinuclear cisterna is clearly visible all round the nucleus, the plane of section must pass fairly close to the equator of the nucleus, since otherwise oblique sectioning effects would have blurred its outline. The Golgi system typically lies close to the nucleus, forming a pale zone on light microscopy, since it fails to stain with the marked basophilia of the surrounding granular reticulum. Multiple groups of Golgi membranes are seen (G) with numerous associated vesicles.

The granular reticulum (R) fills most of the remainder of the cytoplasm. The cisternae are of variable width, and contain granular material of intermediate density, the newly synthesised immunoglobulin. In places, the membranes turn almost parallel to the plane of section (*), revealing the pattern of the polyribosomes attached to the face of the membrane. Notice the typical dense clumped heterochromatin around the inner nuclear membrane, with pale euchromatin channels leading to nuclear pores.

G   Golgi membrane stacks
R   Granular endoplasmic reticulum
*   Indicates obliquely sectioned membranes with face view of polyribosomes

Tissue   Human nasal polyp, plasma cell. Glutaraldehyde and osmium fixation. Uranyl and lead staining.

Magnification   19 000 ×

Refer to   Plates 3, 14, 16, 68, 80b
                Sections 4.2, 4.4, 5.1, 7.2

Micrograph by courtesy of I. A. R. More

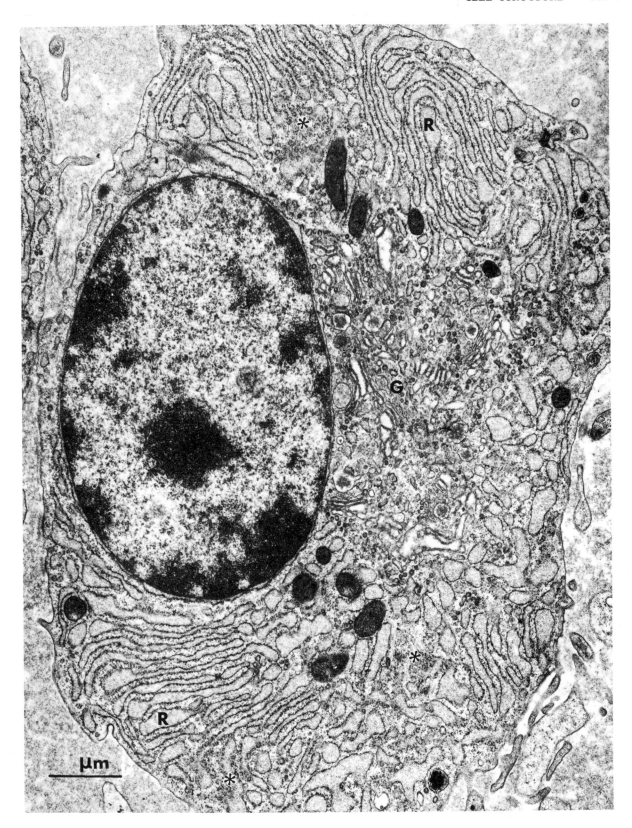

Plate 16a   *Granular endoplasmic reticulum*

This micrograph shows part of a plasma cell, the function of which is to produce and secrete antibody molecules. The cell is, therefore, adapted for large-scale protein synthesis, and is characterised by an elaborate granular endoplasmic reticulum. The cisternae (R) shown here contain finely granular material, which represents newly synthesised immunoglobulin. A few dense intracisternal granules are also seen. These presumably represent condensations of secretion material. Such granules are not uncommon in various types of protein-secreting cells. The plasma cell, however, releases its product not as granules, but at the molecular level.

Close examination of the cisternae of the granular reticulum reveals that the membranes are studded on their external surfaces by numerous ribosomes. In some areas, the membranes of the cisternae are turned obliquely to the plane of section. When this occurs (*) the membrane is no longer seen as a clearly defined line, but becomes blurred. Instead, the arrangement of ribosomes on the face of the membrane becomes apparent. Plane of section effects of a similar type are seen in some mitochondria (M). Where the limiting membrane is not clearly visualised, this may often be due to oblique sectioning effects, as seen also at X in the nuclear envelope.

Notice the particular prominence of the dense heterochromatin blocks (H) separated by pale euchromatin channels (E). This is so characteristic that the term 'clock face' or 'cartwheel' nucleus is often used in describing plasma cells. A final feature is the presence of a faint discontinuous external lamina (arrow) at one point. This is not usually seen in association with a lymphoid cell, but such laminae have been described around plasma cells on some occasions. The significance of this finding is not clear.

E   Euchromatin channel
H   Heterochromatin
M   Mitochondrion showing blurred limiting membrane, due to plane of section
R   Cisternae of granular endoplasmic reticulum
X   Obliquely sectioned nuclear envelope
Arrow   Indicates faint external lamina
*   Indicates areas of obliquely sectioned mem-

brane of GER, showing ribosomes in face view

Tissue   Human nasal polyp. Glutaraldehyde and osmium fixation. Uranyl and lead staining.

Magnification   22 000 ×

Refer to   Plates 3, 14, 15, 16, 17, 68
             Sections 4.2, 5.1, 7.2

Micrograph by courtesy of I. A. R. More

Plate 16b   *Granular endoplasmic reticulum*

This micrograph shows a higher power view of closely packed parallel cisternae of granular endoplasmic reticulum. The cisternae (C) are filled with slightly flocculent material, the stored product of cellular synthesis. The ribosomes are attached to the outer, cytoplasmic surfaces of the membranes. In the lower left hand portion of the micrograph the cisternae are sectioned obliquely. Their membranes are thus not clearly defined, but blurred, and the polyribosomes attached to their surface are seen in face view (*).

C   Cisternal lumen of granular endoplasmic reticulum
*   Indicates face view of polyribosomes attached to membranes obliquely sectioned

Tissue   Magnum, hen oviduct, glutaraldehyde and osmium fixation, uranyl and lead staining.

Magnification   51 000 ×

Refer to   Plates 3, 14, 15, 16, 17, 68
             Sections 4.1, 4.2, 5.1

Micrograph by courtesy of H. S. Johnston

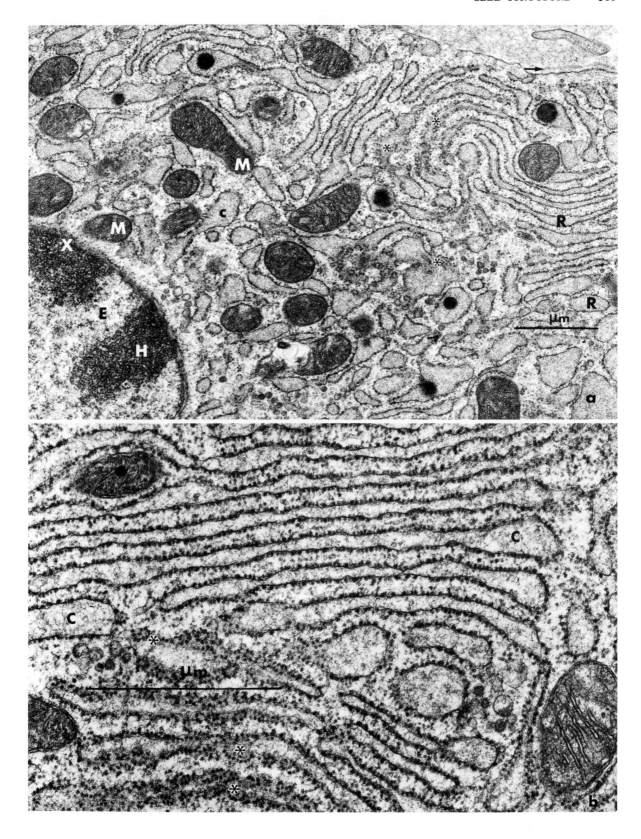

Plate 17a *Intracisternal granules in pancreas*

This micrograph shows part of an exocrine zymogenic cell from a guinea pig, in which the nucleus (N) and several mitochondria (M) are seen. The numerous dense granules present in this cell are smaller than the typical secretion granules and are situated within the cisternae of the granular endoplasmic reticulum. These are not true secretion granules, but represent the accumulation of newly synthesised protein, presumably prior to further processing in the Golgi apparatus, where the final granule is formed. Many cisternae contain multiple granules. These granules are not destined for release in their present form. They have no individual close-fitting limiting membranes such as surround true secretion granules, formed in the Golgi apparatus. The phenomenon of intracisternal granule formation is uncommon. It occurs consistently in the pancreas of certain species, including the guinea pig.

M   Mitochondrion
N   Nucleus of pancreatic zymogenic cell

Tissue Exocrine pancreas. Guinea pig. Glutaraldehyde and osmium fixation, uranium and lead staining.

Magnification   17 000 ×

Refer to   Plates 3, 14, 15, 16, 39, 68
             Sections 4.1, 4.2, 5.1

Micrograph by courtesy of H. S. Johnston

Plate 17b *Intracisternal crystals in pancreas*

This micrograph shows another form of intracisternal accumulation, this time of a crystalline nature. The newly synthesised protein has presumably been sufficiently concentrated and pure to form spontaneously into a crystalline pattern within the cavities of the granular reticulum cisternae. This unusual occurrence may reflect some hold-up in the normal physiological processes of intracellular transport. Notice that the lattice pattern is seen more clearly in some places than in others. This is accounted for by variations in the orientation of different crystals in relation to the plane of section. Between cisternae the cytoplasmic ribosomes are prominent.

Note also that the nucleus is just seen at one margin of the micrograph (N).

N   Nucleus

Tissue Baboon pancreas. Glutaraldehyde and osmium fixation. Uranyl and lead staining.

Magnification   56 000 ×

Refer to   Plates 3, 14, 15, 16, 68
             Sections 4.1, 4.2, 5.1

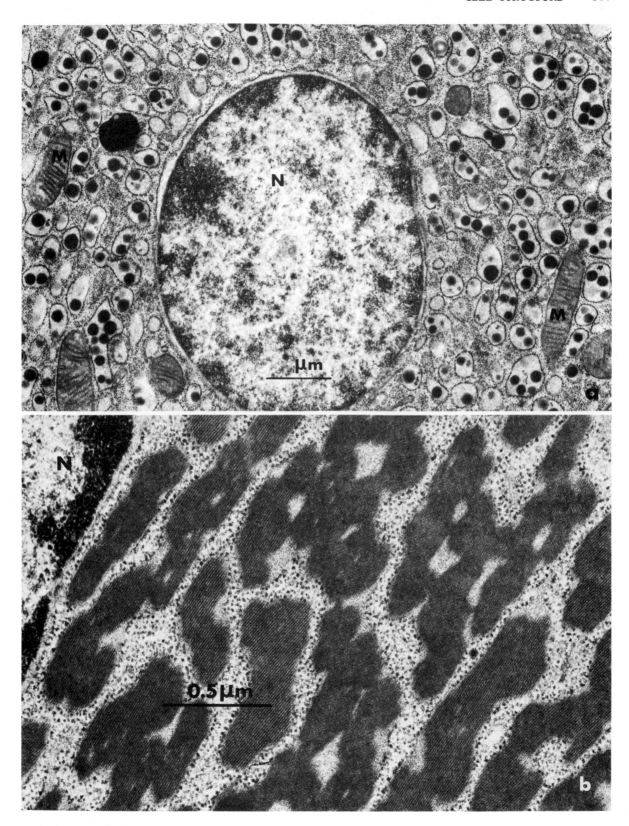

Plate 18a *Agranular or smooth endoplasmic reticulum*

In this steroid-secreting endocrine gland cell, the principal cytoplasmic membrane system is the agranular or smooth endoplasmic reticulum, so called because it lacks the attached ribosomes which characterise the granular or rough endoplasmic reticulum. The cisternal units of this smooth-surfaced membrane system are much more tortuous and interwoven than the structures of the granular reticulum. As a result, the full extent of continuity is not appreciated from a thin section, in which these twisted tubules pass in and out of the field of view. The tubules probably form substantial interconnections, providing for an integration of the functions of the system. Notice the variable diameter of units of the system and the typical pale content of the cisternae. The other typical steroid cell feature is the presence of tubular mitochondrial cristae, but these are not well seen in this micrograph. The scattered dense particles lying free in the intercisternal cytoplasm are glycogen, not ribosomes. Note that these particles are not actually attached to the membranes. These solitary particles of glycogen are known as beta glycogen.

Tissue   Mouse testis, interstitial cell of Leydig. Glutaraldehyde and osmium fixation, uranyl and lead staining.

Magnification   23 000 ×

Refer to   Plates 14, 21, 23a, 40, 45, 46
Sections 4.2, 5.3

Plate 18b *Agranular or smooth endoplasmic reticulum*

This micrograph shows an area of cytoplasm from a liver cell of a rat which had received phenobarbitone for several weeks. This procedure leads to an increase in the amount of smooth endoplasmic reticulum and most of the membranes shown here are of this type. The pale areas are the cavities of the cisternae (C), between which the denser background cytoplasm appears. In contrast, a few typical cisternae of the granular endoplasmic reticulum (G) are seen at one point, with typical ribosomes attached (arrow). The other dense particles which are numerous in this cell are glycogen particles. These are not attached to the nearby smooth membranes but are closely associated with them, perhaps reflecting some involvement of these membranes in carbohydrate metabolism. The glycogen particles (P) are larger than ribosomes and are grouped into rosettes and clusters of considerable size. Rosettes of glycogen such as these are

known as alpha glycogen. Two typical liver mitochondria, with their scattered sparse cristae, appear in the upper part of the plate (M).

C   Cavities of smooth-surfaced cisternae
G   Granular reticulum cisternae
M   Mitochondrion
P   Glycogen particles
Arrow   Indicates ribosomes attached to granular cisternae

Tissue   Rat liver following phenobarbitone administration. Glutaraldehyde and osmium fixation, uranium and lead staining.

Magnification   60 000 ×

Refer to   Plates 14, 21, 23a, 40, 45, 46
Sections 4.2, 5.4

Plate 18c *Annulate lamellae*

A small group of membrane-limited cisternae is present in this testicular germ cell. Their distinctive feature is the presence of fenestrations, or pores (arrows) interrupting the continuity of the smooth-surfaced membranes. These pores have all of the appearances of nuclear pores, including the diaphragm which bridges the aperture of the pore, and the collar, seen here on both sides of the pore apparatus. Other features of the cell include only scanty ribosomes, various smooth-surfaced membrane profiles, a lipid droplet (D) and two mitochondria (M), the configuration of which is very typical of the testicular germ cells. Although annulate lamellae can be seen occasionally in many cell types, they are especially common in ovarian and testicular cells, and in many tumours.

D   Lipid droplet
M   Mitochondria
Arrows   Indicate fenestrations in annulate lamellae

Tissue   Mouse testis. Glutaraldehyde and osmium fixation, uranyl and lead staining.

Magnification   33 000 ×

Refer to   Plates 12a, 26a, 28a, 95c
Section 4.3

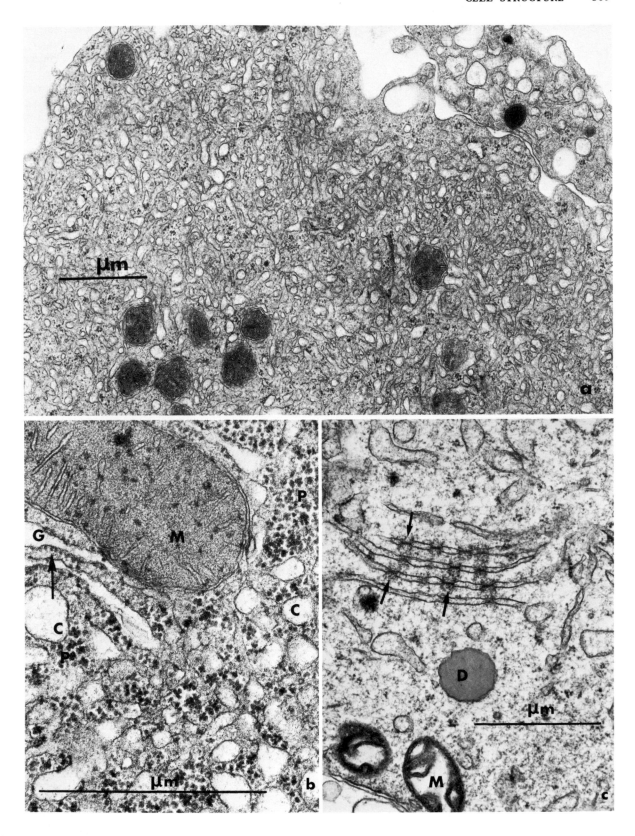

Plate 19 *Golgi apparatus, neutrophil polymorph*

The characteristic features of the neutrophil polymorph are seen here, including the multiple lobes of the nucleus, apparently unconnected to one another, and the presence of several types of cytoplasmic granule. The small Golgi apparatus (G) lies in the centre of the cell, forming two membrane systems in this plane of section. The compact arrangement of the cisternae is entirely characteristic.

The nucleus is of a distinctive pattern, with almost unbroken heterochromatin around the inner aspect of the nuclear envelope (H). The perinuclear cisterna is clearly seen in places, but blurred in others (*) indicating an oblique plane of section.

The granules contain several populations, including the specific granules of the neutrophil and the numerous dense lysosomes. All granules are safely surrounded by an intact limiting membrane, which prevents inappropriate activation of their potentially harmful enzyme apparatus. These granules take part in the complex process of intracellular killing and digestion of microorganisms by the polymorph. In the background cytoplasm many solitary glycogen particles are found and there are a few smooth-surfaced membranes. Glycogen in this solitary form is known as beta glycogen

The polymorph is the cell involved in acute inflammation, a basic response of tissues to injury.

G   Golgi apparatus
H   Heterochromatin
*   Indicates areas where the perinuclear cisterna
            becomes blurred owing to the plane of section

Tissue   Human neutrophil polymorph. Glutaraldehyde and osmium fixation, uranyl and lead staining.

Magnification   52 000 ×

Refer to   Plates, 3, 15, 20, 26a, 30, 37, 52, 63
           Sections 4.4, 4.6, 7.1

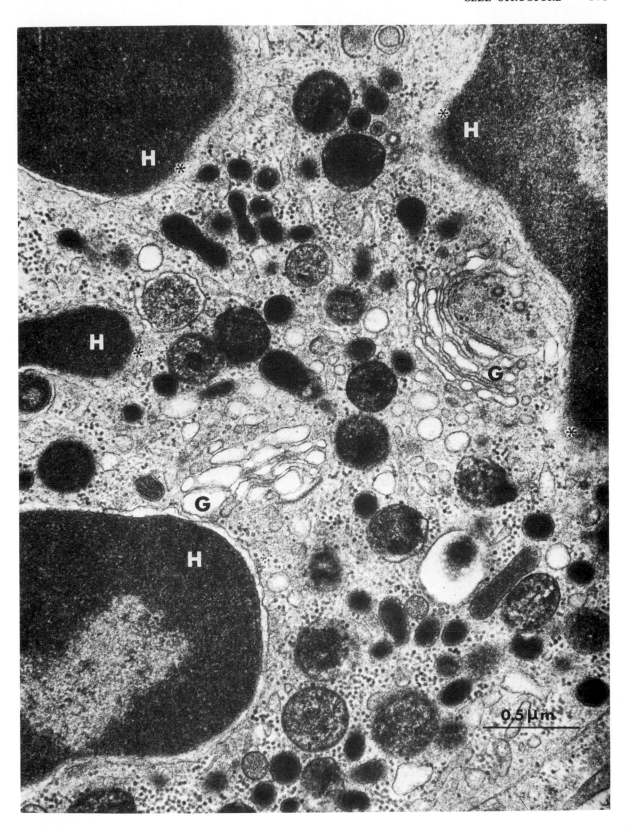

Plate 20a *Golgi apparatus, tumour cell*

This micrograph shows part of a mucus-secreting tumour cell in which the Golgi apparatus is clearly seen. The parallel membrane-limited sacs or lamellae (L) are dilated in places to form vacuoles and numerous small Golgi vesicles (V) lie around the apparatus. Several coated or fuzzy vesicles (X) are also seen, although the significance of these thick-walled structures is not clear. Several irregular dense granules of mucus lie nearby, each limited by a surrounding close-fitting membrane (G). Several large mitochondria (M) are seen and bundles of fine cytoplasmic filaments (F) appear in places. Elongated cisternae of the granular endoplasmic reticulum, with associated ribosomes, come into close relationship with the Golgi apparatus. At one point a transport vesicle appears to be forming from one of these cisternae (arrow). The cell nucleus (N) shows no unusual features.

F	Cytoplasmic filaments
G	Mucus granules
L	Golgi lamellae
M	Mitochondrion
N	Nucleus
X	Fuzzy vesicle
Arrow	Indicates transport vesicle forming from a cisterna of the endoplasmic reticulum

Tissue   Human breast carcinoma. Glutaraldehyde and osmium fixation, uranium and lead staining.

Magnification   35 000 ×

Refer to   Plates 3, 15, 19, 26, 37
Sections 4.4, 4.8

Plate 20b *Golgi apparatus, goblet cell*

This plate shows the immediately supranuclear portion of a typical mature goblet cell from the large intestine. The nucleus lies towards the bottom of the micrograph (N) with the supranuclear cytoplasm above. Small portions of adjacent cells are seen on either side of the goblet cell (X). These adjacent cells interdigitate with one another (I) while at one point a small desmosome is seen (D).

The Golgi apparatus (G) of the goblet cell is particularly elaborate. The membrane cisternae form multiple layers which are grouped into complex horseshoe shapes. Within the concavity of these membranes there lie several large pale vacuoles containing finely flocculent material. These represent the newly synthesised mucus granules about to be released from the Golgi apparatus. These pale vacuoles, (C) known as condensing vacuoles, form at the mature face of the apparatus. Every few minutes, one of these condensing vacuoles will become detached from the Golgi apparatus and pass to the storage area in the apex of the cell.

Interspersed between the Golgi systems there lie closely packed parallel cisternae of the granular endoplasmic reticulum (R). The synthesis of the protein component of the mucus takes place here. It is thought that material passes from the cisternae to the Golgi apparatus by forming small transport vesicles. Notice how the Golgi membranes at two points (*) lose the clear definition seen elsewhere. The membranes at these points lie oblique to the plane of section, leading to blurring of their outline.

C	Condensing vacuoles
D	Desmosome
G	Golgi apparatus
I	Interdigitations between goblet cell and neighbouring cell
N	Nucleus
R	Granular endoplasmic reticulum
X	Adjacent cells
★	Indicate obliquely sectioned Golgi membranes

Tissue   Mouse colon. Glutaraldehyde and osmium fixation, uranyl and lead staining.

Magnification   15 000 ×

Refer to   Plates 3, 15, 19, 34, 38, 49
Sections 4.4, 5.1

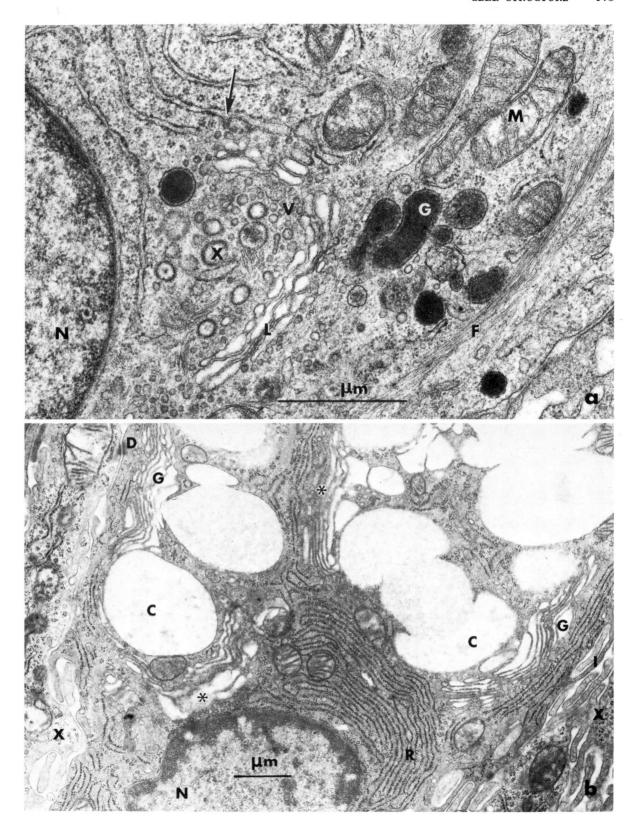

CELL ORGANELLES

Plate 21 *Mitochondria, gastric parietal cell*

This shows the basal part of a large gastric parietal cell, the acid secreting cell of the stomach. A small part of the nucleus (N) lies to one side and the underlying connective tissue (T) is on the other side. At two points the plane of section passes through the intracellular canaliculus, a tubular invagination of the cell surface which branches deep into the cytoplasm (C). This is lined by elaborate microvilli. A basal lamina (L) separates the base of the cell from the connective tissue. Note that the basal surface of the cell is thrown into complex infoldings (F).

Numerous mitochondria (M) are present in the cytoplasm. These make up nearly one third of the total volume of cytoplasm. If individual mitochondria are examined closely (arrows) the close packing of the mitochondrial cristae can be made out. These features indicate a very high rate of oxidative metabolism within the parietal cell. The energy produced by the mitochondria is needed to drive the ion pumps which are responsible for the secretion of gastric acid.

The cytoplasm has few other notable specialisations. A few small Golgi profiles are seen (G) and there are scattered cisternae of granular endoplasmic reticulum. Smooth-surfaced tubulovesicles are only present in very small numbers, not clearly seen at this magnification.

C	Lumen of canaliculus lined by elaborate microvilli
F	Complex interdigitating basal infoldings of the cell membrane
G	Small Golgi profiles
L	Basal lamina
M	Mitochondria
N	Nucleus of gastric parietal cell
T	Connective tissue
Arrows	Indicate mitochondria in which close-packed parallel cristae can be clearly seen

Tissue Human stomach. Glutaraldehyde and osmium fixation, uranyl and lead staining.

Magnification   15 000 ×

Refer to   Plates 4, 22, 40, 41, 51, 70c, 89
Sections 4.5, 5.2

Micrograph by courtesy of H. S. Johnston

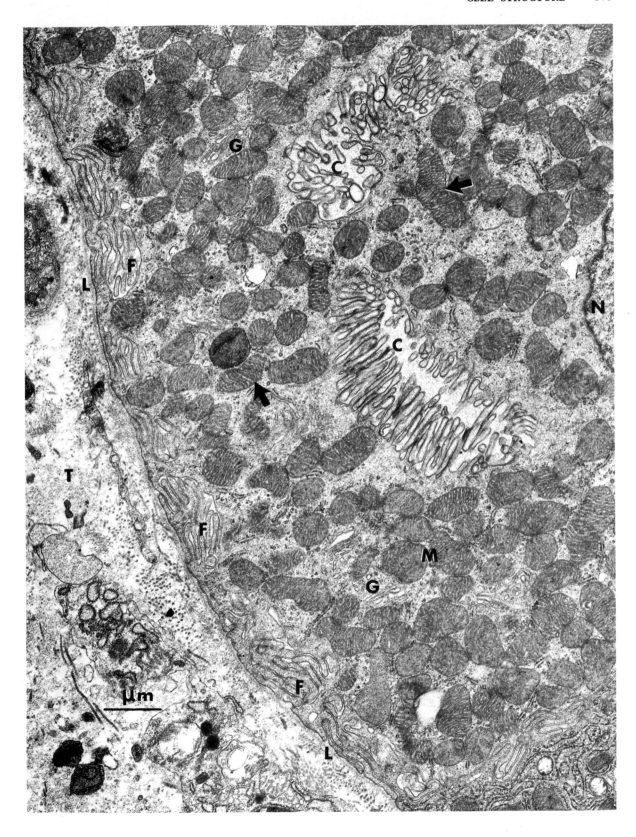

Plate 22a *Mitochondria, liver cells*

The mitochondria in this micrograph of liver are quite numerous (M) but their cristae are rather scanty. This is in contrast to the close-packed cristae seen in the previous plate in association with an ion-secreting function.

The boundary between two cells runs diagonally across the plate. The cell membranes run relatively straight and parallel to each other (C). The nucleus of one cell (N) can be seen. The other cytoplasmic organelles which can be clearly seen in this micrograph include well organised cisternae of granular endoplasmic reticulum (G) and areas of smooth endoplasmic reticulum which are rather poorly defined at this magnification (S). The granular reticulum cisternae are associated with numerous ribosomes, while the smooth reticulum is intermingled with dense aggregates of particulate glycogen. These have taken up the stain particularly intensely. It is important to note that the glycogen is not attached to the smooth-surfaced membranes, whereas the ribosomes are firmly associated with the outer surfaces of the granular reticulum. The glycogen in the liver cell forms coarse aggregates of particles, known as alpha glycogen.

C   Parallel cell membranes in contact with each other
G   Parallel cisternae of granular endoplasmic reticulum
M   Mitochondria of typical liver cell pattern, with sparse randomly orientated cristae
N   Nucleus of liver cell
S   Area containing smooth or agranular endoplasmic reticulum. Dense masses of glycogen have taken up the stain particularly intensely

Tissue Hepatic parenchymal cell. Glutaraldehyde and osmium fixation, uranyl and lead staining.

Magnification   16 000 ×

Refer to   Plates 14, 18b, 32, 33, 43, 44, 45, 46, 85a, 101b
          Sections 4.2, 4.5, 4.11, 5.4

Micrograph by courtesy of G. Bullock

Plate 22b *Mitochondria, gastric parietal cell*

This higher magnification view of the basal part of a gastric parietal cell shows the features of the mitochondria more clearly than the lower magnification view in Plate 21. The basal infoldings of the cell membrane are seen (F), beneath which the continuous basal lamina (L) forms the partition between the parietal cell and the surrounding connective tissue. The background cytoplasm of the parietal cell contains scattered ribosomes in small clusters, along with a few small cisternae of granular endoplasmic reticulum (G).

The mitochondria occupy much of the cytoplasm. Notice their large size, the density of their matrix, and the close packing of the cristae. These are all features of a particularly active mitochondrion. Some mitochondria, however, seem not to show these features (*). This effect results from transverse sectioning of the mitochondrion as opposed to longitudinal sectioning. The transverse section cuts in a plane roughly parallel to the cristae, while the longitudinal section cuts at right angles to the plane of the cristae.

F   Basal infoldings of the cell membrane
G   Solitary cisterna of granular endoplasmic reticulum
L   Basal lamina
*   Indicates transversely sectioned mitochondria in which cristae are not clearly made out

Tissue Human stomach, gastric parietal cell. Glutaraldehyde and osmium fixation, uranyl and lead staining.

Magnification   22 000 ×

Refer to   Plates 21, 40, 51
         Sections 4.5, 5.2

Micrograph by courtesy of H. S. Johnston

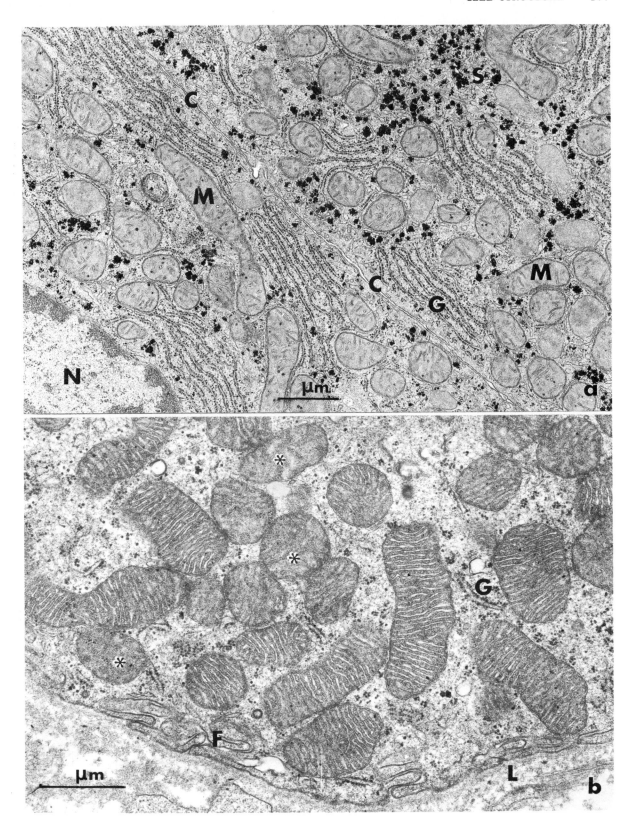

Plate 23a *Mitochondria, adrenal cortex*

Two distinctive specialisations are seen in this field. The smooth endoplasmic reticulum (S) has a predominantly tubular pattern. The mitochondria, which are numerous and large (M), have tubular cristae which present circular cross-sections, instead of the more usual shelf-like structures. These are two common specialisations of steroid-secreting endocrine cells. Some components of the granular endoplasmic reticulum are also found in the cytoplasm (G). At one point an oblique sectioning effect has so obscured the outline of a mitochondrion that its presence can barely be recognised (O), but an area of blurring corresponding to the tangentially sectioned mitochondrial membrane, confirms its presence.

G　　　　Granular endoplasmic reticulum
M　　　　Mitochondrion with tubular cristae
O　　　　Oblique section of mitochondrion
S　　　　Smooth endoplasmic reticulum

Tissue Rat adrenal cortex. Glutaraldehyde and osmium fixation, uranium and lead staining.

Magnification 57 000 ×

Refer to Plates 18, 22
　　　　　Sections 4.2, 4.5, 5.3

Plate 23b *Mitochondria, isolated*

This preparation is unlike most of the others in this book, since it is not an intact cell or tissue. The mitochondria shown here have been isolated from rat liver by disruption of the cells and centrifugation of the resultant homogenate. They were then processed and sectioned to produce this electron micrograph. The reversed contrast configuration of the mitochondria is unusual for liver, reflecting the sensitivity of mitochondria to environmental change. The concentration of ATP and other metabolites can determine isolated mitochondrial morphology. The inner membrane of one of these mitochondria has shrunk away from the outer membrane, emphasising the independence of the outer and inner membranes. Notice how oblique sectioning of the membranes causes blurring of their contour (arrows). Notice also the scattered ribosomes (R) and small membrane vesicles (*), which contaminate the mitochondrial fraction. These are part of the microsome fraction of the cell, derived from the endoplasmic reticulum. Electron microscopy can be usefully employed to assess the purity and state of preservation of isolated subcellular organelles as part of the quality control of experiments in cell biology and biochemistry.

R　　　　　Ribosomes
Arrows　　Indicate obliquely sectioned mitochondrial
　　　　　　membranes
*　　　　　Indicates microsomal vesicles

Tissue Rat liver homogenate, mitochondrial fraction. Glutaraldehyde and osmium fixation, uranyl and lead staining.

Magnification 48 000 ×

Refer to Plates 22a, 45, 46
　　　　　Sections 4.2, 4.5

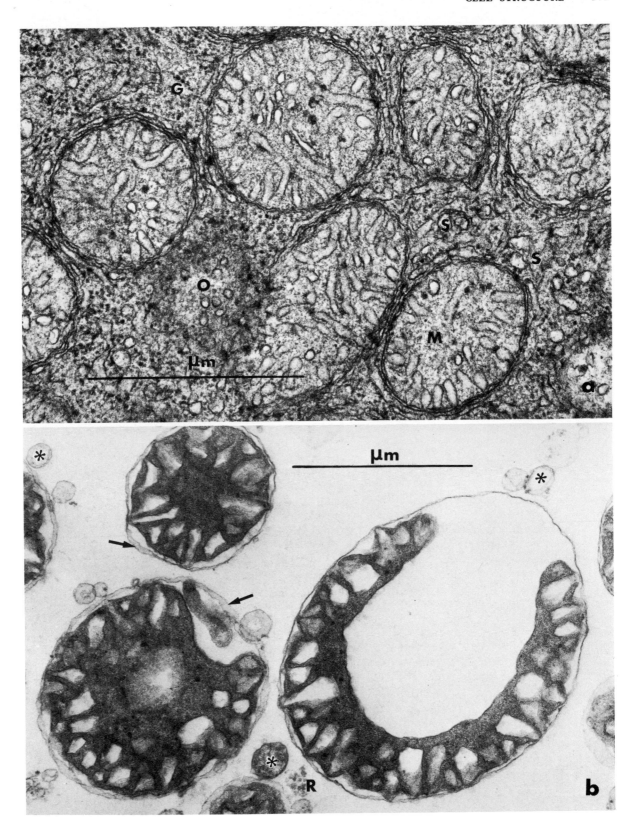

Plate 24a *Lysosomes, Kupffer cell*

This is a micrograph of a liver sinusoid showing a Kupffer cell, a fixed macrophage, bathed in the plasma of the circulating blood. An adjacent lymphocyte is also seen (X). The Kupffer cell is a member of the mononuclear phagocyte series, its function being the uptake of foreign material from the hepatic circulation. As with macrophages elsewhere, the cell is distinguished by surface folds, flaps and ruffles (R) which give an irregular contour on thin section. These surface processes are, of course, dynamic structures, constantly changing as the cell pursues its phagocytic functions. The cell forms phagocytic or pinocytotic vacuoles (P_1) for bulk uptake, by the activity of these surface flaps. In addition, small coated vesicles are formed, indicating selective micropinocytosis, (arrows), as opposed to the relatively non-specific bulk uptake process.

The second typical feature of the macrophage is the presence of numerous membrane-limited primary lysosomes. These are often small dense structures (L), of variable shape. When a phagocytic vacuole is formed at the cell surface and interiorised, (P_1), the primary lysosomes are released into the vacuole to form a secondary lysosome or phagolysosome (P_2). The metabolites released from the digestion process are re-cycled, but the eventual fate of any indigestible residue is to remain as a residual body, or telolysosome, (T) within the cell.

L	Primary lysosomes
P_1	Phagocytic vacuole
P_2	Secondary lysosome, or phagolysosome
R	Surface ruffles
T	Telolysosome, or residual body
X	Lymphocyte
Arrows	Indicate coated vesicles

Tissue   Human liver, Kupffer cell. Glutaraldehyde and osmium fixation, uranyl and lead staining.

Magnification   14 000 ×

Refer to   Plates 14, 19, 25, 26, 32, 51a, 63, 64, 65, 95
              Sections 2.2, 4.6, 7.1

Plate 24b *Lysosomes, pulmonary macrophage*

This section shows part of a macrophage in the connective tissues of an alveolar septum in the lung. A few small primary lysosomes (L) can be seen, but the dominant feature is a large complex telolysosome or residual body (T). This shows areas of variable texture and appearance, including dense particles, granular debris and lamellar structures. These represent the indigestible residues of past phagocytic activity. In the lung, exposed to the external environment, macrophages perform a vital role in clearing from the airways inhaled particulate material, some of which will obviously be resistant to lysosomal hydrolysis. Such residues are often pigmented. In lung this often appears black under light microscopy, on account of the abundance of carbon particles, especially in city dwellers' lungs. In other sites not exposed to airborne particles, the residual bodies or telolysosomes are often golden brown in colour, owing to the presence of lipofuscin pigment, a fatty substance of variable composition. Note a small coated micropinocytotic vesicle forming at the cell surface (arrow), close to a few collagen fibrils (C).

C	Collagen fibrils in transverse section
L	Primary lysosomes
T	Telolysosome or residual body
Arrow	Indicates coated or fuzzy micropinocytotic vesicle

Tissue   Mouse lung, glutaraldehyde and osmium fixation, uranyl staining.

Magnification   46 000 ×

Refer to   Plates 14, 19, 25, 26, 32, 51a, 63, 64, 65, 95
              Sections 2.2, 4.6, 7.1

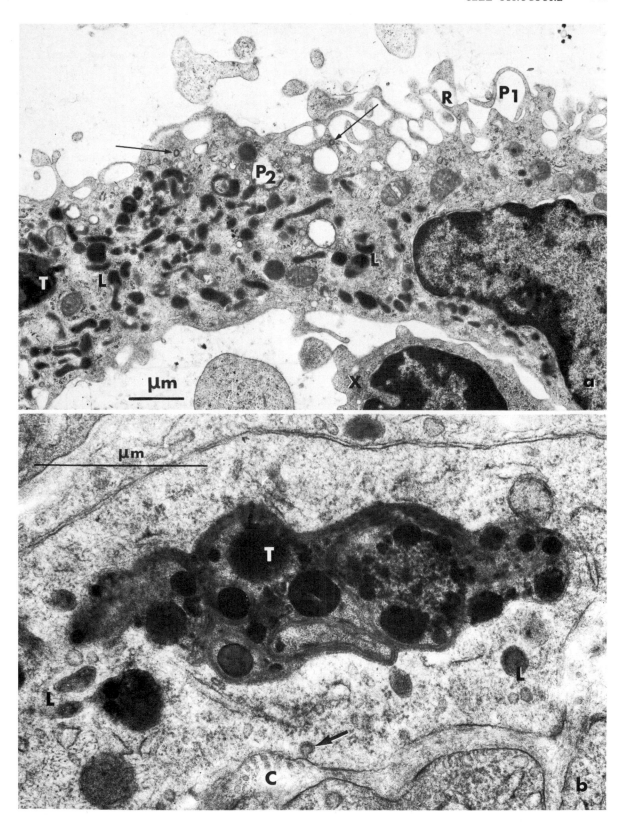

Plate 25a *Lysosomes, autophagocytosis*

This micrograph illustrates the role of lysosomes in the turnover of damaged cell organelles. The cells are crypt cells of the intestine, which are very sensitive to radiation. The animal was exposed to irradiation shortly before this specimen of intestine was taken for electron microscopy. The crypt cells contained many complex structures such as this vacuole, with its partially disrupted cell components. Note that a thin limiting membrane (arrows) demarcates the vacuole from the surrounding normal cytoplasm. This segregation of the damaged area is an essential part of the response to injury. The damaged organelles include a still recognisable mitochondrion (M) and dense granular material (G) suggestive perhaps of fragments of the nucleus.

Into this vacuole are discharged lysosomal enzymes, with the function of digesting the damaged organelles and recycling their constituent molecules. This type of secondary lysosome is known sometimes as a cytolysosome, or autolysosome, reflecting the role of autophagocytosis which the lysosomes of any cell can play in the event of injury.

The intestinal crypt cell is a stem cell, which divides repeatedly to produce replacements for the constantly renewing intestinal epithelium. The crypt cells, like stem cells elsewhere, are structurally undifferentiated, showing few of the specialised features of the intestinal absorptive cell type. They contain numerous free ribosomes but few cytoplasmic membrane systems.

G   Granular material
M   Mitochondrion, partially disrupted
Arrows   Indicate limiting membrane of segregated vacuole

Tissue   Mouse small intestinal crypt cell after irradia-

tion. Glutaraldehyde and osmium fixation, uranyl and lead staining.

Magnification   50 000 ×

Refer to   Plates 14, 19, 24, 26, 32, 51a, 63, 64, 65, 95
           Sections 2.2, 4.6, 7.1

Plate 25b *Lysosomes, autophagocytosis*

As in the above micrograph, this shows the reaction of radiosensitive crypt cells to ionising radiation. The segregation of damaged cytoplasmic fragments has advanced further than in the previous figure, resulting in the formation of several well-defined circular secondary lysosomes, through the process of autophagocytosis. Fragments suggestive of nuclear debris (N), several mitochondria (M) and other recognisable cell components are present in these vacuoles. The largest of these structures might possibly represent the phagocytosis of a dying crypt cell by its neighbour, a phenomenon reported to occur under such circumstances. Note, as above, the predominance of free ribosomes in the cytoplasm of these crypt cells.

M   Mitochondria within lysosomes
N   Material suggesting nuclear origin

Tissue   Mouse intestinal crypt cell after irradiation. Glutaraldehyde and osmium fixation, uranyl and lead staining.

Magnification   46 000 ×

Refer to   Plates 14, 19, 24, 26, 32, 51a, 63, 64, 65, 95
           Sections 2.2, 4.6, 7.1

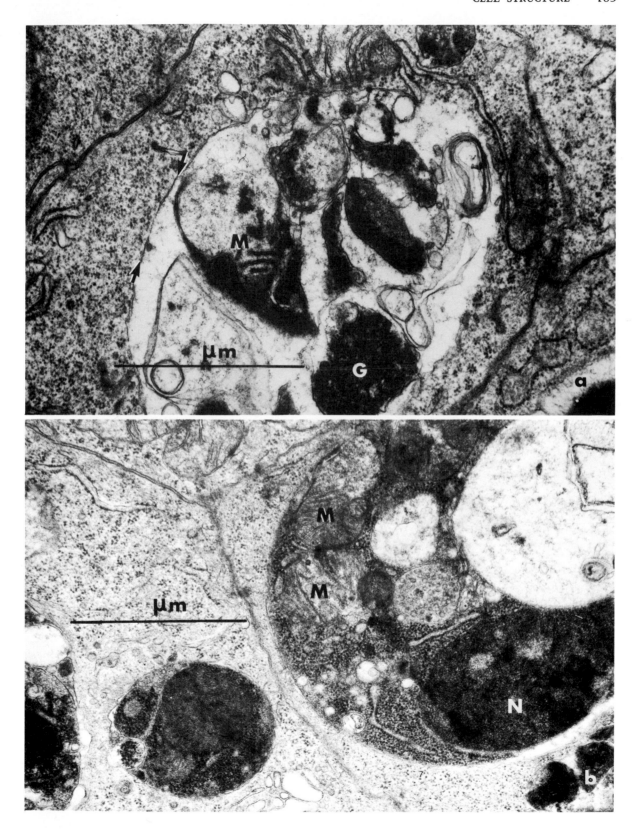

Plate 26a *Lysosomes, acrosome of spermatid*

The testicular germ cells have numerous highly distinctive specialisations related to their unique function. The acrosome is one of these specialisations. The Golgi apparatus of the maturing germ cell (G) is highly complex, with numerous lamellae. The product of this elaborate system accumulates in a cytoplasmic vesicle known as the acrosome, which is applied to the anterior pole of the nucleus (A). This often contains a condensation of acrosomal material, the acrosome granule. As the germ cell matures, this structure fills with denser material and becomes progressively flattened against the anterior nuclear pole.

The acrosome is believed to be a specialised form of lysosome, with a role in the process of penetration of the zone pellucida of the ovum prior to fertilisation. Notice the distinctive mitochondria of the male germ cell (M). An obliquely sectioned sperm tail (T) lies in an intercellular space.

A   Acrosome
G   Golgi apparatus
M   Mitochondria
T   Obliquely sectioned sperm tail

Tissue   Mouse testis. Glutaraldehyde and osmium fixation, uranyl and lead staining.

Magnification   24 000 ×

Refer to   Plates 19, 24, 25, 32, 63, 64, 65, 95
            Sections 2.2, 4.6, 7.1, 10.2

Micrograph by courtesy of H. S. Johnston

Plate 26b *Lysosomes, acid phosphatase reaction*

The concept of the lysosome is based on biochemical evidence of acid hydrolytic enzyme activity in specific membrane-limited cell organelles. As the preceding plates have shown, lysosomes can be recognised by electron microscopy as dense bodies of various types. Morphology alone, however, cannot positively confirm lysosomal identity, without evidence of the appropriate enzyme activity.

This micrograph shows the result of a cytochemical procedure to demonstrate acid phosphatase, a standard lysosomal marker enzyme. The tissue is left unstained apart from the cytochemical reaction, which deposits lead salts at the sites of enzyme activity. Lead, which is electron-dense, thus stains only the lysosomes (L) and related structures. For example, the Golgi apparatus (G) shows a faint deposit of reaction product.

L   Lysosome stained by acid phosphatase reaction
G   Golgi apparatus showing faint reaction product

Tissue   Human tumour cell stained with the acid phosphatase method.

Magnification   18 000 ×

Refer to   Plates 19, 24, 25, 32, 63, 64, 65, 95
            Sections 2.2, 4.6, 7.1, 10.2, 14.6

Plate 26c *Microbody, hepatocyte*

Two microbodies are seen in this plate, one sectioned through the middle (M) and one tangentially (T). A close association with a cisterna of granular endoplasmic reticulum can be seen (C), and an adjacent mitochondrion (X) helps to indicate the relative scale. The microbody contains a granular matrix with a structured nucleoid, consisting of roughly parallel bundles of dense material. The presence or absence of nucleoids and their detailed morphology is species dependent and correlates with urate oxidase activity. The function of the microbody however is obscure, although it seems to play a part in oxidative metabolism.

C   Closely related cisterna of granular reticulum
M   Microbody sectioned through its equator, showing dense paracrystalline nucleoid
T   Microbody tangentially sectioned
X   Mitochondrion

Tissue   Mouse liver. Glutaraldehyde and osmium fixation, uranyl and lead staining.

Magnification   90 000 ×

Refer to   Plates 22a, 45, 46
            Sections 4.7, 5.4

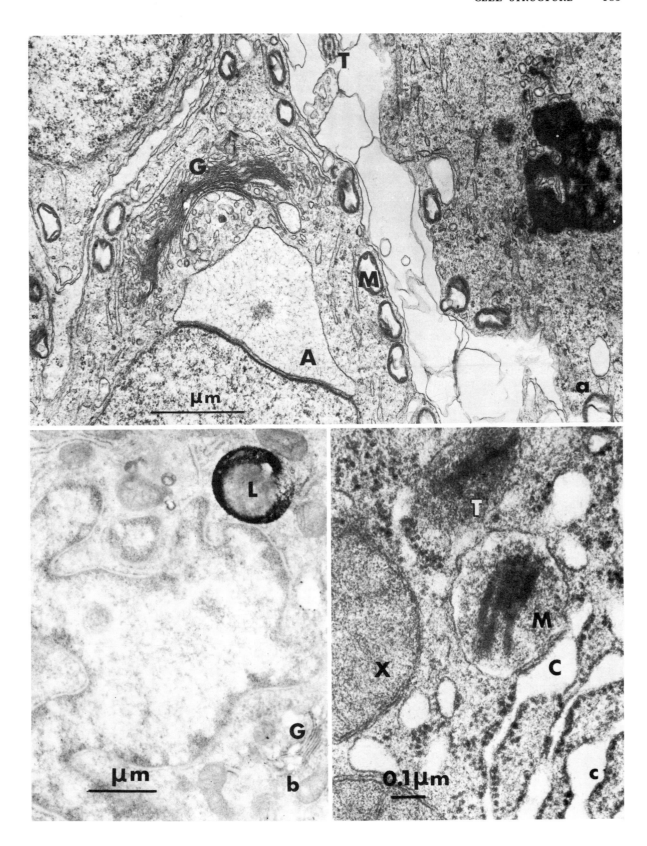

Plate 28a *Microtubules*

This rather complicated plane of section of a developing male germ cell shows a dense bundle of microtubules (T) in close association with the nucleus. The tubules clearly form a highly orientated shell, made visible here by the grazing or tangential cut across the edge of the nucleus. Microtubules, which measure around 23 nm in diameter, are hollow structures of indeterminate length. They are reversibly assembled from tubulin subunits, and are susceptible to environmental influences.

Note also in this micrograph the effect of a tangential plane of section of the nucleus, which reveals closely packed nuclear pores in face view (P). In a pocket of the nucleus (*) a complex centriolar structure forms a basal body, giving rise to what would in due course be the sperm tail.

P	Face view of nuclear pores
T	Microtubules cut at an angle
*	Indicates basal body forming sperm tail

Tissue   Mouse testis. Glutaraldehyde and osmium fixation, uranium and lead staining.

Magnification   40 000 ×

Refer to   Plates 10b, 12, 31a, 93, 94, 95, 96
            Sections 3.3, 4.9, 4.10, 10.2

Plate 28b *Microtubules and microfilaments, nerve*

This cross-section of an unmyelinated nerve bundle containing two axons serves as a good example of the importance of cytoskeletal structures in highly asymmetrical cell shapes. The axon of a nerve is characterised by the presence of longitudinally orientated microtubules, seen here as small empty circles. Interspersed with these are 10 nm filaments, appearing in cross-section as dense dots (arrows). These tubules and filaments play an important part in the maintenance of the shape of the axon, as well as providing a possible basis for the various types of trans-axonal cytoplasmic flow. The other axonal cytoplasmic structures are mostly mitochondria.

Notice the mesaxon (M) suspending the nerve process within its Schwann cell tunnel. The Schwann cell cytoplasm is indicated (S). Note the presence of a well formed basal or external lamina (L) around the Schwann cell, with surrounding cross-sectioned collagen fibres (C).

C	Cross-sectioned collagen fibres
L	Basal or external lamina around Schwann cell
M	Mesaxon
S	Schwann cell cytoplasm
Arrows	Indicate cross-sectioned filaments

Tissue   Human peripheral nerve. Glutaraldehyde and osmium fixation. Uranyl and lead staining.

Magnification   62 000 ×

Refer to   Plates 8, 10b, 93, 94, 98, 100
            Sections 4.8, 4.9, 11.1

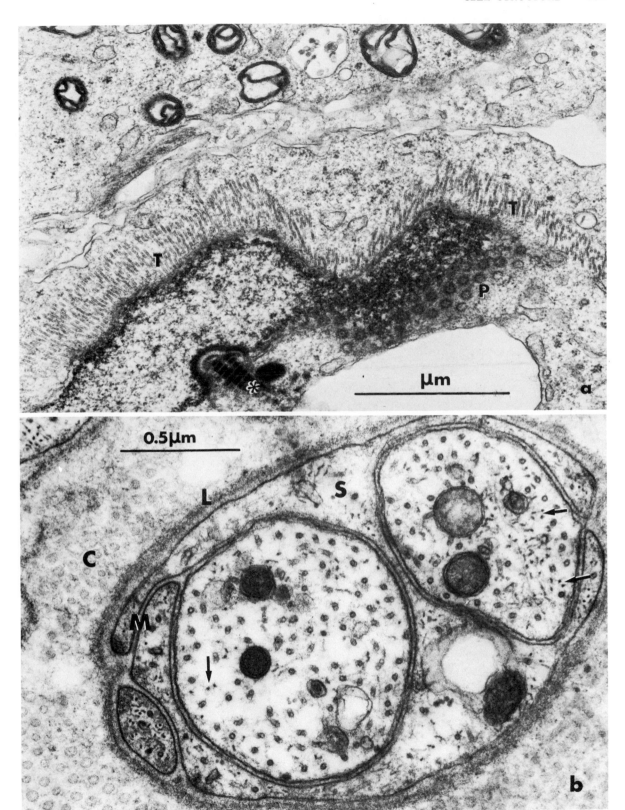

Plate 29 *Microtubules, telophase midbody*

At first glance the dense structure connecting these two tumour cells looks like a large desmosome, but this is not so. On careful examination, this turns out to be a residual intercellular bridge, left at the end of mitosis when the remainder of the cytoplasm has divided into the two daughter cells. The midbody of telophase consists of a densely packed mass of overlapping microtubules, which represent the remains of the central part of the mitotic spindle. The tubules are associated with surrounding dense material. The role of the midbody in mitosis is not clear, but it has been suggested that it is a device to keep the newly formed daughter cells from fusing again in this early unstable phase of their existence.

Tissue   Human tumour cells. Glutaraldehyde and osmium fixation, uranyl and lead staining.

Magnification   39 000 ×

Refer to   Plates 10b, 11
            Sections 3.1, 4.9

Micrograph by courtesy of I. A. R. More

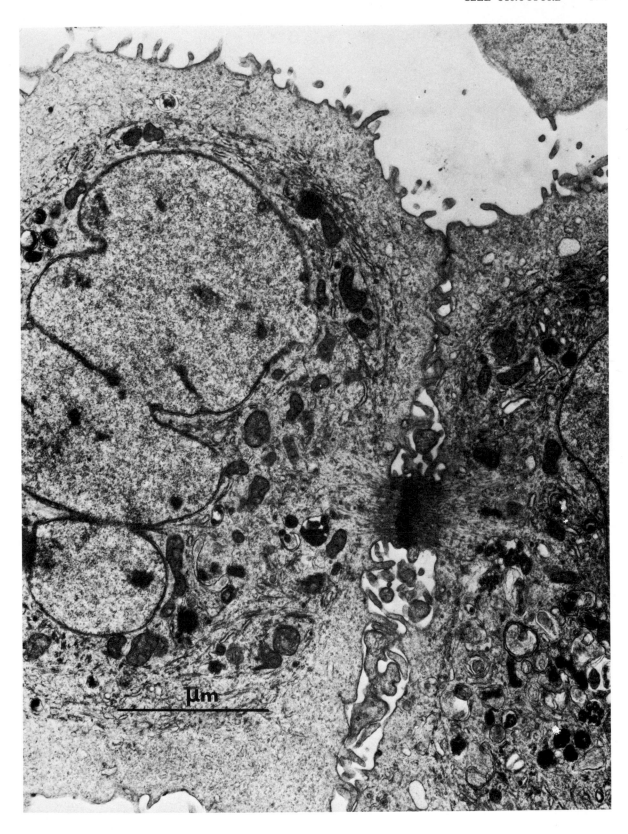

1μm

Plate 30 *Centriole, cell centre*

This high magnification micrograph shows part of the cytoplasm of an endothelial cell. The nucleus (N) is limited by a clearly defined nuclear envelope with a pale perinuclear cisterna (PC). On the inner aspect of the nuclear envelope, dense heterochromatin aggregates are broken at two points where nuclear pores occur (P). The inner aspect of the nuclear pore is usually related to a pale euchromatin channel (E). A centriole in cross-section displays the nine 'triplet' subunits which are characteristic of this organelle, but reveals no central tubular component. Close to the centriole lies the Golgi apparatus (G), part of which is seen in this micrograph. These two cell components together form the region noted by light microscopists as the 'cell centre'. Several mitochondria (M), with poorly organised internal structure, lie between the centriole and the nucleus and there are occasional irregular bundles of filaments in the cytoplasm (F). The effect of oblique sectioning is seen at several points, where mitochondrial membranes appear blurred (O).

C	Centriole
E	Euchromatin channel
F	Cytoplasmic filaments
G	Golgi apparatus
H	Heterochromatin
M	Mitochondrion
N	Nucleus
O	Obliquely sectioned mitochondrial membranes
P	Nuclear pores
PC	Perinuclear cisterna

Tissue   Human endothelial cell. Glutaraldehyde and osmium fixation, uranium and lead staining.

Magnification   76 000 ×

Refer to   Plates 10b, 12, 28a, 31, 32b, 92, 93, 94
Sections 3.3, 4.10, 10.2

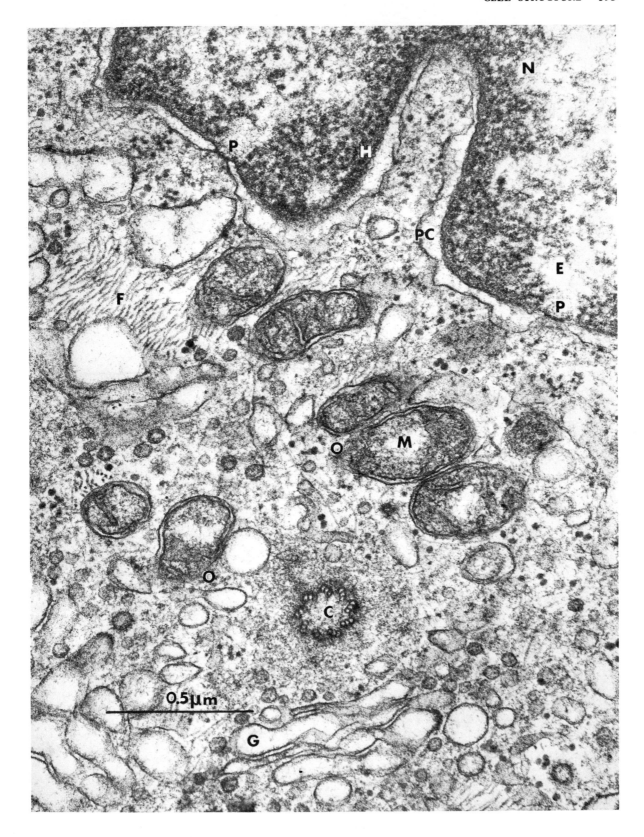

0.5μm

Plate 31a *Centrioles, reproduction*

In this micrograph the centriolar profiles are seen close to the nucleus (N), part of which has been obliquely sectioned (O) displaying several nuclear pores in face view as a result (P). In all, four centrioles can be seen, double the normal number. This micrograph shows the stage of centriolar reduplication which precedes mitosis. The two parent centrioles, one longitudinally sectioned, (L), the other transversely (T) lie, as is common, with their axes at right angles to one another. Closely related to each is a small daughter centriole, cut in longitudinal section in each case. The daughter centriole is, as yet, only quarter of the length of the parent centriole. Once the two new centrioles are formed, the two pairs migrate to opposite sides of the nucleus, were they form the poles of the mitotic spindle.

D   Daughter centriole growing at right angles to the parent
L   Longitudinally sectioned parent centriole
N   Nucleus
O   Obliquely sectioned portion of nucleus, showing pores
P   Nuclear pores in face view
T   Transversely sectioned parent centriole

Tissue Human foetal oesophagus. Glutaraldehyde and osmium fixation, uranium and lead staining.

Magnification 53 000 ×

Refer to Plates 10b, 28a, 30, 32b, 92, 93, 94
          Sections 4.10, 10.2

Micrograph by courtesy of H. S. Johnston

Plate 31b *Centrioles, oblique section*

The two centrioles in this micrograph are obliquely sectioned, one rather more so than the other. They lie close to the nucleus on the one hand (N) and the Golgi structures on the other (G). In the region of the lower centriole several microtubules appear (arrows), but their mode of attachment to the centriole is not clear.

G   Golgi apparatus
N   Nucleus
Arrows   Indicate microtubules

Tissue Human jejunum. Glutaraldehyde and osmium fixation, uranium and lead staining.

Magnification 46 000 ×

Refer to Plates 10b, 28a, 30, 32b, 92, 93, 94
          Sections 4.10, 10.2

Plate 31c *Centriole, satellites*

This centriole is again obliquely sectioned. It shows several lateral projections known as satellites (S), the nature of which is not clear. They may act in some way as points of attachment for microtubules.

S   Satellites of centriole

Tissue Human colon. Glutaraldehyde and osmium fixation, uranium and lead staining.

Magnification 48 000 ×

Refer to Plates 10b, 28a, 30, 32b, 92, 93, 94
          Sections 4.10, 10.2

Plate 31d *Centriole and cilium*

This longitudinally sectioned centriole, one of a pair seen in this cell, has given rise to a single cilium (C) projecting from the cell. Structures such as this are occasionally seen even in cells which do not normally have cilia. They may represent a functionally insignificant expression of centriolar potential, although some single cilia might have other non-motile functions, perhaps as some form of sensory receptor for the cell. During the maturation of true ciliated cells, such as respiratory epithelial cells, repeated centriolar replication gives rise to numerous centrioles which form the basal bodies from which the multiple cilia grow out into the lumen.

C   Base of a single cilium

Tissue Human skin biopsy. Glutaraldehyde and osmium fixation, uranium and lead staining.

Magnification 45 000 ×

Refer to Plates 10b, 28a, 30, 32b, 92, 93, 94
          Sections 4.10, 10.2

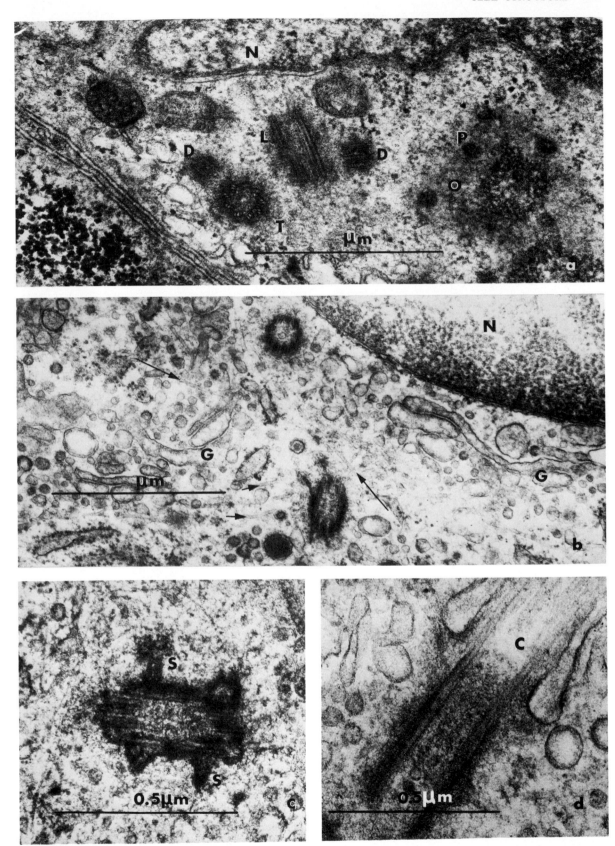

Plates 32a, 32b *Ferritin particles*

A high magnification view of a portion of cytoplasm from a macrophage shows the presence of ferritin. The numerous small dense particles are individual ferritin molecules, rendered visible by virtue of the high concentration of iron in the core of the molecule. The particles lie both free and enclosed within dense granules (Gr), which are probably residual bodies, and which represent a form of iron store. A fine substructure which can be distinguished in some of these ferritin molecules may reflect a supposed octahedral pattern of the iron core of the ferritin molecule. The dimensions and morphology of ferritin are sufficiently characteristic to allow these particles to be distinguished from ribosomes and from particulate glycogen. Ferritin is a smaller particle than either. It can be understood that the ferritin molecules, if attached to antibody as a label, can act as an ultrastructural tracer for immunological reactions.

In the lower power view of the same cell, the dense ferritin-containing granules (Gr), sometimes called siderosomes, are particularly prominent but ferritin molecules are widely scattered through the cytoplasm. They do not appear within the mitochondria (M), the Golgi sacs (G), the endoplasmic reticulum or the nucleus (N), but several particles are present in the hollow centres of the centrioles (C), suggesting that these components are open to the cytoplasm. Both centrioles of this cell are present in this section, the one on the left being obliquely sectioned while that on the right is cut in transverse section, revealing the nine subunit construction as seen in Plate 30. Notice the association between the Golgi apparatus, in the upper part of the plate, and the centrioles.

Mitochondria can also be seen in this section, but one of them has been cut obliquely so that its limiting membranes are blurred and the outlines of the mitochondrion are indistinct. Two portions of the nucleus appear in the plate on the left of the field, the upper portion cut tangentially, showing a surface view of nuclear pores, as indicated by arrows, but obscuring the nuclear envelope.

Several 'coated vesicles' (*), thick-walled structures with a fuzzy lining, are seen in this micrograph, one close to the two nuclear profiles on the left of Plate 33b, another close to a mitochondrion on the right of the picture. These coated vesicles are thought to be formed by selective micropinocytosis from the surface of the cell.

C	Centrioles
G	Golgi apparatus
Gr	Ferritin-containing granules or siderosomes
M	Mitochondrion
N	Two parts of the nucleus included in plane of section
Arrows	Indicate nuclear pores sectioned tangentially showing surrounding annuli
*	Indicate coated vesicles

Tissue   Mouse epididymis. Glutaraldehyde and osmium fixation, lead staining.

Magnification   Plate 32a 90 000 ×
                  Plate 32b 38 000 ×

Refer to   Plates 14, 16, 19, 22a, 24, 25, 26, 33, 51a, 63, 64, 65, 85a, 95, 101b
            Sections 4.6, 4.11, 7.1

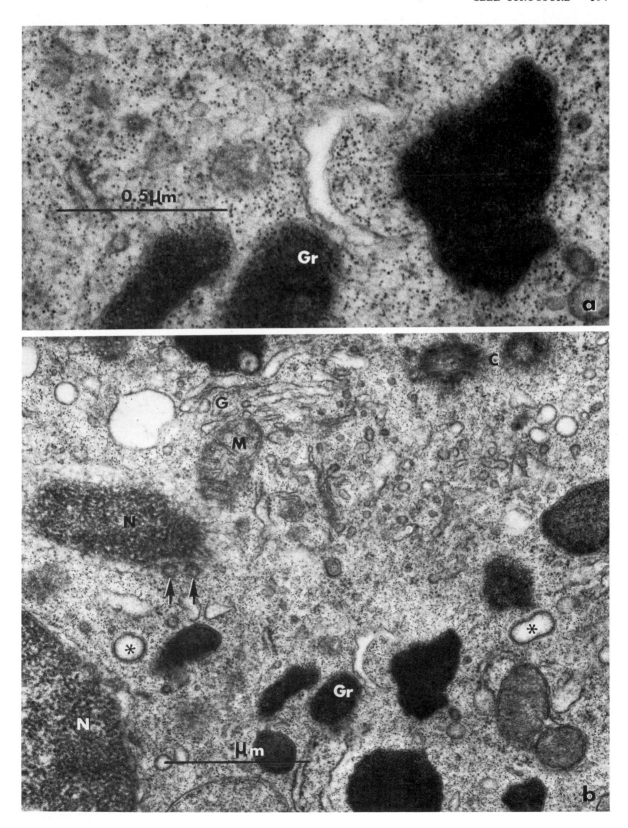

Plate 33a *Glycogen particles*

In the previous plate, ferritin was distinguished by its size from other particulate cell components such as ribosomes and glycogen. In this micrograph the particles are glycogen, in the solitary or beta configuration. Typically, in striated muscle, the glycogen is gathered mainly in the region of the sarcoplasmic reticulum, the smooth-surfaced membrane system which occupies the cytoplasm between the myofibrils. The glycogen serves as a stored fuel reserve for muscle action. The cross-banded sarcomere pattern of the muscle can be made out, with prominent Z lines (Z).

Z   Z lines of striated muscle

Tissue   Skeletal muscle, glutaraldehyde and osmium fixation, uranyl and lead staining.

Magnification   64 000 ×

Refer to   Plates 22a, 32, 45, 46, 85a, 101b, 111
              Sections 4.11, 10.1

Micrograph by courtesy of G. Bullock

Plate 33b *Glycogen and cell coat*

Particulate glycogen is present in many cells, including these superficial squamous cells of the nonkeratinising human oesophagus. The chemical nature of the glycogen can be hinted at by the use of a cytochemical reaction for carbohydrate groupings. This is a variant of the periodic acid-Schiff method, with the final Schiff reagent stage replaced by a reaction with silver proteinate. This constitutes an electron-dense label which visualises the reaction sites for carbohydrates. Note the densely stained particulate beta glycogen and the fine linear reactivity of the cell surface coat, also rich in carbohydrate.

The surface projections of nonkeratinising squamous cells provide a useful lesson in ultrastructural interpretation. The processes extending from the cell surface might reasonably be described as microvilli from their appearances in thin section, but scanning microscopy has shown that they are microridges rather than microvilli.

Tissue   Human oesophageal epithelium. Glutaraldehyde and osmium fixation. Uranyl and lead staining.

Magnification   100 000 ×

Refer to   Plates 7, 22a, 32, 45, 46, 72, 85a, 101b, 115
              Sections 2.2, 4.11, 8.3, 14.6

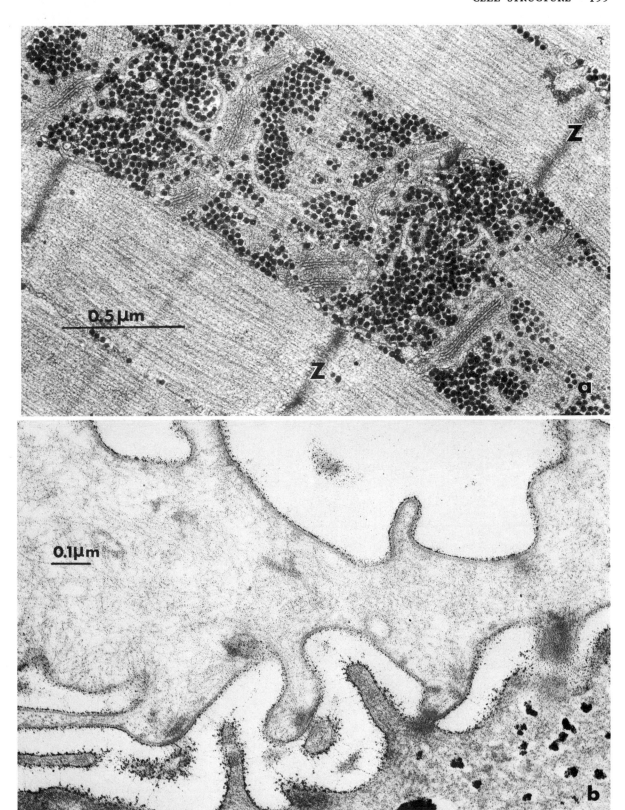

SECRETION

Plates 34a, 34b *Goblet cells, human colon*

These two scanning micrographs are from the same specimen of colonic mucosa. The boxed area of the left hand micrograph is enlarged to show detail of a goblet cell.

Scanning electron microscopy is a very sensitive method for the detailed analysis of topographical detail. At low magnification, the slightly undulating mucosal contours are well seen, with the outlines of individual cells being made out as a polygonal pattern. The frequency and distribution of the goblet cells is easily assessed in such preparations. The higher magnification micrograph shows up the irregular surface detail of the active goblet cell, bulging with undischarged secretion. The pits seen at several points on the cell surface (arrows) probably represent empty vacuoles where a secretion granule has been released. Note the surrounding enterocytes, or absorptive cells, with their surface microvilli, whose even distribution contrasts with the irregularity of the goblet cell surface.

Arrows   Indicate surface pits, possibly caused by granule release.

Tissue Human colonic mucosa, glutaraldehyde and osmium fixation, critical point dried, gold coated.

Magnification   Plate 34b 350 ×
                   Plate 34b 2700 ×

Refer to   Plates 7, 20b, 38, 39b, 49
            Sections 5.1, 14.3

Plate 34c *Secretory cells, bronchus*

This scanning micrograph contrasts the surface features of the two cell types of the bronchial mucosa, the ciliated cell and the secretory cell, also known as the Clara cell. The cilia are closely packed, forming a moving carpet across which moist secretions are constantly being propelled, to trap and eliminate inhaled dust particles. The elongated apical projections of the Clara cell do not have microvilli or other surface features of note. This cellular proboscis is presumably a device to enhance the surface area available for secretion, although the nature of the Clara cell's secretion is not known with certainty.

Tissue Rat lung. Glutaraldehyde and osmium fixation. Critical point dried. Gold coated.

Magnification   7000 ×

Refer to Plates 56, 91, 92, 93, 94
           Sections 10.2, 14.3

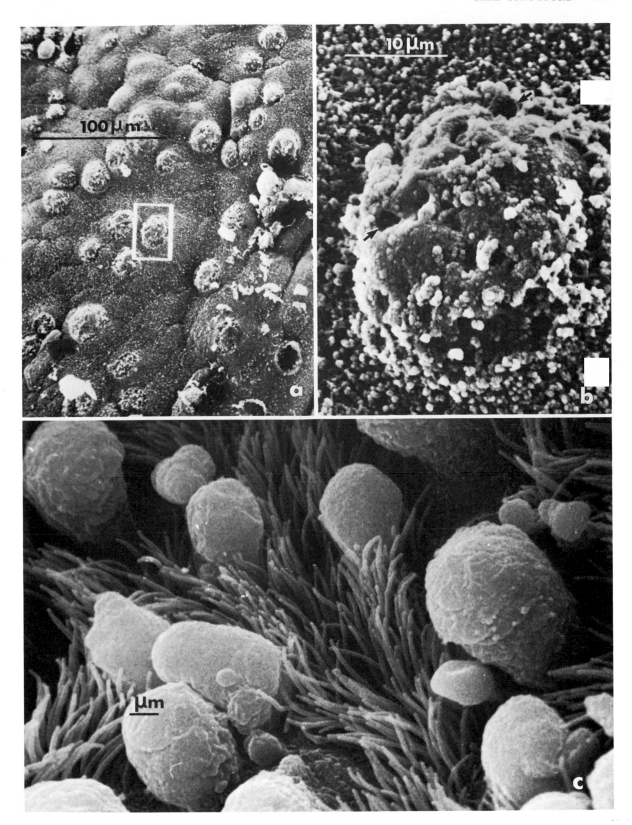

Plate 35 *Secretory cells, gastric gland*

The purpose of this micrograph is to show the relation-ship between the cells of an exocrine gland. The section is taken through a gastric gland in the body of stomach. The gland cells enclose a lumen (L) which contains some cellular debris. Around the outside of the gland is the loose connective tissue which supports the epithelium (CT). Here can be seen scattered cells including plasma cells (P), a normal component of mucosal surfaces.

Detailed examination of the gland shows that it consists of several cell types, all involved in secretory activity of some kind. Most numerous are the gastric parietal cells, which secrete acid. These cells are identified by their prominent intracellular canaliculi (C) and by the numerous closely packed mitochondria, which can be distinguished without difficulty (M). These cells also contain dense residual body structures, which probably arise through autophagocytic turnover of cell organelles, especially the mitochondria. The function of the parietal cell depends on its secretory surface area. Note how limited is the surface area of the narrow gastric gland lumen; the additional 'free' surface provided by the canaliculus greatly enhances the effi-ciency of the function of acid secretion, while avoiding the need for a more elaborate or larger gland system to provide the necessary secretory surface.

Two other main cell types can be seen in this gland, the zymogenic or chief cells, which secrete the gastric digestive enzymes (Z), and occasional mucous neck cells (N), which secrete protective mucus. Cytoplasmic details of these cell types cannot be easily made out at this magnification. A third cell type, not seen here, is the endocrine cell, of which various subtypes are recognised. Generally, the endocrine cells do not reach the gastric gland lumen, but have a broad basal surface exposed to the surrounding vascular connective tissue into which their hormones, such as gastrin, are released.

The gastric gland, therefore, is an assembly of heterogeneous cells pursuing an integrated physiological function. Each gland forms an anatomical entity, separated from its neighbours by delicate vascular connective tissue.

C	Canaliculus of gastric parietal cell
CT	Loose connective tissue
L	Lumen of gastric gland
M	Mitochondria of parietal cell, closely packed
N	Mucous neck cell
P	Plasma cell
Z	Zymogenic cell

Tissue Human gastric gland. Glutaraldehyde and os-mium fixation, uranyl and lead staining.

Magnification   5000 ×

Refer to   Plates 4, 21, 22b, 36a, 39, 40
             Sections 5.1, 5.2

Micrograph by courtesy of H. S. Johnston

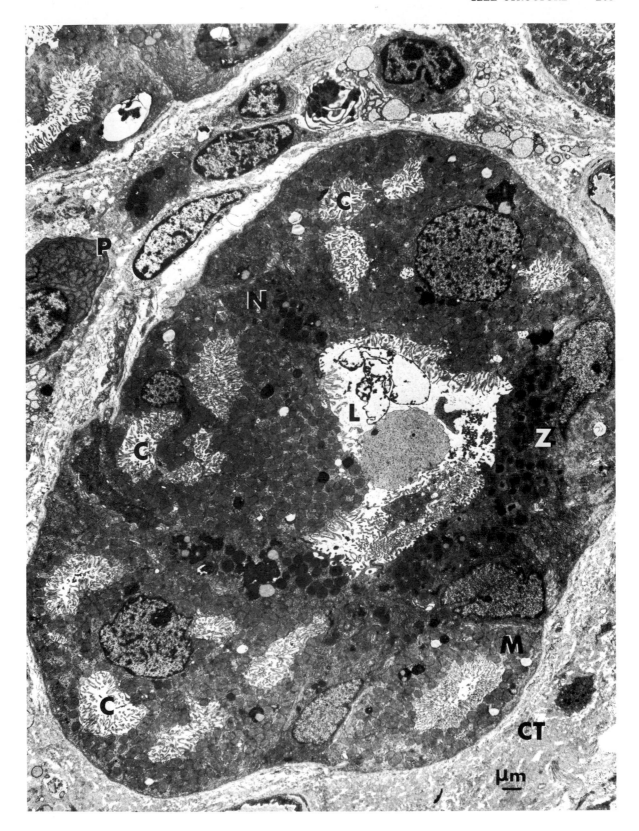

Plate 36a *Exocrine secretion, Paneth cell*

Exocrine function typically involves the release of a secretory product into the lumen of a gland. This micrograph shows the lumen (L) of an intestinal crypt, a simple test-tube shaped gland which communicates with the main lumen of the gut. At the foot of the crypt are found the Paneth cells, which are a useful model of the exocrine protein secretory process although their exact function is still in some doubt. They have an elaborate granular endoplasmic reticulum (R) and Golgi system (G) and their large secretion granules (S) accumulate in the cell apex, to be released into the lumen. The basal parts of these cells contain more granular reticulum, but secretion granules are not found below the nucleus. Thus the ultrastructural specialisation and polarisation of these cells gives a strong indication of their exocrine function. The story would be perfect if we knew what this cell actually does, but it remains a mystery. Recently the granules have been shown to contain lysozyme, an antibacterial enzyme also present in tears, while the cells have been shown to be able to ingest micro-organisms from the lumen. Perhaps, therefore, the Paneth cell is concerned with maintaining the cleanliness of the intestinal crypt.

G   Golgi apparatus of Paneth cell
L   Lumen of intestinal crypt
R   Granular endoplasmic reticulum
S   Secretory granules gathered in cell apex

Tissue   Human small intestine. Glutaraldehyde and osmium fixation, uranyl and lead staining.

Magnification   12 000 ×

Refer to   Plates 6a, 35, 37, 39
          Sections 5.1, 12.1

Plate 36b *Endocrine secretion, intestine*

Most of the epithelial surfaces of the body have their own populations of specialised endocrine cells, no doubt concerned with regulation of function, but sometimes poorly understood. Such cells share many common functional properties, such as polypeptide hormone secretion and amine storage. Typical structural features include a basal location, with the apex of the cell rarely reaching the surface. Such cells also contain marked secretory granulation, with a basal orientation. This is the site of discharge of secretion into the connective tissue and thus into the circulation. The APUD cell concept has been advanced to bring many of these cells under a common developmental neuroendocrine 'umbrella', since some, at least, originate from the embryonic neural crest.

This micrograph shows the typical ultrastructural features of an intestinal endocrine cell. For orientation, note the position of the striated border (B) and the lumen of the gut (L) on the apical side of the granulated cell. The underlying connective tissue (T) on the basal side contains a nearby capillary vessel filled by a red blood corpuscle (R). The proximity of blood vessel and basal secretion granules emphasises that the secretory product of this cell is destined for distribution by the blood stream. The adjacent cells lying parallel to the endocrine cell are absorptive cells (A).

A   Absorptive cells flanking the intestinal endocrine cell
B   Striated border of the intestinal epithelium
L   Lumen of intestine
R   Red blood corpuscle lying within capillary vessel
T   Connective tissue

Tissue   Human small intestine, glutaraldehyde and osmium fixation, uranyl and lead staining.

Magnification   9000 ×

Refer to   Plates 18a, 42, 69c
          Section 5.3

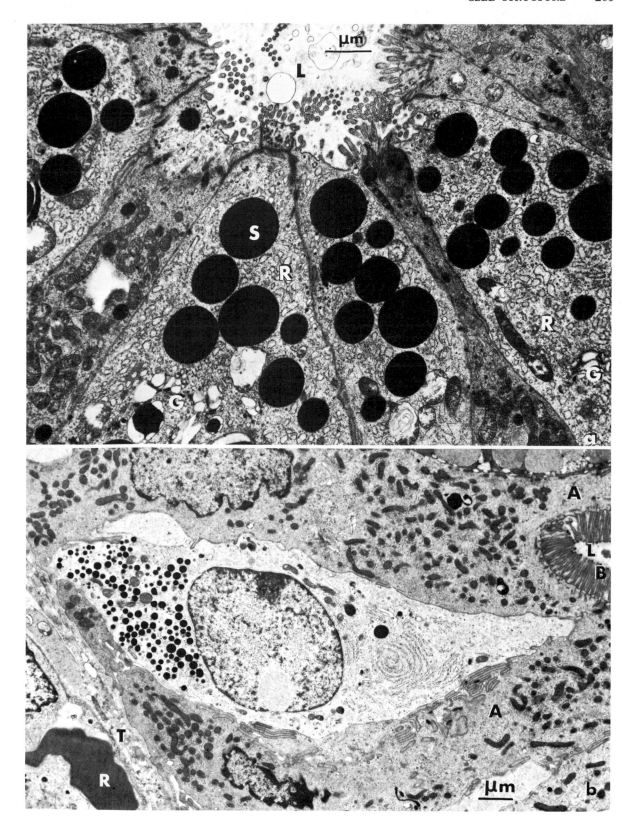

Plate 37 *Protein secretion, Golgi apparatus*

The Golgi apparatus is heavily involved in the secretory process, although its detailed metabolic functions are not yet fully known. It certainly plays a part in granule packaging, as seen in this Paneth cell, the granules of which (G) contain lysozyme, an enzyme with antibacterial properties. It remains uncertain what the function of the Paneth cell may be.

The Golgi system consists of many vesicles and multiple groups of paired membrane lamellae, dilated in places to form Golgi vacuoles (V). Dense material is seen at points in this lamellar system (arrows), being finally segregated into condensing vacuoles (C) which become detached from the apparatus to form new secretion granules. The complexity of the system is remarkable, and the dynamic interpretation of the role of the Golgi membranes, vacuoles and vesicles remains open to discussion.

C   Condensing vacuoles
G   Granules
V   Golgi vacuoles
Arrows   Indicate dense secretion material in the interstices of the Golgi apparatus

Tissue Human small intestine, glutaraldehyde and osmium fixation, uranyl and lead staining.

Magnification   33 000 ×

Refer to   Plates 3, 6a, 15, 36a
              Sections 5.1, 12.1

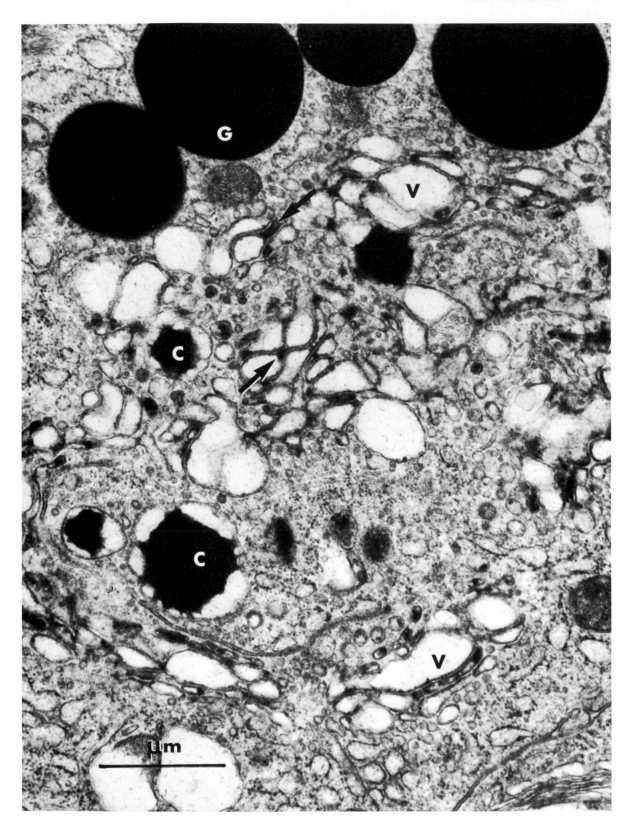

Plate 38 *Mucus secretion, small intestine*

Goblet cells do not always pout out on the mucosal surface. Often the apex of the cell lies slightly below the general surface of the surrounding enterocytes, so that a small dimple is produced through which its secretion trickles out. In this case, notice the relationships between the goblet cell and adjacent enterocytes. They are linked by junctional complexes (J) at their apices, but the margins of the enterocytes overhang the goblet cell mouth, producing the characteristic dimple.

The accumulated mucus mass consists of dense individual granules, mutually distorted into roughly polygonal shapes. A thin shell of goblet cell cytoplasm (C) extends right up to the junctional complex around the mucus mass. A couple of residual microvilli (M) are present at the goblet cell apex at this marginal zone on either side of the orifice, recalling that the goblet cell, in its early differentiation, has many features in common with the enterocyte. Across most of the surface of the goblet cell, the membrane is attenuated and stretched over mucus droplets on the point of release.

Note the dense and elaborately organised cytoplasm of the goblet cell, contrasting with the paler background of the neighbouring absorptive cells. A highly complex granular endoplasmic reticulum is seen. The plane of section has, however, missed ther membranes of the Golgi apparatus.

C   Peripheral thin shell of goblet cell cytoplasm
J   Junctional complexes
M   Residual goblet cell microvilli

Tissue   Mouse small intestine, glutaraldehyde and osmium fixation, uranyl and lead staining.

Magnification   19 000 ×

Refer to   Plates 7, 20b, 34, 39b, 49
             Section 5.1

Micrograph by courtesy of H. S. Johnston

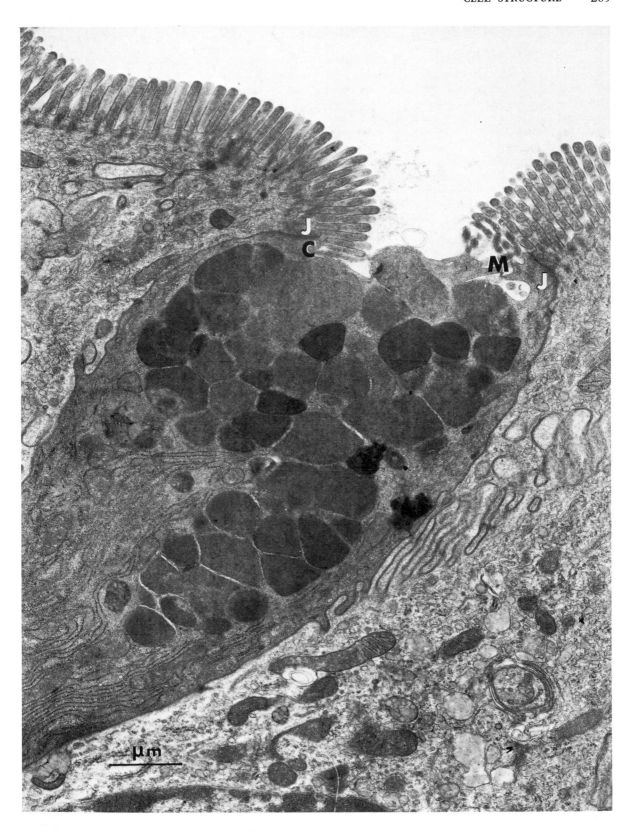

Plate 39a *Zymogen secretion, pancreas*

This micrograph shows the lumen (L) of a pancreatic acinus filled with dense secretion material. Parts of four surrounding acinar cells are seen, with their large dense zymogenic granules (Z). Parts of several other cells (C) are seen in continuity with the acinar epithelium. These are the so-called centro-acinar cells, which represent the first part of the pancreatic ductular system, carrying secretion from the acini towards the main pancreatic duct. There is a striking contrast between the active secretory cells, with complex granular reticulum (R), and stored apical zymogenic granules (Z), and these passive duct lining cells, with little cytoplasmic speci-alisation (C).

Notice that one of the zymogenic cells has an apical concavity (arrow) opening at its surface. This is where a newly discharged zymogenic granule is pouring its contents into the lumen of the gland.

C	Centro-acinar cell with unspecialised cytoplasm
L	Lumen of gland
R	Granular endoplasmic reticulum
Z	Zymogen granules in acinar cells
Arrow	Indicates discharge of secretory granule into gland lumen

Tissue Baboon pancreas, glutaraldehyde and osmium fixation, uranyl and lead staining.

Magnification   17 000 ×

Refer to   Plates 3, 7, 17, 20b, 34, 35, 36a, 38, 49
Section 5.1

Plate 39b *Mucus secretion, goblet cell*

Exocrine secretion involves the release into the lumen of cytoplasmic granules stored in the cell apex. This is normally accomplished by fusion of the limiting mem-brane of the granule with the apical membrane of the cell surface, allowing the release of granule contents without disruption of continuity of the cytoplasmic envelope, a process known as merocrine release. In zymogenic cells, the granule contents are rapidly voided into the lumen, but the sticky mucus granule is slower to discharge and can be more easily 'caught in the act' of secretion.

The goblet cell apex contains masses of partially fused mucus granules (M) and is sandwiched between the apical parts of two adjacent absorptive cells (A) to which it is joined by junctional complexes (J). A thin shell of goblet cell cytoplasm (C) extends to the level of the surrounding cells but the mucus mass has become excavated to a lower level by discharge of some of its content. Much of the discharged mucus (D) remains plugging the orifice of the cell. One granule (G$_1$) is actively streaming out to join this plug, while others (G$_2$) show imminent discharge.

A	Adjacent absorptive cells
C	Thin shell of goblet cell cytoplasm
D	Plug of discharged mucus
G$_1$	Mucus granule pouring out contents
G$_2$	Mucus granule about to discharge
J	Junctional complexes
M	Mass of fused mucus granules in cell apex

Tissue Colonic goblet cell, glutaraldehyde and osmium fixation, uranyl and lead staining.

Magnification   15 000 ×

Refer to   Plates 3, 7, 17, 20b, 34, 35, 36a, 38, 49
Section 5.1

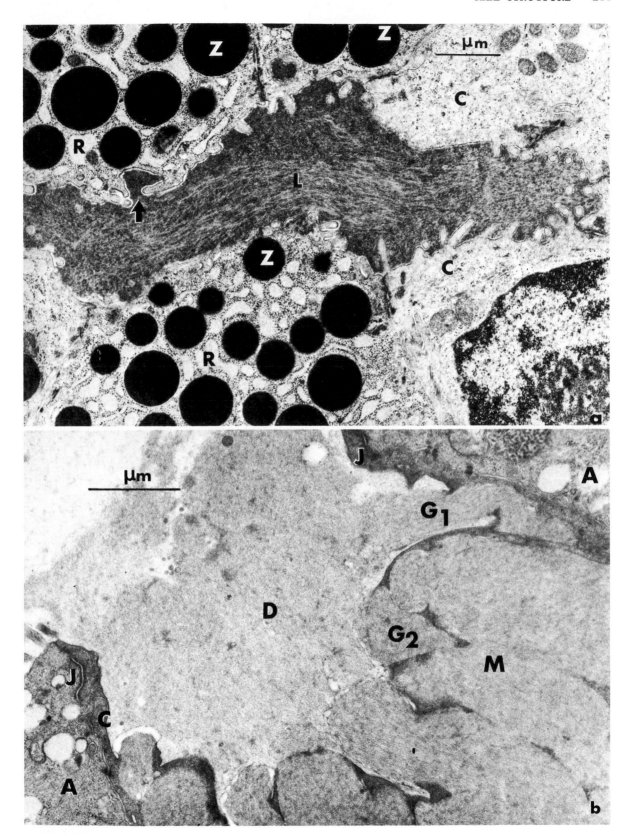

Plate 40a *Ion secretion, gastric parietal cell*

This micrograph shows part of a gastric parietal cell from human stomach. The nucleus is featureless. The base of the cell displays several complex areas of membrane infolding which greatly increase the surface area (F). The underlying connective tissue contains numerous collagen fibrils and a few small fibroblast processes. Within the parietal cell the elaborate intracellular canaliculus (C) is a striking feature. Its lumen is nearly occluded by the presence of closely packed elongated, club-shaped microvilli. The cytoplasm contains many large and elaborate mitochondria (M) which may account for up to a third of the total cytoplasmic volume of the parietal cell. Their presence reflects the importance of a high rate of oxidative metabolism in the function of acid secretion. Several dense lysosome-like structures (L) are present, along with one multivesicular body (B). Between the mitochondria there are a few cisternae of the granular endoplasmic reticulum but smooth-surfaced tubulovesicles are almost entirely absent from this particular cell.

B	Multivesicular body
C	Intracellular canaliculus
F	Basal infolding
L	Lysosome-like dense body
M	Mitochondrion
N	Nucleus

Tissue   Human stomach. Glutaraldehyde fixation, uranium and lead staining.

Magnification   13 000 ×

Refer to   Plates 4, 21, 22b, 35, 41
Sections 4.2, 5.2

Micrograph by courtesy of H. S. Johnston

Plate 40b *Ion secretion, tubulovesicles*

This high magnification micrograph shows an area of cytoplasm from a different parietal cell in which numerous tubulovesicles are found (T). These appear as closely packed winding tubular structures which at times turn at right angles to the plane of section to give circular profiles. A few glycogen particles are associated with them. Part of the canalicular lumen (L) is present and a large and almost empty multivesicular body is also seen (B). These tubulovesicles appear in some species to make contact with the cell surface, but the part they play in acid secretion is not yet clearly defined. They are now regarded as distinct from the smooth endoplasmic reticulum.

B	Multivesicular body
L	Lumen of intracellular canaliculus
T	Tubulovesicles

Tissue   Human stomach. Glutaraldehyde fixation, uranium and lead staining.

Magnification   37 000 ×

Refer to   Plates 4, 21, 22b, 35, 41
Sections 4.2, 5.2

Micrograph by courtesy of H. S. Johnston

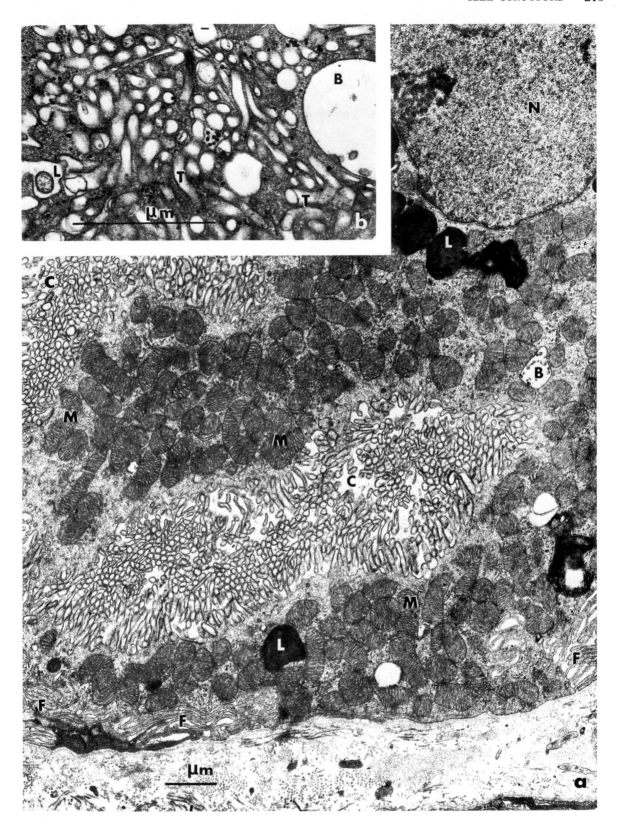

Plate 41a *Ion secretion, salivary duct*

The columnar cells of the striated ducts of salivary glands have distinctive basal ultrastructural specialisations. The appearance of infranuclear striation is due to the presence of complex infoldings and interdigitations of the cell membrane, associated with accumulated mitochondria. In this plate the cells have been obliquely cut across the infranuclear area. The lower pole of one nucleus is included in the section (N). The closely packed mitochondria are the most striking cytoplasmic feature, but the elaborately interdigitated and infolded cell membranes of the lateral basal parts of the cells can be well seen (X). The association of such membrane specialisations and numerous mitochondria is a reliable pointer to a localised ion transportation function.

N   Nucleus
X   Interdigitated infoldings of cell membrane

Tissue   Rat salivary gland, glutaraldehyde and osmium fixation, uranyl and lead staining.

Magnification   17 000 ×

Refer to   Plates 4, 21, 22b, 35, 40, 50, 51, 60
             Sections 4.5, 5.2

Plate 41b *Ion secretion, salivary duct*

This micrograph shows, at higher magnification, the complex membrane specialisations associated with ion secretion. The surface area of this region of the cell is greatly increased by the formation of these interdigitating flaps of cytoplasm. Note that the cell membrane at many places (*) is not clearly defined. This blurring is the characteristic sign of obliquity of the plane of section in relation to well-orientated membranes. The mitochondria of these cells are closely packed in the basal region (M). These provide the energy resources for ion transportation.

M   Mitochondria
*   Indicate areas of obliquely sectioned membrane infoldings

Tissue   Rat salivary gland, glutaraldehyde and osmium fixation, uranyl and lead staining.

Magnification   45 000 ×

Refer to   Plates 4, 21, 22b, 35, 40, 50, 51, 60
           Sections 4.5, 5.2

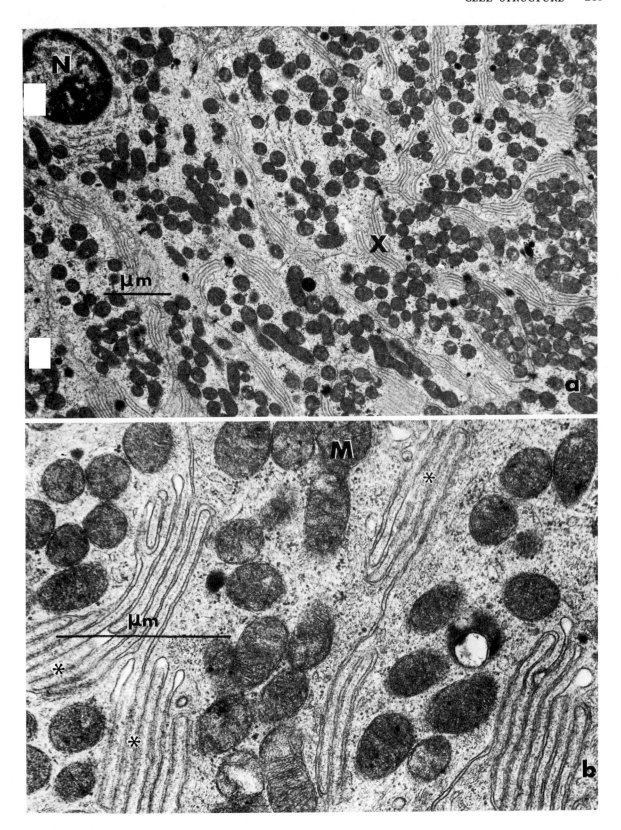

Plate 42a *Endocrine secretion, small intestine*

Although most of the granules of the intestinal endocrine cell are stored in the basal part of the cell, below the nucleus, the apparatus for their manufacture is found largely in the apical region. This micrograph shows the supranuclear cytoplasm of such a cell, with part of the nearby nucleus included (N). The adjacent cells are both of columnar absorptive type, with prominent Golgi membrane systems (G_1).

The endocrine cell also has a clearly seen Golgi system (G_2), in association with which can be seen several secretion granules in the course of formation. A few mature granules released from the Golgi system have not yet made their way past the nucleus to join the main mass of stored secretion granules.

In addition to the Golgi membranes, other cytoplasmic membrane systems are seen, including some cisternae of granular endoplasmic reticulum (R), and various smooth-surfaced membranes and vesicles. Mitochondria are present, showing the usual variations in outline associated with tangential sectioning (*). A single centriole is seen in longitudinal section (C). Some glycogen granules are seen in small groups (arrows).

Various distinct types of intestinal endocrine cell can be identified on the basis of immunocytochemical stains and even by granule morphology. These cells secrete polypeptide hormones and other substances. Although some of these cells have quite well-defined regulatory functions, such as the secretion of cholecystokinin, the role of other types remains to be clarified.

C	Centriole in longitudinal section
G_1	Golgi apparatus of absorptive cell
G_2	Golgi apparatus of endocrine cell showing associated secretion granules
N	Nucleus of endocrine cell
R	Granular endoplasmic reticulum
*	Indicate tangential sectioning effects on mitochondria
Arrows	Indicate groups of glycogen particles

Tissue Human small intestine, glutaraldehyde and osmium fixation, uranyl and lead staining.

Magnification 37 000 ×

Refer to Plates 18a, 36, 69c
Section 5.3

Plate 42b *Endocrine secretion, pancreatic beta cell*

In many of the polypeptide hormone-secreting endocrine cells, granule morphology can contribute to the precise identification of the cell type. For example, all of the gastric and intestinal endocrine cell types have their own distinct ranges of granule shape, size and appearance. This is particularly true of the pancreatic beta cell, which secretes the hormone insulin. This cell is characterised by the frequent occurrence of polycrystalline granules, a pattern not seen normally in any other endocrine cell type.

The micrograph shows part of a human beta cell. There are several small mitochondria (M), some strands of granular endoplasmic reticulum (R) and groups of ribosomes. Although a few granules contain amorphous dense material (A) the majority in this cell are of the typical crystalline pattern.

A	Amorphous secretion granule
M	Mitochondrion
R	Granular endoplasmic reticulum

Tissue Human pancreatic islet, beta cell. Glutaraldehyde and osmium fixation, uranyl and lead staining.

Magnification 34 000 ×

Refer to Plates 18a, 36, 39, 69c
Section 5.3

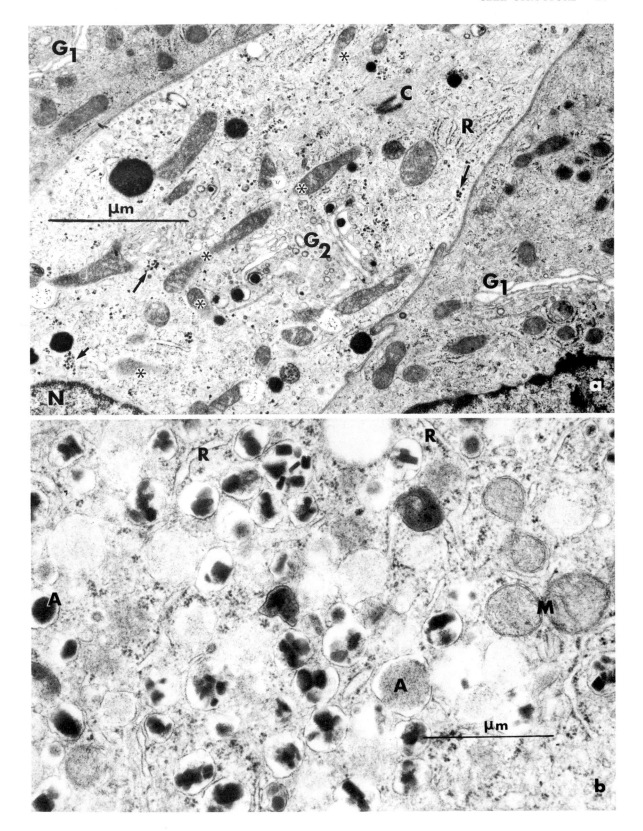

Plate 43a   *Liver plates, bile canaliculi*

This scanning electron micrograph was made from the fractured surface of the liver parenchyma. It displays clearly the interconnecting plates of liver cells (H). The exocrine function of liver is reflected in the presence of the bile canaliculi (C) running along the liver plate in the plane of the fracture, which has revealed the cell surface. These ducts are lined by small microvilli, which can just be made out. The endocrine function of liver is reflected in the abundant vascular channels (V) which alternate with the plates of hepatocytes. These are the sinusoids of the liver, which are normally filled with blood.

C   Linear biliary canaliculi
H   Plates of hepatocytes
V   Vascular channels or sinusoids

Tissue   Liver tissue. Perfusion fixation. Fractured and coated for SEM.

Refer to   Plates 9, 18b, 22a, 23b, 24a, 26c, 44, 46, 55a, 67a, 71, 111
Sections 5.4, 6.4

Micrograph by courtesy of P. Motta

Plate 43b   *Liver sinusoid, space of Disse*

This scanning micrograph shows the endothelial lining of the hepatic sinusoid. Numerous small fenestrations or pores (P) are grouped into aggregates known as sieve plates. In addition to these numerous small gaps, there are also large discontinuities (D) in the endothelial lining. Through these gaps can be seen the underlying microvilli of the hepatocyte, which project into and fill the space of Disse. It is clear from such a graphic view of the endothelium that it must have an unusually high level of permeability to the plasma flowing through the sinusoid. In fact the sinusoidal surfaces of the hepatocytes exposed to the space of Disse are bathed in plasma, facilitating the numerous metabolic exchanges between the blood and the hepatocytes.

D   Large endothelial discontinuities underlain by microvilli of the space of Disse
P   Small endothelial pores grouped in sieve plates

Tissue   Liver tissue. Perfusion fixation, fractured and coated for SEM.

Magnification

Refer to   Plates 9, 18b, 22a, 23b, 24a, 26c, 44, 46, 55a, 67a, 71, 111
Sections 5.4, 6.4

Micrograph by courtesy of P. Motta

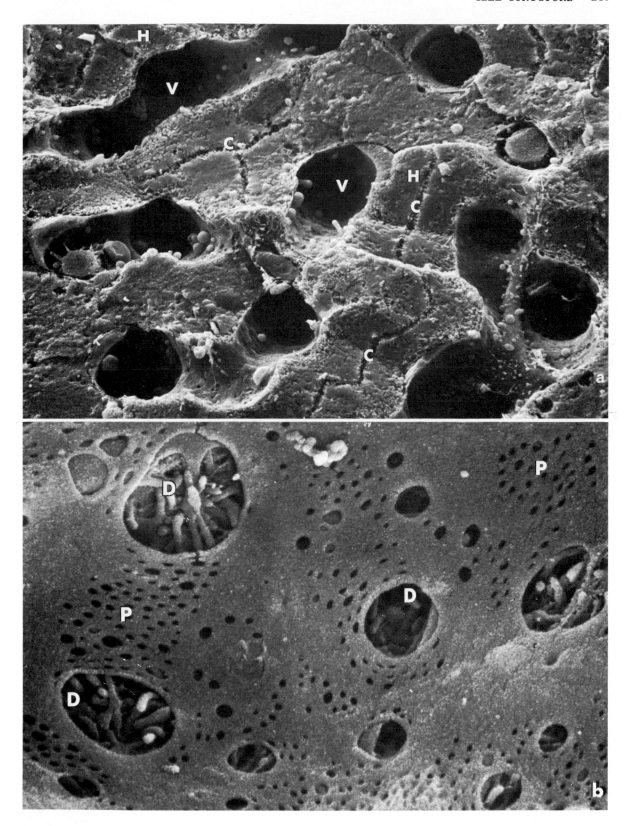

Plate 44 *Isolated hepatocyte*

This scanning electron micrograph shows a single isolated polygonal liver cell, with its various faces exposed. The anastomosing network of biliary canaliculi (C) runs across the contact surfaces of the cell, whereas the microvilli of the space of Disse (D) define the sinusoidal aspects of the hepatocyte. The bile canaliculi are also lined by stubby microvilli. Micrographs such as this give an instant appreciation of the three-dimensional reality of the cell which can only be inferred indirectly from a thin section micrograph.

C   Canaliculi lined by stubby microvilli
D   Microvilli of space of Disse defining the sinusoidal surface of the cell

Tissue   Isolated hepatocyte, metal-coated and studied by SEM.

Refer to   Plates 9, 18b, 22a, 23b, 24a, 26c, 43, 46, 55a, 67a, 71, 111
             Section 5.4

Micrograph by courtesy of P. Motta

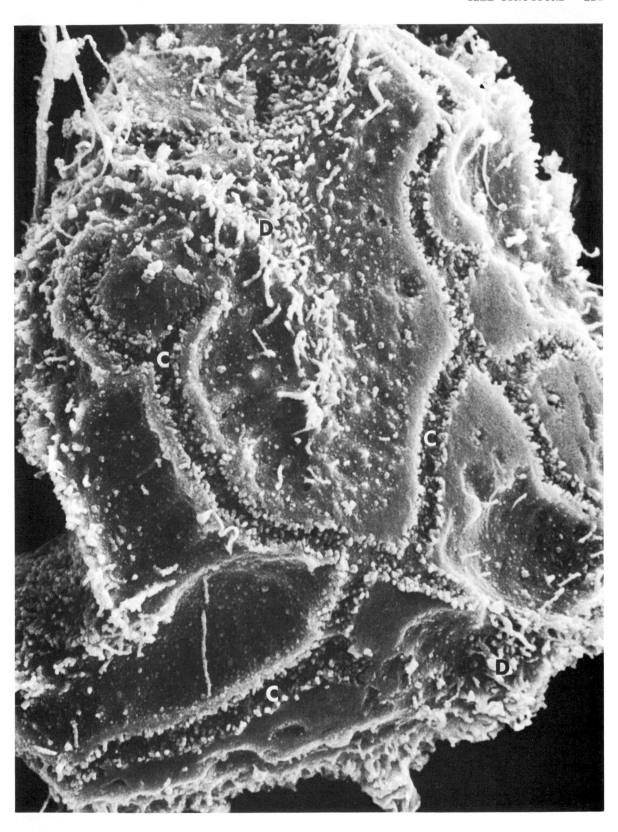

Plate 45 *Liver cells*

Parts of three liver cells appear in this low magnification micrograph. In one corner the hepatic sinusoid is seen, with parts of two red corpuscles (R) appearing as dense homogeneous structures of irregular shape. The sinusoid lining cells present a discontinuous barrier to the plasma (E) and no underlying basal lamina can be seen. The plasma thus has direct access to the sinusoidal surface of the liver cell, with its numerous irregular microvilli (MV). The space between the liver cell and the sinusoid lining cell, known as the space of Disse, is largely filled by these microvilli.

Between adjacent liver cells there appear tubular channels known as bile canaliculi (B). These are also lined by microvilli projecting from the surfaces of the liver cells. The lumen of the canaliculus is sealed off from the intercellular space by junctional specialisations which resist free passage or interchange between the space of Disse and the canalicular lumen (J). Thus the only route by which material normally reaches the bile canaliculus is through the cytoplasm of the liver cell. Most of the common cytoplasmic organelles are found in the liver cell. The mitochondria are typically large (M) but rather poorly organised. The pericanalicular distribution of the liver lysosomes (L) is well seen. These are large, dense and of variable appearance. Microbodies can be distinguished from the lysosomes by the presence of a central dense nucleoid within an otherwise homogeneous background (MB). The microbodies are less dense and are slightly smaller than lysosomes. Numerous glycogen particles are scattered in rosettes and clusters through the cytoplasm (P). The Golgi apparatus (G) appears at several points, and shows the typical features. In this field the Golgi apparatus lies close to the bile canaliculus. Cisternae of the granular endoplasmic reticulum (GER) and areas of smooth endoplasmic reticulum (SER) can be found. Two large empty spaces represent fat removed during processing (F).

B	Bile canaliculi
E	Sinusoidal lining cells
F	Fat droplets
G	Golgi apparatus
GER	Granular endoplasmic reticulum
J	Junctional complex
L	Lysosome
M	Mitochondrion
MB	Microbody
MV	Microvilli
P	Glycogen particles
R	Red blood corpuscle
SER	Smooth endoplasmic reticulum

Tissue   Mouse liver. Glutaraldehyde and osmium fixation, uranium and lead staining.

Magnification   13 000 ×

Refer to   9, 18b, 22a, 23b, 24a, 26c, 43, 44, 46, 55a, 67a, 71, 111
Sections 4.11, 5.4, 6.4, 8.1

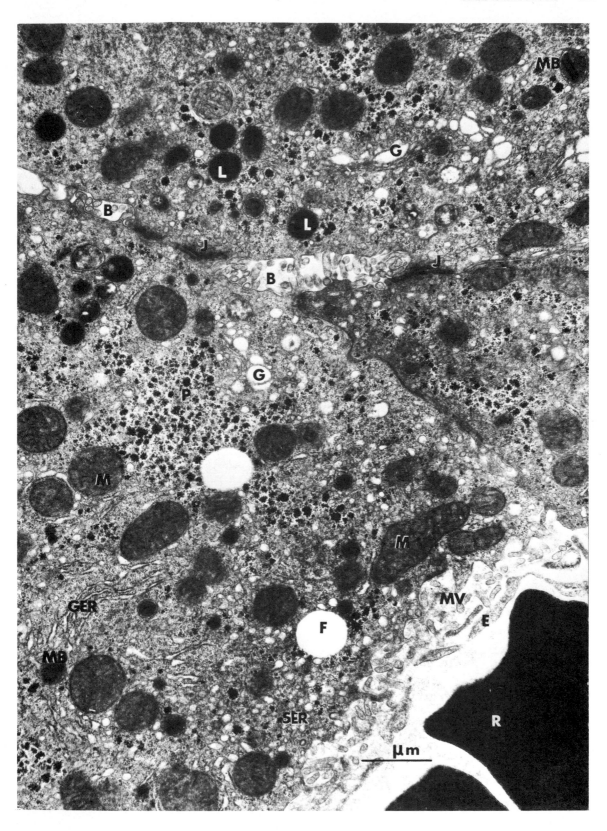

Plate 46 *Bile secretion*

The principal exocrine secretory function of the liver is the production of bile, although to some extent this can be seen also as an excretory function. The hepatocytes are arranged in cords, between which run the delicate tubules of the biliary canaliculi. These tiny ducts, formed out of matching tubular grooves on the contact surfaces of adjacent liver cells, run out along the liver cords to the portal tracts, where they drain into true duct structures which carry the bile out of the liver to the gall bladder and to the gut lumen.

The micrograph shows the canalicular lumen (L) into which small microvilli project from the adjacent hepatocyte surfaces. The margins of the canalicular lumen are sealed off by junctional complexes (J) from continuity with the narrow extracellular space. Within the two hepatocytes are seen mitochondria (M), granular reticulum (R) and smooth or agranular reticulum (*), with plentiful glycogen in the alpha configuration, appearing as rosettes (arrows).

J	Junctional complex
L	Lumen of biliary canaliculus
M	Mitochondrion
R	Granular endoplasmic reticulum
*	Indicate elements of the smooth endoplasmic reticulum
Arrows	Indicate glycogen rosettes

Tissue   Baboon liver. Glutaraldehyde and osmium fixation. Uranyl and lead staining.

Magnification   40 000 ×

Refer to   Plates 9, 18b, 22a, 23b, 24a, 26c, 43, 44, 45, 55a, 67a, 71, 111
Sections 4.2, 4.11

Micrograph by courtesy of H. S. Johnston

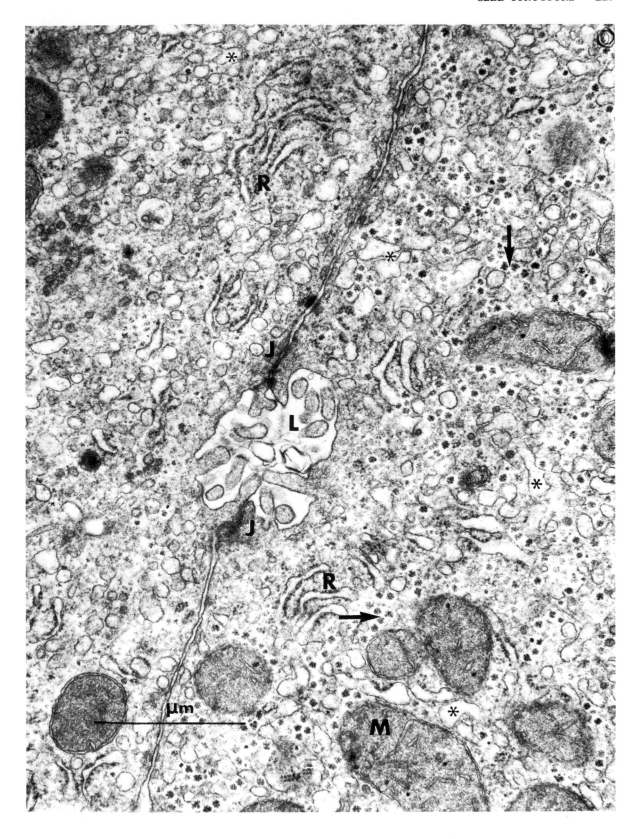

ABSORPTION AND PERMEABILITY

Plate 47 *Small intestine, surface morphology*

The familiar villi of the small intestinal mucosa are generally presented to students in two dimensions, in histological sections, leaving the observer to integrate the image into a three-dimensional concept through the imagination. The scanning electron microscope is an ideal tool for visualising such surface details, since it combines high resolving power with great depth of focus. The three-dimensional pattern of the mucosa can then be instantly appreciated, without the risk of error in extrapolating from two-dimensional sections.

In this scanning micrograph, the variation in duodenal mucosal morphology is clearly seen. Some villi (A) are finger-shaped, some are broader (B) like leaves or tongues, while others are convoluted (C). Individual villi have slightly creased, irregular surfaces. The effect of such mucosal complexity is to increase by a substantial amount the surface area available for intestinal absorption.

A Finger-shaped villi
B Leaf- or tongue-shaped villi
C Convoluted villi

Tissue Human duodenum. Glutaraldehyde and osmium fixation, air dried, carbon-platinum coated.

Magnification 110 ×

Refer to Plates 7a, 10a, 14, 27a, 38, 48, 49, 50, 113
Sections 2.2, 6.1

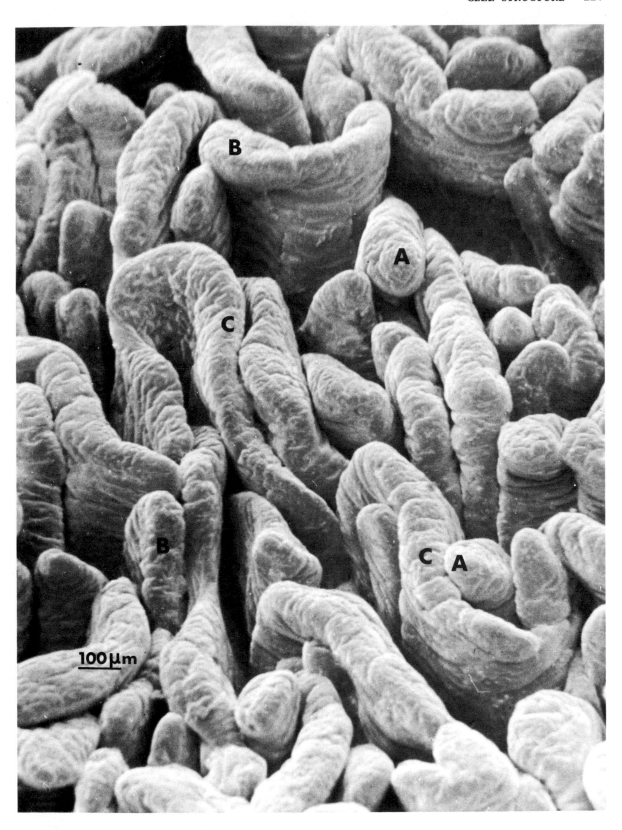

Plates 48a, 48b *Intestinal epithelial cell surfaces*

These higher magnification scanning electron micrographs show the pattern produced on the intestinal epithelial surface by the closely packed columnar absorptive cells. Individual cell outlines are seen at the lower magnification forming polygonal territories, held tightly to their neighbours at the apical junctional complexes, but bulging as a slight dome into the lumen. A fine surface granularity is just seen, indicative of the pattern of microvilli, barely resolved at this magnification.

The higher magnification scanning electron micrograph shows the pattern of individual microvilli on the intestinal epithelial cell surface. A rough hexagonal stacking exists, but the pattern is not precisely geometrical.

Tissue   Mouse small intestinal epithelium. Glutaraldehyde and osmium fixation, critical point dried, gold coated.

Magnification   Plate 48a 3500 ×
                              Plate 48b 42 000 ×

Refer to   Plates 10a, 14, 27a, 38, 47, 49, 50, 113
                   Sections 2.2, 6.1

Plate 48c *Intestinal microvilli*

This is a thin section viewed by transmission electron microscopy, showing the intestinal microvilli in transverse section. The individual finger-like processes are seen as circular profiles with central cores of microfilaments, which have been identified as actin. The cell membrane of each microvillus is recognised by its trilaminar structure. From the outer surface of the membrane, fine fuzzy projections represent the cell coat or glycocalyx.

Tissue   Mouse small intestine, glutaraldehyde and osmium fixation, uranyl and lead staining.

Magnification   90 000 ×

Refer to   Plates 14, 27, 38, 47, 113
                   Sections 2.2, 6.1

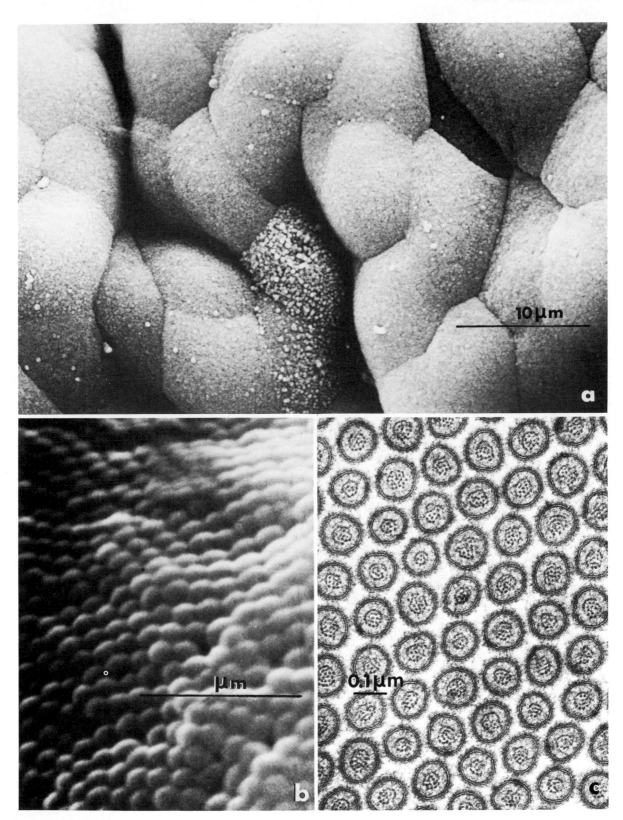

Plates 49a, 49b *Large intestine, surface morphology*

The small intestine and the large intestine share a general function in relation to absorption, although the role of the large bowel is much more restricted than that of the small bowel. The colonic mucosa is mainly concerned with the reabsorption of water from the bowel contents, since most of the absorption related to general nutritional needs has been completed by the time the food material leaves the small intestine. An elaborately expanded absorptive surface is, therefore, less necessary in the colon than in the small bowel

This functional difference is reflected in the surface pattern of the colonic mucosa. At low magnification, there is a total absence of surface projections. The intestinal crypts open on to the general mucosal surface, where they are visible as small holes (arrows). Around each crypt, a roughly circular territory of mucosa is marked off by shallow furrows. Two deep furrows (F) represent part of a general contouring of the mucosa.

In the higher magnification micrograph, the orifice of a single crypt (C) is seen as a narrow slit. A strand of mucus is seen within the lumen. The individual enterocyte territories are less clearly defined than in the small intestine. The microvilli can just be made out as a fine stubble across the epithelial surface. The mouths of numerous goblet cells (G) are also seen, reflecting the much greater importance of lubricant mucus secretion in the colon than in the small intestine, where the bowel content is still fluid.

C   Slit-shaped crypt orifice
F   Mucosal furrow
G   Goblet cells
Arrows   Indicate crypt orifices in low magnification
         micrograph

Tissue Human colonic mucosa, glutaraldehyde and osmium fixation, critical point dried, gold coated.

Magnification   Plate 49a 290 ×
                  Plate 49b 2600 ×

Refer to   Plates 47, 48, 114
             Sections 2.2, 6.1

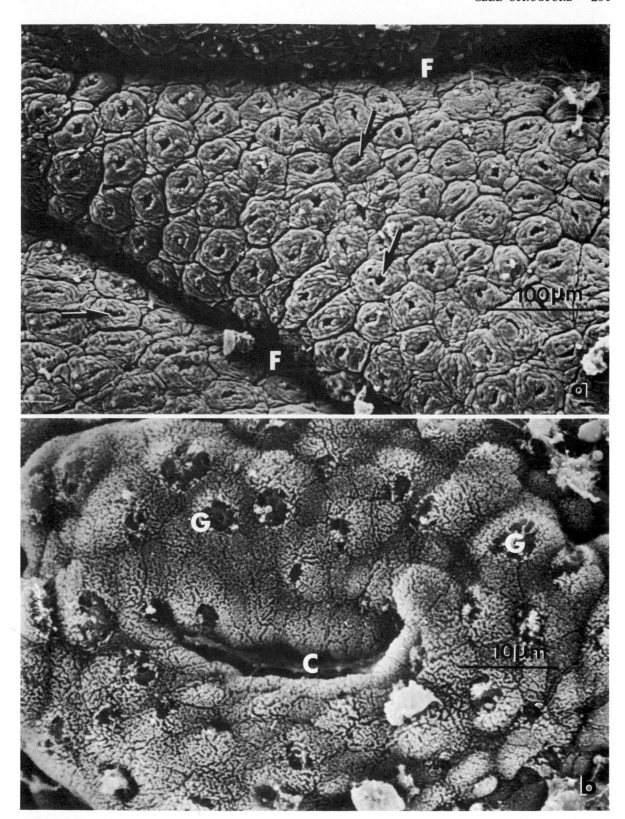

Plates 50a, 50b *Gall bladder, surface morphology*

These two scanning electron micrographs show the elaborate surface contours of the gall bladder mucosa, which reabsorbs water and thus concentrates the bile. At low magnification, individual cells appear as small polygonal units on the mucosal surface, which is thrown as a whole into complex folds. The higher magnification micrograph shows the polygonal pattern of cells in greater detail. Individual microvilli can just be made out. These surface specialisations are typical of an absorptive epithelium.

Tissue Human gall bladder mucosa, glutaraldehyde and osmium fixation, critical point dried, gold coated.

Magnification   Plate 50a 300 ×
                   Plate 50b 5000 ×

Refer to   Plates 14, 47, 48, 49
              Sections 6.1, 6.2

Micrographs by courtesy of A. E. Williams

Plate 50c *Gall bladder epithelium*

A distinctive feature of gall bladder epithelium is the presence of basal intercellular spaces of variable extent, the appearance of which is dependent on physiological function. In the resting gall bladder these spaces are relatively inconspicuous, while during active fluid resorption they become widely patent.

As shown here, these basal intercellular spaces (*) are occupied by complex flap-like cytoplasmic projections which interdigitate with each other to form an elaborate labyrinth. When the intercellular spaces dilate, the interdigitations retain contact across the gap between the cells, preserving the labyrinthine nature of the intercellular space. This configuration of the cell surface is believed to be important in establishing trans-epithelial fluid flow during active resorption.

Note the basal portion of an epithelial cell nucleus (N). Note also the presence of the basal lamina (L) and underlying collagen fibrils (C).

C   Collagen fibrils in connective tissue
L   Basal lamina
N   Nucleus of gall bladder epithelial cell
*   Indicate labyrinthine intercellular space crossed by interdigitating cytoplasmic processes

Tissue Mouse gall bladder epithelium, glutaraldehyde and osmium fixation, uranyl and lead staining.

Magnification   21 000 ×

Refer to   Plates 4, 21, 22b, 40, 41, 51, 60
              Sections 6.1, 6.2

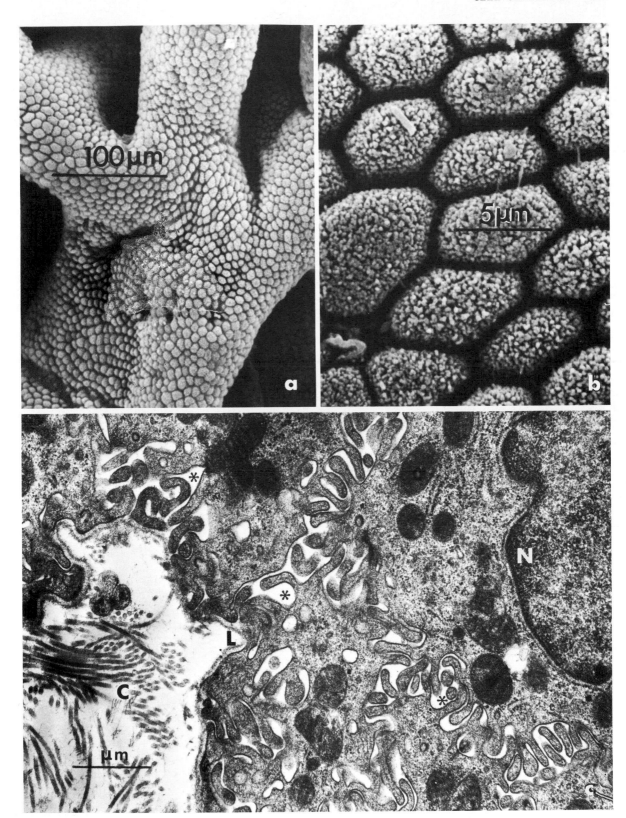

Plate 51a *Renal tubular epithelium*

The epithelium of the renal tubules is active in many aspects of the absorption and transport of metabolites, including ions. Its distinctive specialisations include abundant mitochondria (M) of great size and complexity, which generate the energy resources for its transport functions. The cell membrane at the apex, bordering the tubular lumen, is thrown into closely packed, elongated microvilli (MV), between which active micropinocytosis takes place (*), resulting in the formation of many apical vesicles. The cell membrane at the base of the tubule cell is infolded (arrows), the infoldings running up between the vertically orientated parallel mitochondria. The extent of cell uptake and transport is reflected in the many secondary lysosomes and residual bodies seen in the cytoplasm (L).

L   Lysosomal structures of various types
M   Elaborate mitochondria
MV Microvilli
*   Indicate basal infoldings of cell membrane

Tissue Renal tubular epithelium, glutaraldehyde and osmium fixation, uranyl and lead staining.

Magnification 11 000 ×

Refer to Plates 21, 40, 41, 50, 60
    Section 6.3

Micrograph by courtesy of G. Bullock

Plate 51b *Renal tubule, basal infoldings*

This shows the basal part of an epithelial cell from the convoluted tubule of the kidney, resting on a continuous basal lamina (L). Immediately adjacent to the tubule is the fenestrated endothelium of a capillary vessel (V) containing a red blood corpuscle (R). Note the narrow gap between the circulation and the ion-transporting basal surface membrane of the epithelial cell. The base of the epithelial cell is deeply infolded (arrows) producing pale clefts of extracellular space (*) which separate the resultant tongues of basal cytoplasm. Numerous large mitochondria (M) are found in these cytoplasmic processes, sandwiched between the layers of infolded membrane.

L   Basal lamina of epithelial cell
M   Mitochondrion sandwiched between basal infoldings
R   Red blood corpuscle
V   Capillary vessel with fenestrated endothelium
*   Indicate clefts of extracellular space produced by basal infolding
Arrows Indicate basal infoldings of cell membrane

Tissue Human kidney. Glutaraldehyde and osmium fixation, uranyl and lead staining.

Magnification 25 000 ×

Refer to Plates 21, 40, 41, 50, 60
    Section 6.3

Micrograph by courtesy of A. L. C. McLay

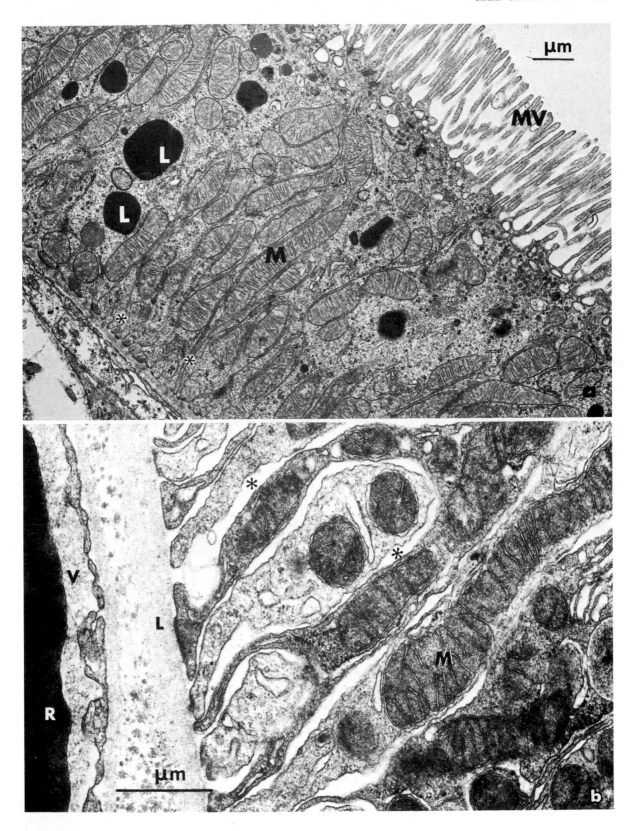

Plate 52 *Capillary endothelium*

This micrograph shows a thin-walled blood capillary in the exocrine pancreas. Within the lumen (L) lies a neutrophil polymorph, with its typical convoluted nucleus (N_1). Notice the different chromatin patterns of the polymorph and the nearby acinar cell nucleus (N_2). The endothelial lining of the vessel is very delicate. A few areas show small fenestrations (F) but these sieve plate areas are in the minority. Elsewhere the endothelium is continuous, but shows the formation of micropinocytotic vesicles (arrows). At one point an intercellular adhesion zone is seen (*).

Note the complexity of the basal cytoplasm of the adjacent acinar cells. The granular endoplasmic reticulum (R) is particularly elaborate in protein-secreting cells such as these pancreatic acinar cells. Zymogenic granules (Z) and mitochondria (M) are also seen.

F	Fenestrations forming sieve-plates
L	Lumen of capillary
M	Mitochondria
N_1	Nucleus of neutrophil polymorph
N_2	Nucleus of pancreatic acinar cell
R	Granular endoplasmic reticulum
Z	Zymogen granule
*	Indicate endothelial cell junction
Arrows	Indicate micropinocytotic vesicles

Tissue Baboon pancreas, glutaraldehyde and osmium fixation, uranyl and lead staining.

Magnification   12 000 ×

Refer to   Plates 43b, 45, 53, 54, 55, 57, 58, 69, 84, 89, 99
Sections 5.1, 6.4

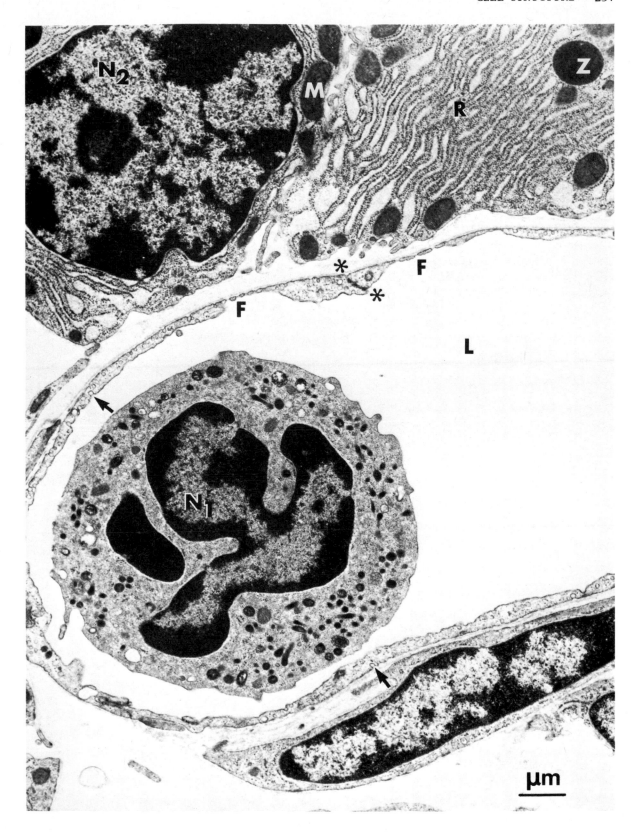

μm

Plate 53a *Capillary endothelium, fenestrated*

This micrograph shows a capillary vessel from the exocrine pancreas of the guinea pig. The bases of nearby acinar cells contain elaborate cisternae of granular endoplasmic reticulum (ER) in which a few intracisternal granules appear, a feature of pancreas in this species. The basal lamina of the epithelium is tenuous, but can be made out here and there. The capillary external lamina is equally diffuse. The endothelial cell nucleus is convoluted and indented, perhaps partly on account of contraction of the vessel on fixation. Numerous caveolae (C) appear at the external surface of the cell close to the nucleus (N). The lumen (L) appears pale. At one point there is an intercellular junction (J) which is accompanied by a marginal fold, or projection of the endothelial cell into the lumen. Much of the circumference of the vessel consists of an attenuated layer of endothelial cytoplasm, perforated by pores or fenestrations (F), each apparently bridged by a thin diaphragm. This type of capillary is found in sites characterised by marked permeability. Two thin processes of cytoplasm probably represent sections of a capillary pericyte, a contractile cell, which is related to smooth muscle and is contained within the capillary basal lamina (X).

C	Cavelolae or micropinocytotic vesicles
ER	Granular endoplasmic reticulum
F	Capillary fenestrations
J	Intercellular junction
L	Capillary lumen
N	Nucleus
X	Pericyte

Tissue Guinea pig pancreas. Glutaraldehyde and osmium fixation, uranium and lead staining.

Magnification 20 000 ×

Refer to   43b, 45, 52, 54, 55, 57, 58, 69, 84, 89, 99
Section 6.4

Plate 53b *Capillary endothelium, non-fenestrated*

This plate shows part of a vessel in a connective tissue septum in muscle. One Z line (Z) and a group of glycogen particles (G) can be seen close to the surface of a skeletal muscle cell. An endothelial cell junction is seen, with associated marginal folds (F) which project into the lumen, coming into contact with a red blood corpuscle (R). Oblique sectioning effects account for the blurring of the cell membranes at points along the endothelial junction (O). A number of micropinocytotic vesicles (V) can be seen within the thickness of the endothelial cell, some communicating freely with the cell exterior. In muscle the endothelium is slightly less permeable than in some other sites. This is reflected in the relative thickness of the endothelial layer and the absence of fenestrations.

F	Marginal folds
G	Glycogen particles
L	Capillary lumen
O	Obliquely sectioned membrane
R	Red blood corpuscle
V	Micropinocytotic vesicles
Z	Z line

Tissue Human skeletal muscle. Glutaraldehyde and osmium fixation, uranium and lead staining.

Magnification 80 000 ×

Refer to   43b, 45, 52, 54, 55, 57, 58, 69, 84, 89, 99
Section 6.4

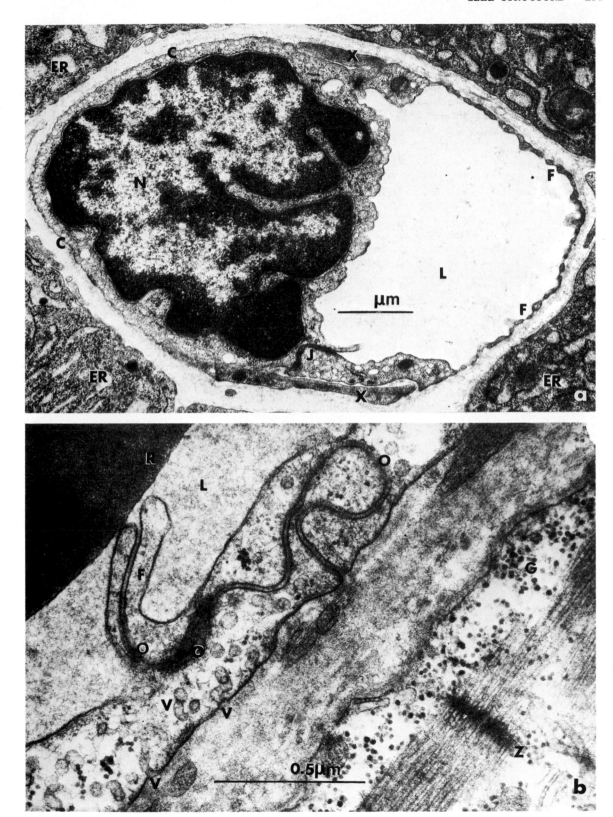

Plate 54a *Fenestrated endothelium*

This micrograph shows part of the wall of a fenestrated capillary. The connective tissue (C) around the vessel contains scattered collagen fibrils, with a well-defined basal lamina (B) applied closely to the endothelial cell. The endothelium is perforated by many fenestrations (*) each bridged by a thin diaphragm. A single endothelial cell granule (G) is present.

Within the lumen (X) notice the circulating mono-nuclear cell, probably a blood monocyte. Its nucleus (N) is tangentially sectioned (T), with consequent blurring of the envelope. There are two nuclear pores seen in face view (arrows). The cytoplasm of this cell contains primary lysosomes (L), scattered membranes of the endoplasmic reticulum and small mitochondria. Notice that the margin of this cell is blurred, indicating again an oblique sectioning effect (O).

B	Basal lamina
C	Connective tissue containing collagen fibrils
G	Endothelial cell granule
L	Monocyte lysosomes
N	Nucleus of monocyte
O	Oblique section of cell surface, with blurred membrane
T	Tangentially sectioned nuclear envelo 213
X	Lumen of capillary 214
*	endothelial fenestrations
Arrows	Indicate nuclear pores in face view

Tissue Human small intestine, glutaraldehyde and osmium fixation, uranyl and lead staining.

Magnification 38 000 ×

Refer to Plates 45, 52, 53, 55, 57, 58, 69, 84, 89, 99
Section 6.4

Plate 54b *Endothelial fenestration*

This high-magnification micrograph shows a single endothelial fenestration. The lumen of the capillary (L) lies on one side, the extravascular connective tissue space (C) on the other. The basal lamina (B) is separated from the endothelial cell by a pale interspace. The endothelial cell cytoplasm (E) is featureless.

B	Basal lamina
C	Connective tissue
E	Endothelial cell cytoplasm
L	Lumen of capillary

Tissue Rat pituitary gland. Osmium fixation. Uranyl staining.

Magnification 190 000 ×

Refer to Plates 45, 52, 53, 55, 57, 58, 69, 84, 89, 99
Section 6.4

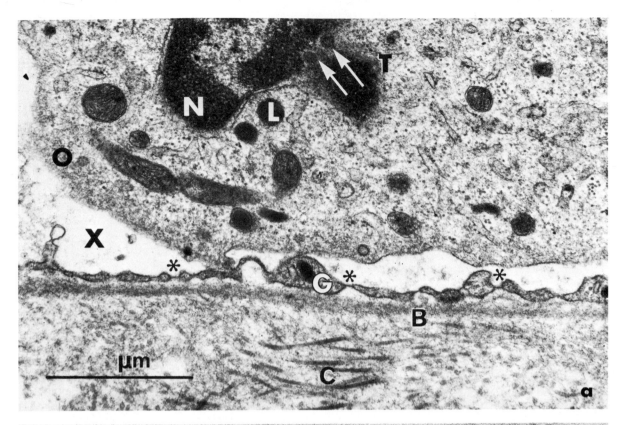

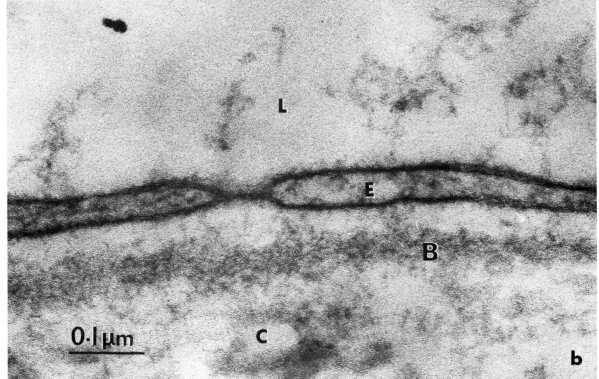

Plate 55a *Hepatic sinusoid*

The sinusoid in liver is the most permeable variety of blood vessel, a fact readily appreciated from the fine structural features of its lining. The lumen (L) is separated from the adjacent hepatocytes (H) by a discontinuous and partially overlapping endothelial lining (E) with wide gaps (*) allowing free access for the blood plasma to the underlying perivascular space of Disse (D). The space of Disse, lined by surface projections of the hepatocyte, provides the vital interface between blood and liver for the numerous metabolic functions of that organ. Also present in the perivascular space is the delicate intralobular connective tissue framework consisting of the small scattered bundles of collagen fibrils (C) which appear as the framework of the liver cords in reticulin-stained preparations. Within the hepatocyte there are mitochondria (M), smooth endoplasmic reticulum (S) and abundant alpha glycogen (G).

C   Collagen fibrils in cross-section
D   Space of Disse lined by projections from hepatocyte surface
E   Endothelial cell cytoplasm
G   Rosettes of alpha glycogen
H   Hepatocyte
L   Lumen of hepatic sinusoid
M   Mitochondria of hepatocyte
S   Smooth endoplasmic reticulum
*   Indicate wide gaps in the endothelial lining of the sinusoid

Tissue Human liver, glutaraldehyde and osmium fixation, uranyl and lead staining.

Magnification   20 000 ×

Refer to   Plates 43, 44, 45, 46, 52, 53, 54, 57, 58
            Sections 4.11, 5.4, 6.4

Plate 55b *Endothelial granules*

A characteristic inclusion of the endothelial cell is the specific endothelial granule, or Weibel-Palade granule. These membrane-limited structures are present in variable numbers in different parts of the vascular system. They are typically rod-shaped, with an internal structure consisting of fine tubules, closely packed, which lie parallel to the long axis. Two of these granules are seen here in longitudinal section (*) and several others in transverse or oblique section. The nature of the content of these granules is uncertain and their function unknown. Perhaps they represent part of the body's system of defence against accidental vascular blockage by thrombosis.

Note also in this micrograph the presence of a red blood corpuscle (R). Its margins, and the inner margins of the endothelial cell, are obliquely sectioned (arrows), causing blurring of the membrane image. Part of an endothelial nucleus (N) and a mitochondrion (M) are also seen.

M   Mitochondrion of endothelial cell
N   Nucleus of endothelial cell
R   Red blood corpuscle in vascular lumen
*   Endothelial Weibel-Palade granule in longitudinal section

Tissue Human vascular endothelium. Glutaraldehyde and osmium fixation, uranyl and lead staining.

Magnification   48 000 ×

Refer to   Plates 54, 69
            Section 6.4

Plate 55c *Endothelial cytoplasm*

This micrograph shows a portion of the wall of a larger vessel, probably a small venule. The lumen (L) is to the left. The endothelial cell (E) contains a centriole (C) cut in longitudinal section and a small Golgi apparatus (G). Note the presence, in relation to the centrioles, of a distinct cross-striated rootlet (R). Such structures are sometimes also seen attached to the basal bodies of cilia, perhaps serving in some way as an anchor within the cytoplasm. Several micropinocytotic vesicles can be seen forming at the endothelial cell surface (arrows). There is an underlying basal lamina (B) between the endothelium and the adjacent flattened cells of the deeper layers of the vessel wall.

B   Basal lamina
C   Centriole in longitudinal section
E   Endothelial cell cytoplasm
G   Golgi apparatus
L   Lumen of vessel
R   Rootlet of centriole, showing cross-striations
Arrows Indicate micropinocytotic vesicles forming at the surface of the endothelial cell

Tissue Mouse testis. Glutaraldehyde and osmium fixation, uranyl and lead staining.

Magnification   26 000 ×

Refer to   Plates 30, 31, 93
            Sections 2.2, 4.10, 6.4

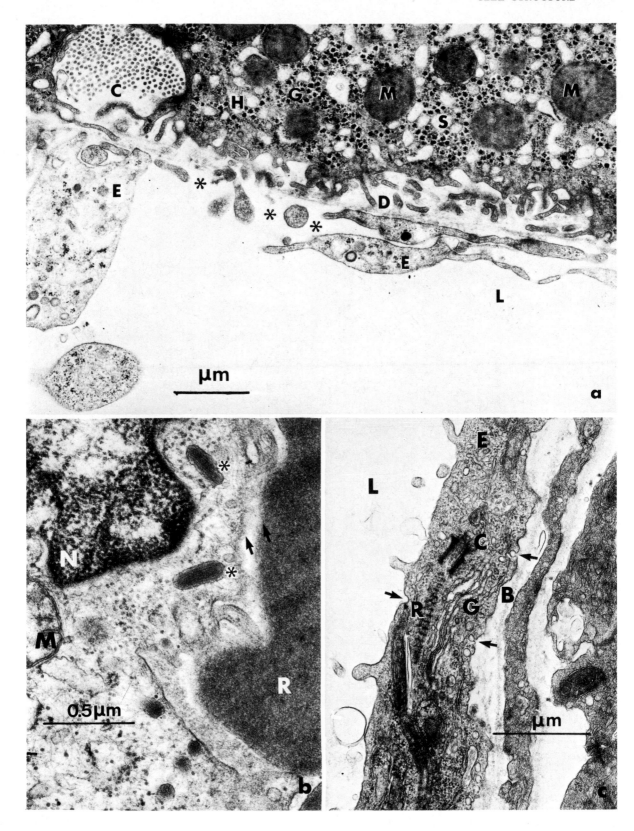

Plates 56a, 56b *Lung, pulmonary alveoli*

These are scanning electron micrographs of the cut surface of rat lung. Although SEM is still most often used for looking at tissue surfaces, these surfaces may be exposed by cutting, sectioning, or microdissection. The low-power micrograph in Plate 56a shows the open sponge-like appearance of the alveoli (A) in the gaseous exchange parenchymal component of the lung, contrasted with the components of the tracheobronchial tree (B), seen cut obliquely in two places. From their diameter, it can be estimated that these are bronchi or large bronchioles. At higher magnification in Plate 34c, secretory cells and cilia can be seen lining the inner aspect of these bronchi. The accompanying branches of the pulmonary blood vessels can be identified (V) with red blood cells filling the lumina.

Most of the micrograph, however, shows alveoli, cut in various planes of section. These are seen at higher power in Plate 56b, where the details of the alveolar wall can be made out. The significant feature here is the extreme thinness of the wall, consisting of flattened type 1 pneumocytes, covering the underlying capillary network. The capillaries make up the bulk of the alveolar septum and bulge into the alveoli. The outlines of red blood cells (R) in the capillaries can be made out clearly through the alveolar wall. Many such profiles can be seen in this one alveolar wall, underlining the importance of surface area in the function of gaseous exchange.

Two alveolar pores can also be seen (P). These are openings in the alveolar wall, connecting one alveolus with its neighbour through the shared alveolar septum. The function of these pores is still in doubt.

A	Sponge-like texture of pulmonary alveoli
B	Bronchus or large bronchiole
P	Alveolar pore
R	Outline of red blood corpuscle in capillary seen through the thin alveolar lining layer
V	Branch of pulmonary vessel containing red corpuscles

Tissue Rat lung, inflated by fixative. Glutaraldehyde and osmium fixation, uranyl and lead staining.

Magnification   Plate 56a 90 ×
                  Plate 56b 1800 ×

Refer to   Plates 57, 58, 59
             Section 6.5

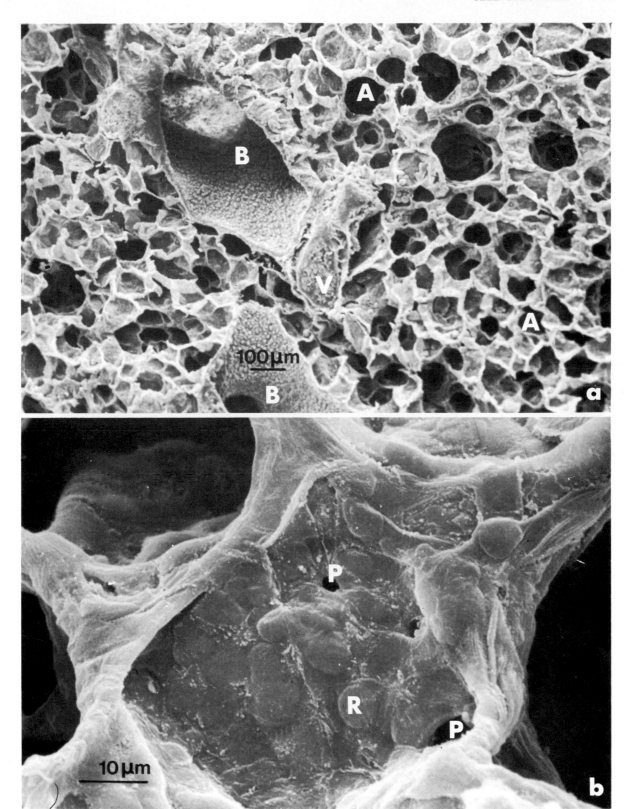

Plate 57 *Pulmonary alveolar lining*

This is perhaps the most remarkable physiological interface of the body, the boundary between the air that is breathed in (A) and the circulating blood (C). Note that there is a finely granular precipitate of blood plasma in the capillary lumen, whereas the alveolar space appears completely empty. The interface consists of the capillary endothelial cell (E), the flat alveolar epithelial type 1 lining cell (L_1) and the intestinal connective tissue space (T) with the two basal laminae (*) associated with the endothelial and epithelial layers. The large, plump cell (L_2) which is continuous with the flattened alveolar lining cell and is joined to it by the junctional complex (J), is a type 2 alveolar epithelial lining cell or great alveolar cell. It is characterised by the presence of large secretory inclusions (S) containing dense laminated material. This is phospholipid-rich surfactant, secreted by the type 2 cell. Its function is to reduce the surface tension of the broad flat expanses of alveolar wall, to avoid the hazard of pulmonary collapse. The strength of the normal surface tension of wet layers is such that the force of the respiratory movements would be unable to re-open collapsed alveoli. Absence of the surface-tension reducing effect of surfactant is thought to be a factor in the acute respiratory distress syndrome of the premature infant.

The capillary wall is of the non-fenestrated type, but shows abundant micropinocytotic activity. An endo-thelial nucleus (N) is present, and an endothelial cell junction is seen (arrow).

A	Alveolar space containing respired air
C	Capillary lumen containing precipitated plasma
E	Capillary endothelial cytoplasm showing numerous micropinocytotic vesicles.
J	Junctional complex joining type 1 and type 2 cells
L_1	Type 1 alveolar epithelial lining cell
L_2	Type 2 alveolar epithelial lining cell, or great alveolar cell
N	Nucleus of endothelial cell
S	Surfactant-containing secretory granules of type 2 cell
T	Connective tissue space
*	Indicate basal laminae
Arrow	Indicates endothelial cell junction

Tissue   Dog lung. Glutaraldehyde and osmium fixation, uranyl and lead staining.

Magnification   29 000 ×

Refer to   Plates 56, 58, 59
              Section 6.5

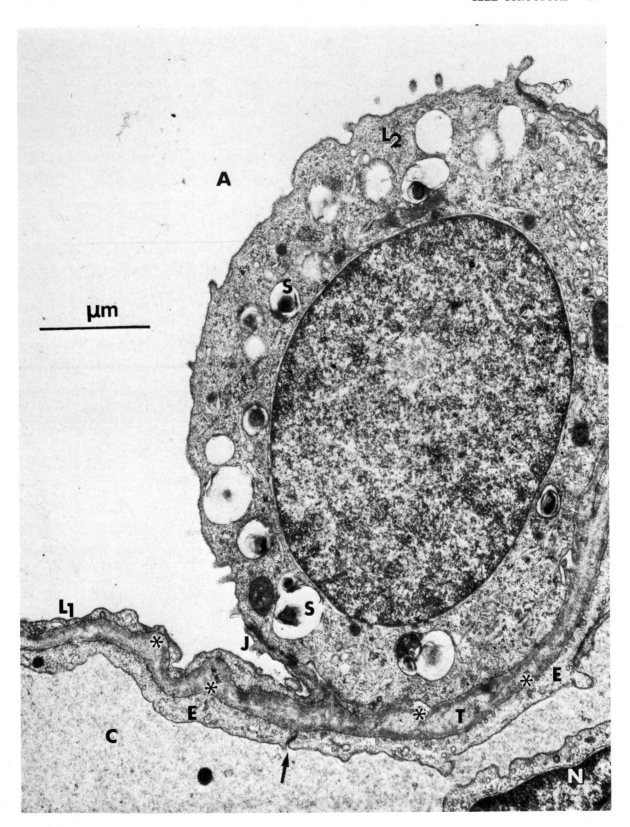

Plate 58 *Blood-air barrier*

This high magnification micrograph shows in detail the narrow interface between blood and air. The airspace of the pulmonary alveolus (PA) is to the right of the plate; the dense area on the left is part of a red blood corpuscle (RBC). The red corpuscle comes very close to the endothelial wall of the capillary (E). The narrow space between the surface membrane of the red corpuscle and the cell membrane of the endothelial cell is indicated by arrows. This space is occupied by plasma. The endothelial wall is less than 20 nm thick at points.

The basal laminae of endothelium and epithelium are fused in this region, no true connective tissue space remaining between them. The amorphous appearance of the material composing the basal lamina is typical of this structure (BL).

The alveolar lining is formed by the alveolar epithelial cell, which is itself less than 100 nm in thickness and is relatively featureless (Epi). In this area of the lung, the total width of the tissue separating blood from air is in places as little as 0.1 μm, thus allowing rapid passage of carbon dioxide and oxygen across the barrier in accordance with the physical constants which determine diffusion rates. In view of the unusual delicacy of this barrier, with the scantiness of support tissues over large areas, it may seem surprising that the alveolar walls are so well able to resist the mechanical stresses of respiration.

BL   Basal lamina or lamina densa
E   Endothelial cell
Epi   Alveolar epithelium
PA   Pulmonary alveolus, air space
RBC   Red blood corpuscle
Arrows   Indicate the narrow gap between red corpuscle and endothelial cell

Tissue   Mouse lung. Glutaraldehyde and osmium fixation, uranium staining.

Magnification   170 000 ×

Refer to   Plates 56, 57, 59, 69
             Sections 6.5, 8.1

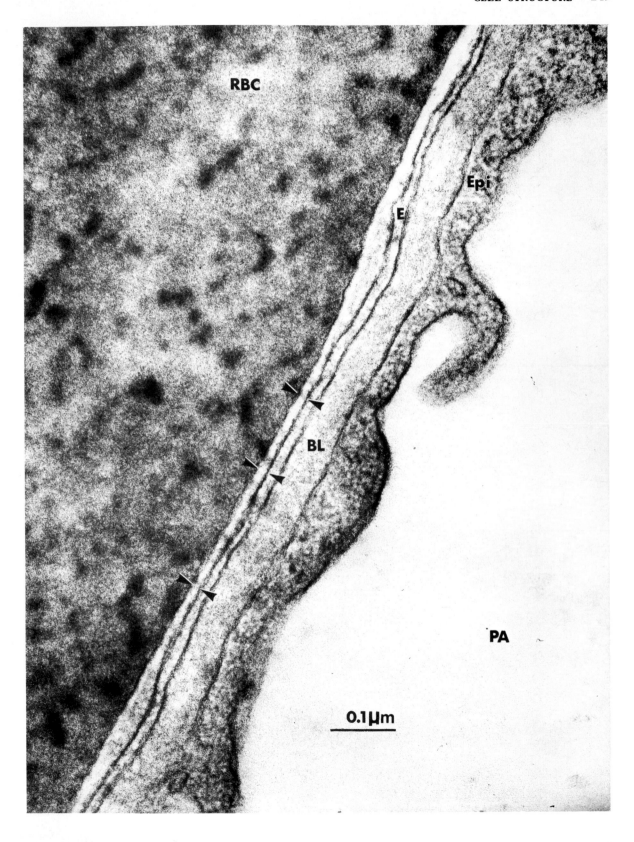

0.1μm

Plates 59a, 59b, 59c *Renal glomerulus, podocytes*

This series of three scanning electron micrographs at increasing magnifications demonstrates different structural features of the surfaces of the renal glomerulus. Plate 57a shows the outer aspect of the whole glomerulus, revealed by removal of the parietal layer of Bowman's capsule, thus opening into the urinary space. The tortuous arrangement of the capillaries (C) can be seen through the cells of the visceral epithelial layer of Bowman's capsule (B), the podocytes which ensheath and support the individual capillary loops.

At the higher magnification of Plate 57b, a podocyte cell body (B) lies in the hollow of a capillary loop, extending its broad major foot processes (*) across the vessel's surface. Note, however, that some of these foot processes do not appear to originate from this podocyte, but come from another podocyte behind the vessel. The minor foot processes, or pedicels (P), of the two cells thus interdigitate intimately, clasping the capillary between them.

The higher power view of Plate 57c shows in greater detail the interlocking network of minor foot processes or pedicels (P), like the fingers of clasped hands. The glomerular filtrate trickles between these delicate foot processes into the urinary space. Presumably the podocyte plays an important mechanical role in the support of the capillary loops of the glomerulus.

B   Cell body of glomerular podocyte
C   Capillary loops of glomerular tuft
P   Minor foot processes, or pedicels, of the glomerular podocyte
*   Indicate major foot processes of glomerular podocyte

Tissue   Rat kidney, glutaraldehyde and osmium fixation, critical point dried, gold coated.

Magnification   Plate 59a 800 ×
                  Plate 59b 3900 ×
                  Plate 59c 7000 ×

Refer to   Plates 56, 60, 62, 109, 110
             Sections 6.6, 13.2

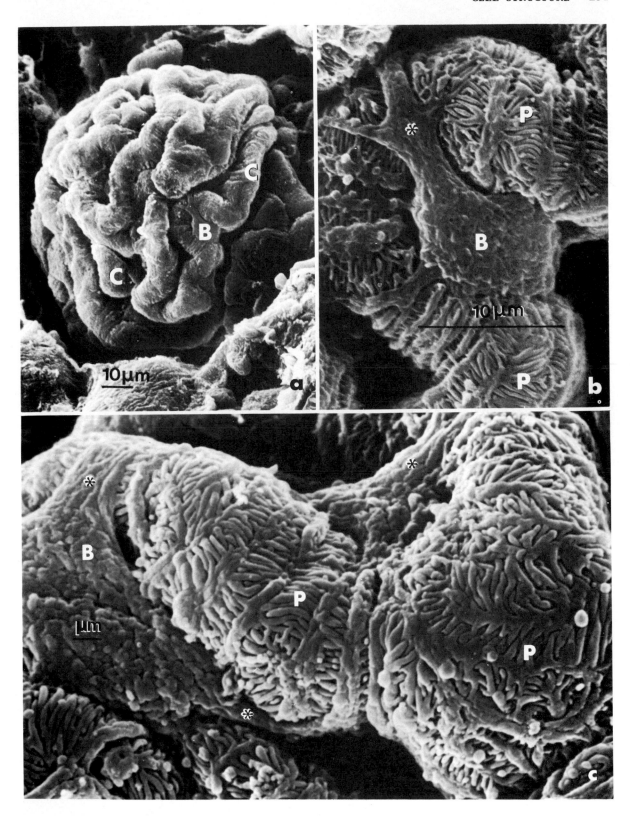

Plate 60 *Renal glomerulus, Bowman's capsule*

This survey view shows a peripheral segment of the renal glomerulus, with adjacent tubules. The capsule of the glomerulus consists of the elaborate multi-layered basal lamina (B) lined by the parietal epithelial cells of Bowman's capsule (P). This interface separates the urinary space of the glomerulus (U) from the surrounding connective tissue space of the kidney (C), in which a large thin-walled capillary is seen. Notice the red corpuscles in the lumen of this vessel, and of a glomerular capillary (R).

Parts of several glomerular capillary loops are seen, their vascular lumina (L) filled with coagulated plasma. The urinary space, by contrast, which contains the almost protein-free glomerular filtrate, appears empty. For identification purposes, two capillary endothelial nuclei are labelled (N) but the endothelial cytoplasmic layer is so thin that it can barely be seen at this low magnification. Outside the capillaries, the visceral epithelial cells, or podocytes, of Bowman's capsule are seen (V). These cells have foot processes which are applied to the outer surfaces of the capillary loops.

Adjacent to the glomerulus lies a proximal convoluted tubule (T). The obliquely sectioned lumen is filled with the tightly packed microvilli of these epithelial cells (*). Complex basal infoldings (arrows) envelope numerous large dense mitochondria (M). Large dense lysosomes are plentiful (D). Notice the close association between a segment of the base of this tubule and the adjacent interstitial capillary.

B	Thick basal lamina of Bowman's capsule
C	Interstitial connective tissue space
D	Dense lysosomes of tubular epithelial cell
L	Lumen of glomerular capillary containing precipitated protein
M	Mitochondria of renal tubular epithelial cell
N	Nucleus of glomerular endothelial cell
P	Parietal epithelial cells of Bowman's capsule
R	Red blood corpuscles
T	Proximal convoluted tubule
U	Urinary space of glomerulus
V	Visceral epithelium of Bowman's capsule, or podocytes
*	Microvilli of tubule cell, filling tubular lumen

Tissue Rat renal cortex. Glutaraldehyde and osmium fixation, uranyl and lead staining.

Magnification 5100 ×

Refer to Plates 51, 59, 61, 62, 109, 110
Sections 6.6, 13.2

Micrograph by courtesy of H. S. Johnston

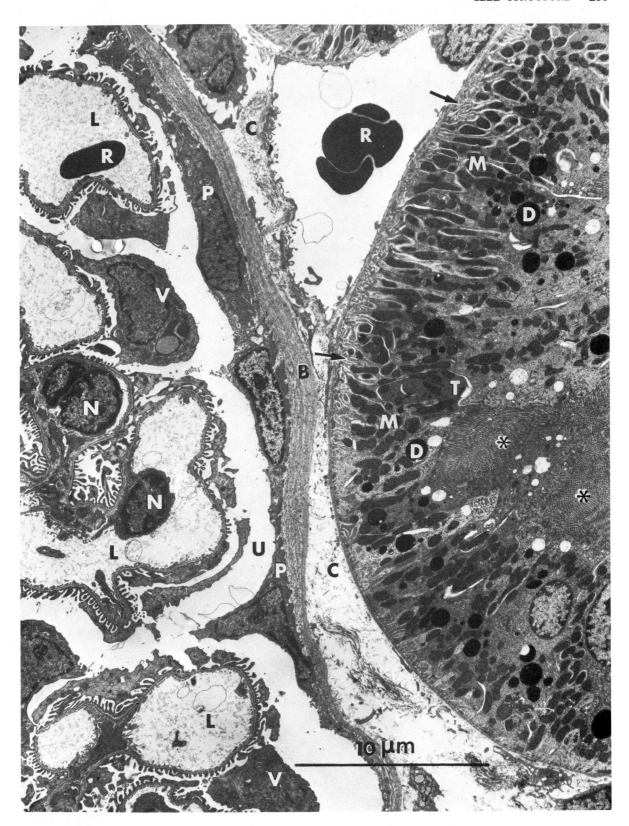

10 μm

Plate 61a *Renal glomerulus capillaries*

This picture is easily interpreted once a few basic landmarks are recognised. The location of the fenestrated endothelium (E) of the glomerular capillary allows identification of the vascular lumen (L). The glomerular basement membrane lies immediately outside the endothelium, sandwiched between endothelial and glomerular visceral epithelial cells, or podocytes, the foot processes of which are easily seen (P_1). Note how these minor foot processes arise from the broad major foot processes (P_2) which extend out from the bodies of the podocytes. It is now possible to identify the urinary space (U). The nucleus (N) of an endothelial cell is seen. Nearby, the lining of the capillary has been tangentially sectioned, revealing in face view the closely arranged endothelial cell fenestrations (F).

E	Fenestrated endothelium of the glomerular capillary
F	Fenestrations seen in face view, due to tangential sectioning of capillary loop
L	Vascular lumen of glomerular capillaries
N	Nucleus of endothelial cell
P_1	Minor foot processes
P_2	Major foot processes
U	Urinary space of Bowman's capsule

Tissue   Rabbit renal glomerulus. Glutaraldehyde and osmium fixation, uranyl and lead staining.

Magnification   10 000 ×

Refer to   Plates 43b, 51, 59, 60, 62, 109, 110
Sections 6.6, 13.2

Micrograph by courtesy of J. S. Kennedy

Plate 61b *Renal glomerulus, filtration inferface*

This higher magnification view of the filtration interface of the kidney shows the pale urinary space (U) containing glomerular filtrate, and the coagulated protein content of the vascular lumen (L), with part of an adjacent red corpuscle (R). The fenestrated endothelial cells (E) and the closely arranged minor podocyte foot processes, or pedicels (P), sandwich between them their shared thick basal lamina, known by histologists as the glomerular basement membrane or filtration membrane (★). This is subdivided into three components, the central lamina densa, and two pale zones on either side. The lamina rara interna is the pale zone just beneath the endothelial cell; the lamina rara externa is the equivalent zone just beneath the foot processes of the podocyte. The recognition of pathological changes affecting various parts of this filtration interface becomes of great importance in the accurate diagnosis of some types of renal disease.

E	Endothelial cells showing many fenestrations
L	Lumen of glomerular capillary, containing protein coagulum
P	Pedicels, or minor podocyte foot processes
R	Red blood corpuscle
U	Urinary space
★	Lamina densa of glomerular filtration membrane

Tissue   Rat kidney. Glutaraldehyde and osmium fixation, uranyl and lead staining.

Magnification   38 000 ×

Refer to   Plates 43b, 51, 59, 60, 62, 109, 110
Sections 6.6, 13.2

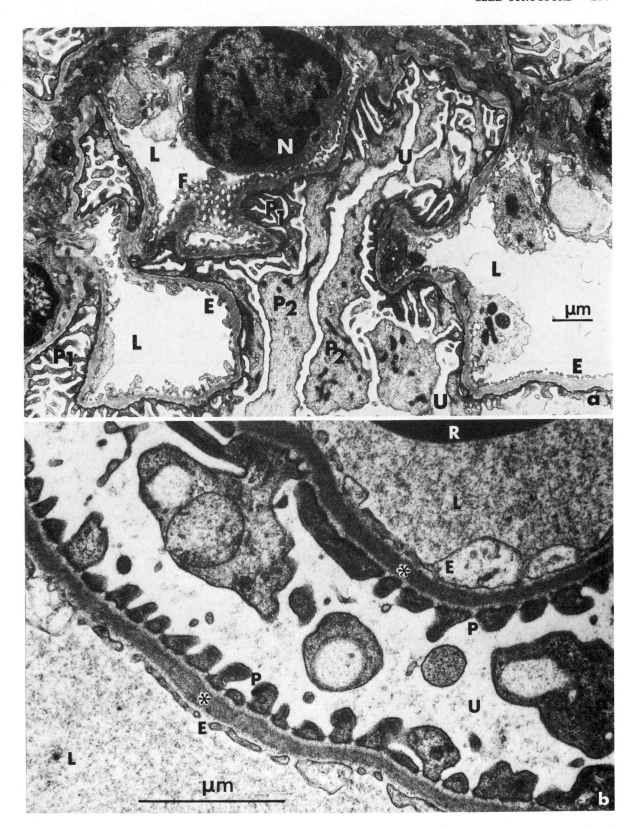

Plate 62 *Renal glomerulus, filtration inferface*

This is a high magnification micrograph showing the barrier between the lumen of the glomerular capillary (C) and the urinary space of Bowman's capsule (U).

The endothelial cell which lines the capillary (En) has pores (P) or fenestrations in its thin wall. The basal lamina is a moderately dense, almost homogeneous layer, in which the only detail seen is a suggestion of a fibrillar or filamentous network (BL). The basal lamina separates the endothelial and the epithelial components of the renal glomerulus. Each podocyte foot process (F), or pedicel, is covered by the podocyte surface membrane which has the typical trilaminar membrane structure shown elsewhere. The filtration slit membrane, indicated by an arrow, the final barrier between blood and urine, links adjacent foot processes. It does not have a trilaminar substructure.

When the glomerulus is the site of disease, the foot processes may become fused and thickened and dense deposits may produce irregular thickening of the basal lamina. These changes are seen in Plates 109 and 110. In some cases such changes may be reversible on treatment.

BL	Basal lamina
C	Lumen of glomerular capillary
En	Endothelium
Ep	Glomerular epithelial cell; podocyte
F	Minor podocyte foot processes or pedicels
P	Endothelial pore
U	Urinary space
Arrow	Indicates filtration slit membrane between adjacent podocyte foot processes

Tissue Mouse kidney. Glutaraldehyde and osmium fixation, uranium staining.

Magnification   58 000 ×

Refer to   Plates 51, 59, 60, 61, 109, 110
Section 6.6

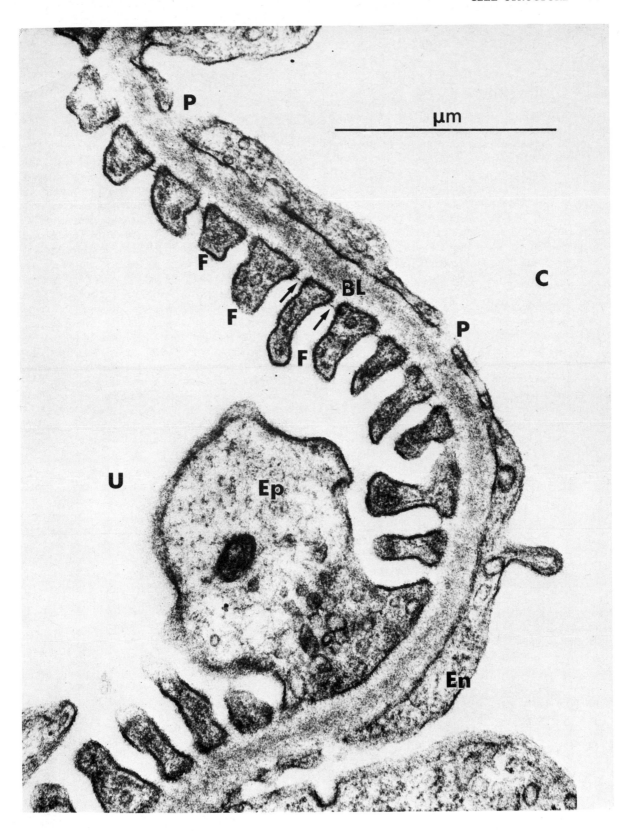

DEFENCE

Plate 63a *Phagocytosis, neutrophil polymorphs*

This cell is taken from an in vitro experiment designed to test the ability of the polymorph to take up bacteria by phagocytosis. The polymorphs were separated from the red blood corpuscles and were incubated in culture medium in the laboratory. Bacteria were then added to the culture fluid. One hour later, the polymorphs were examined to observe their phagocytic efficiency.

This micrograph shows a single polymorph. Only one nuclear lobe (N) is seen in this plane of section. The cytoplasmic granules are less numerous than in normal control cells since some have already been used up. Several bacterial cells can be seen. They are staphylococci, small dense round organisms. One of the bacteria is still outside the cell (B_1) but is becoming enveloped by cytoplasmic flaps prior to engulfment. Two others are seen within phagocytic vacuoles within the cytoplasm (B_2). Several other large phagocytic vacuoles (V) contain no clearly identifiable bacterial cells, but do contain variable amounts of dense debris, some of which might represent fragments of destroyed bacteria. Alternatively, the plane of section through these vacuoles may simply have missed a contained organism. This micrograph shows that the polymorph is capable of engulfing the test organisms and of discharging its lysosomes and specific granules into the phagocytic vacuoles which are formed in the process.

B_1 Bacterial cell outside polymorph, but on the point of being engulfed
B_2 Bacterial cells within phagocytic vacuoles in polymorph
N Nucleus of polymorph
V Phagocytic vacuoles containing debris but no recognisable organism

Tissue Human neutrophil polymorphs incubated in vitro with staphylococci. Glutaraldehyde and osmium fixation, uranyl and lead staining.

Magnification 20 000 ×

Refer to Plates 19, 24, 52, 64, 65
Sections 4.6, 7.1

Plate 63b *Phagocytosis, neutrophil polymorphs*

This higher magnification micrograph shows part of another cell from the same preparation as the one shown above. There are parts of three nuclear lobes visible (N). Three phagocytic vacuoles all contain staphylococci (B), along with dense granular material. At two points (*) it is particularly obvious that the little aggregates of material within the phagocytic vacuole represent the material from polymorph granules, released into the vacuole, but not yet fully dispersed in the contents of the vacuole. Other granules seen close to the vacuoles are presumably also destined to add themselves to the digestive effort. Notice that each phagocytic vacuole is surrounded entirely by a limiting membrane, which partitions the potentially lethal vacuole contents from the cytoplasm of the still viable polymorph.

Although the bacteria shown here were present one hour after the start of this experiment, no identifiable bacteria remained at 24 hours, proving that the polymorphs could not only take up the bacteria, but could also effectively kill them. In a parallel experiment with cells from a patient with defective polymorphs, similar bacterial uptake was seen at one hour, but viable bacteria remained at 24 hours, showing that ingestion was still effective, but that the killing mechanism was defective.

B Bacteria within phagocytic vacuoles
N Nuclear lobes of polymorph
* Indicate incompletely dispersed polymorph granules released recently into the vacuole

Tissue Human neutrophil polymorphs incubated in vitro with staphylococci. Glutaraldehyde and osmium fixation, uranyl and lead staining.

Magnification 25 000 ×

Refer to Plates 19, 24, 52, 64, 65
Sections 4.6, 7.1

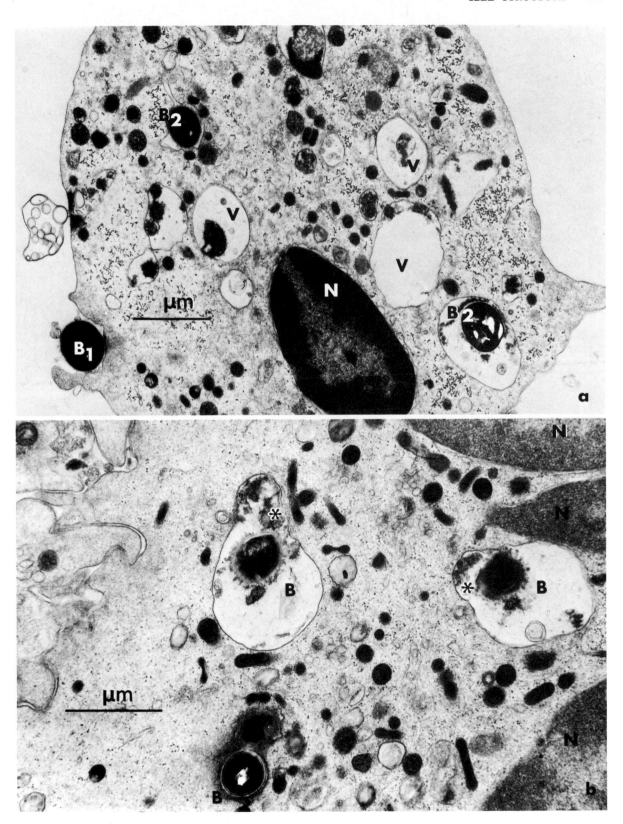

Plate 64a *Eosinophil polymorph*

The eosinophil cytoplasm contains numerous characteristic dense granules, each of which is limited by a membrane. These are specialised lysosomes. The eosinophil granule (E) commonly contains a dense crystalline inclusion. Some of the granules shown here have multiple inclusions of this type (X). In many instances the elliptical configuration of the eosinophil granule appears to be caused by the distorting effect of an elongated crystal. The nucleus of the eosinophil is rather dense in appearance with little euchromatin and a pronounced peripheral heterochromatin aggregation (N). Apart from the distinctive cytoplasmic granules, the eosinophil has few other notable features.

E   Elliptical eosinophil granule with crystal
N   Eosinophil nucleus
X   Multiple crystalline inclusions in eosinophil granules

Tissue   Rat colon. Glutaraldehyde and osmium fixation, uranium and lead staining.

Magnification   32 000 ×

Refer to   Plates 19, 24, 52, 63, 65, 69c
Sections 7.1, 9.1

Plate 64b *Mast cell*

The functions of the mast cell are not fully understood, but it is a relatively common cell in connective tissue. Its granules show a wide range of structural features. In this case they are of variable pattern, with homogeneous areas interspersed by crystalline (C) and lamellar (L) components. Cisternae of the granular endoplasmic reticulum are scattered between the mast cell granules (ER). Mast cell granules generally contain heparin and histamine.

The mast cell has surface receptors for the IgE class of immunoglobulins. When sensitised individuals are exposed again to the antigens of such immunoglobulins, the mast cell to which the IgE molecules are bound undergoes explosive degranulation. The release of the powerful pharmacological agents contained in the mast cell granules can cause serious illness, such as asthma and anaphylactic shock.

C   Crystalline areas
ER   Granular endoplasmic reticulum
L   Lamellar areas of mast cell granules

Tissue   Human mast cell. Glutaraldehyde and osmium fixation, uranium and lead staining.

Magnification   68 000 ×

Refer to   Plates 19, 24, 52, 63, 65, 69c
Sections 7.1, 9.1

Plate 64c *Mast cell*

Human mast cell granules may contain characteristic tubular scroll-like inclusions with a fine periodicity. These are shown in longitudinal section (L) and in transverse section (T). The cytoplasm between the granules contains many fine filaments.

L   Longitudinally sectioned scroll-like inclusions
T   Transversely sectioned scroll-like inclusions

Tissue   Human small intestine. Glutaraldehyde and osmium fixation, uranium and lead staining.

Magnification   90 000 ×

Refer to   Plates 19, 24, 52, 63, 65, 69c
Sections 7.1, 9.1

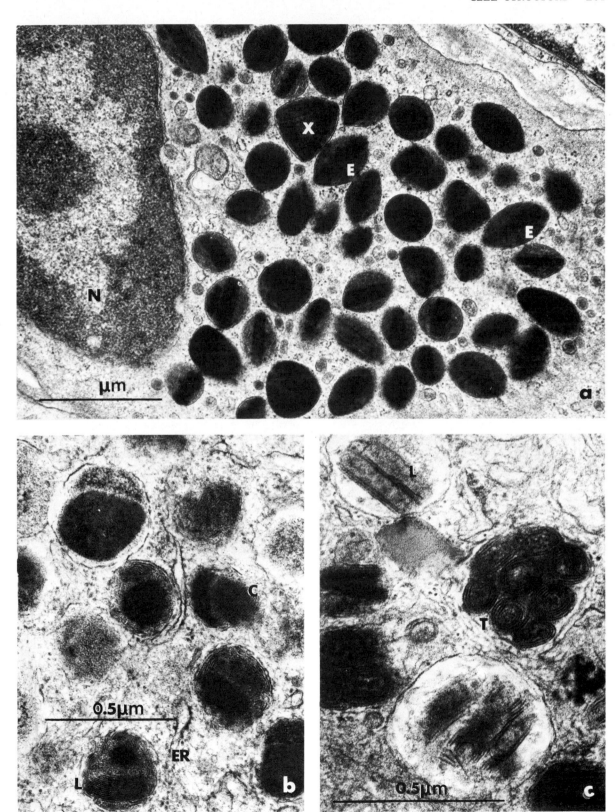

Plate 65a *Monocyte arriving at work*

This micrograph shows a monocyte close to a group of mature macrophages (M) in an experimental granuloma, a type of chronic inflammation. The mature cells have features indicating phagocytic activity, such as the presence of surface processes, appearing as flaps and ruffles (F). The monocyte shows its immaturity by its relative lack of these features. Its peripheral location in the clump of cells suggests that it has just arrived by emigrating from the blood stream and crossing the connective tissue space to join its colleagues at work. It has been shown that most of the macrophages in lesions of this type originate in this way from the monocytes of the blood stream.

The monocyte nucleus (N) shows well defined peripheral heterochromatin and central euchromatin masses. Two tangentially sectioned nuclear pores are seen (arrows). The cytoplasm of the monocyte contains moderate numbers of small dense primary lysosomes (*), but no obvious secondary lysosomes or phagocytic vacuoles. The membrane systems of the cytoplasm are becoming developed in preparation for lysosome synthesis, a necessary preliminary to active phagocytosis.

F	Surface flaps and ruffles of mature macrophages
M	Mature macrophages adjacent to newly arrived monocyte
N	Monocyte nucleus
*	Indicate primary lysosomes in monocyte cytoplasm
Arrows	Indicate two tangentially cut nuclear pores

Tissue Experimental granuloma in liver. Glutaraldehyde and osmium fixation. Uranyl and lead staining.

Magnification 11 000 ×

Refer to Plates 10a, 12, 19, 24, 63, 66, 67, 68
Section 7.1

Specimen by courtesy of W. D. Thompson

Plate 65b *Macrophage hard at work*

This rather complicated micrograph shows an area of human lung tissue seven days after the patient had accidentally swallowed some paraquat weedkiller. This chemical destroys the alveolar lining cells and subsequently produces fatal lung fibrosis. A large active macrophage (M) lies in the alveolar space (S), in contact with remnants of the alveolar lining epithelial cells (E). The basal lamina of the alveolus is indicated (arrows). Notice that the epithelial cells are irregularly detached from their basal lamina. In the connective tissue space, an active fibroblast (F) contains well-developed granular endoplasmic reticulum (G) and a mast cell (X) is present.

The macrophage is actively responding to the abnormal conditions in the alveolus. It was, no doubt, on the point of enveloping the epithelial remnants and removing them from the alveolar wall, since fragments of dead cells are offensive to a macrophage. The cell contains numerous small dense primary lysosomes and a few larger residual bodies. Notice the elaborate ruffles on both of the surfaces of the macrophage (R).

Unfortunately, although the macrophage can tidy up the mess, it cannot renew the devastated epithelial lining of the alveolus. The fibroblasts move into the fluid exudate which fills the alveoli in an attempt to heal the injured tissues by making collagen. This, however, only worsens the already disastrous condition of the lungs, and leads rapidly to death through respiratory failure.

E	Epithelial cells of alveolus, becoming detached from their basal lamina
F	Active fibroblast
G	Granular endoplasmic reticulum in fibroblast
M	Macrophage in alveolar space
R	Ruffles on macrophage surface
S	Alveolar air space
X	Mast cell
Arrows	Indicate alveolar epithelial basal lamina

Tissue Human lung biopsy, seven days after paraquat poisoning. Glutaraldehyde and osmium fixation, uranyl and lead staining.

Magnification 12 000 ×

Refer to Plates 10a, 12, 19, 24, 63, 66, 67, 68
Section 7.1

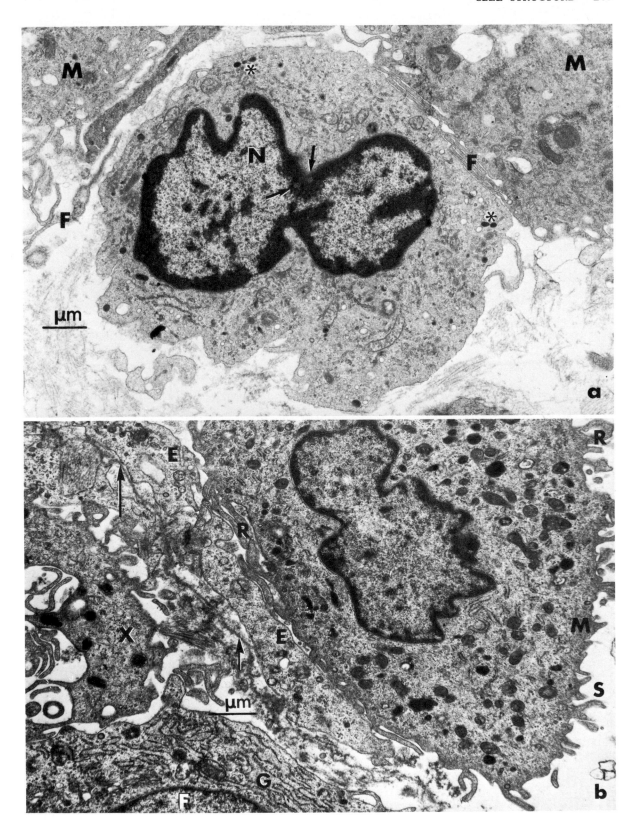

Plate 66a *Active macrophage in vitro*

This scanning micrograph shows a cell surrounded by bacteria, which have stimulated it to great phagocytic activity. The bacterial cells are seen on the substrate, surrounding the macrophage. The macrophage surface is thrown into elaborate folds and ruffles extending out at different angles. One single elongated process extends from the macrophage to contact a clump of organisms (arrowed). Surface irregularity of this degree indicates a particularly active physiological state, implying avid phagocytosis.

Arrow   Indicates a group of organisms in contact with an elongated cytoplasmic process

Tissue   Macrophage in vitro, bacterial culture.

Magnification   7000 ×

Refer to   Plates 45, 46, 65, 67, 68
            Section 7.1

Micrograph by courtesy of D. A. Powell and K. A. Muse

Plate 66b *Peyer's patch*

This scanning electron micrograph shows the intestinal mucosa at low magnification. The typical nodules of the Peyer's patches have a distinctive surface pattern, consisting of circular domes surrounded by intestinal villi. The dome is covered by epithelial cells, but has a pattern which differs from that of villous epithelium.

Tissue   Rat intestinal mucosa

Magnification   70 ×

Refer to   Plates 14, 27a, 38, 47, 48
            Section 6.1

Micrograph by courtesy of L.-G. Friberg

Plate 66c *Peyer's patch*

This scanning micrograph shows the surface of a single nodule of a Peyer's patch. The surface is rather more wrinkled towards the centre of the nodule, which may be an extrusion zone, where worn-out cells are shed at the end of their active lives. The entire surface of the dome is studded with small indentations or pockmarks. Although the magnification is too low to distinguish their individual surface features, the pockmarks correspond to the surfaces of intestinal M cells. These numerous specialised absorptive cells appear to play a role in antigen uptake from the lumen, thus acting as the first component of an intestinal immunological sensory system which may be of great importance in defence mechanisms.

Tissue   Rat intestinal mucosa

Magnification   290 ×

Refer to   Plates 14, 27a, 38, 47, 48
            Section 6.1

Micrograph by courtesy of L.-G. Friberg

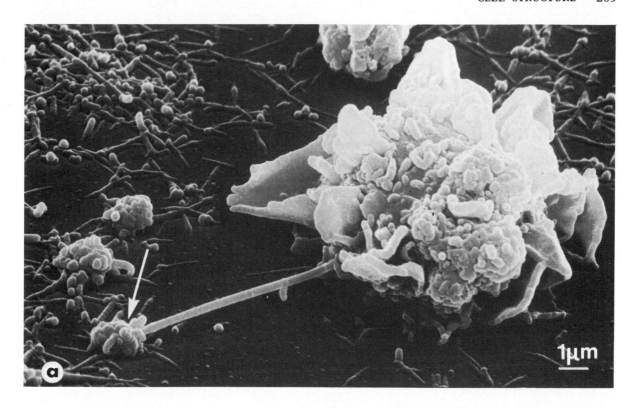

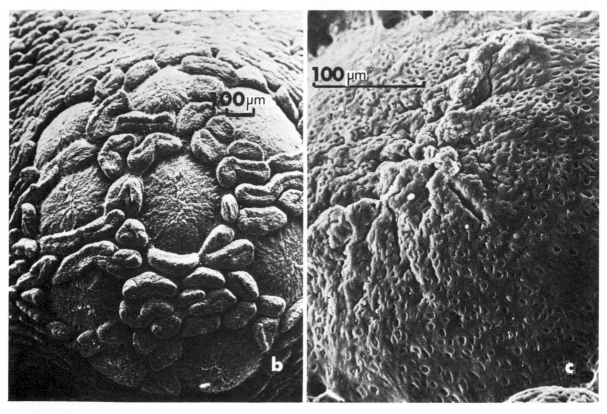

Plate 67a *Macrophages, injured liver*

The peculiar structure in the middle of this micrograph is a dying liver cell (L). Most of its cytoplasm is vacuolated, with residual clumps of cytoplasmic organelles and occasional pale lipid droplets (D). The injury was produced by the experimental administration of carbon tetrachloride. Around this dying cell are macrophages (M). These are spreading along the surfaces of the hepatocyte, which they will eventually surround completely and engulf, forming a small inflammatory lesion recognisable by a pathologist as a granuloma. Notice the elaborate surface contours of the large macrophage on the left, which already contains several large secondary lysosomes (*).

D	Lipid droplets in injured hepatocyte
L	Vacuolated cytoplasm of dying liver cell
M	Macrophages around the dying cell
*	Indicate secondary lysosomes

Tissue   Experimental granuloma in liver, glutaraldehyde and osmium fixation, uranyl and lead staining.

Magnification   11 000 ×

Refer to   Plates 45, 46, 65, 68
         Section 7.1

Specimen by courtesy of W. D. Thompson

Plate 67b *Macrophage, parasitic infection*

This tissue is from a hamster experimentally infected with a protozoal parasite *Leishmania donovani*, similar to that which causes a serious human disease in tropical countries. The parasite is taken up by macrophages, but is not killed. It lives and multiplies in the host cells, causing various serious ill-effects including the production of an abnormal glycoprotein which is deposited as filamentous material known as amyloid (A).

Most of this plate is occupied by a single macrophage, with an indented nucleus (N). The numerous small rounded structures within this macrophage are the individual parasites (P), each an independent cell. Each parasite has a nucleus (X) and various cytoplasmic organelles including flagella (F). The structure of the parasite is maintained by a sub-surface basketwork of microtubules, cut at various angles (arrows). The cell membrane of the parasite is intact, since the macrophage is unable to injure it.

A	Amyloid filaments deposited in the extracellular connective tissues
F	Flagella in cross-section, showing axial filament complex
N	Nucleus of host macrophage
P	Protozoal parasites within host macrophage
X	Nucleus of parasite
Arrows	Indicate subsurface microtubules of parasite

Tissue   Hamster experimentally infected with *Leishmania donovani*. Glutaraldehyde and osmium fixation, uranyl and lead staining.

Magnification   15 000 ×

Refer to   Plates 24, 63, 65, 106, 112
         Section 7.1

Micrograph by courtesy of J. D. Anderson

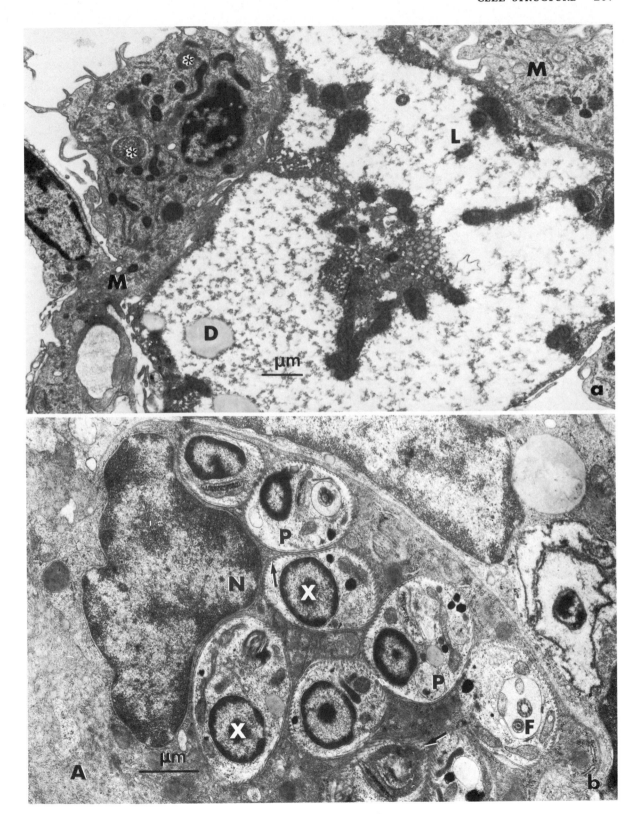

Plate 68a *Lymphocyte*

This active-looking lymphocyte (L) is in the process of pushing its way either into or out of the intestinal epithelium. Unfortunately electron microscopy is not a good way of answering questions about movement, since all movement has been 'frozen' by fixation. The surrounding cells which are being indented by the lymphocyte's cytoplasmic processes are intestinal absorptive cells (A). The connective tissue of the lamina propria lies to the lower right (C). The basal lamina of the epithelium (arrows) would normally run right across in a straight line, but it becomes discontinuous on either side of the lymphocyte, indicating that the passage of this wandering cell has disrupted its continuity. The cytoplasm of the lymphocyte is almost totally unspecialised, containing only a few free ribosomes. The nucleus is the only prominent feature.

All epithelia, but especially those of the intestine, have lymphoid cells constantly wandering through them. It was once thought that the lymphocytes were heading for the lumen, but it is now known that they come and go, in and out of the lamina propria, as part of their general job of searching for foreign antigens. When a lymphocyte meets the antigen for which it is primed, it is able to proliferate and produce other cells, which together can combat the antigen. Each lymphocyte, therefore, must simply remember one antigen, and be able to move around from place to place. This rather basic job description does not need a complex cytoplasmic apparatus for its fulfilment, which explains the unspecialised nature of the cytoplasm.

A   Intestinal absorptive cells
C   Connective tissue of lamina propria
L   Lymphocyte migrating through epithelium
Arrows   Indicate the basal lamina, broken on either
          side of the lymphocyte

Tissue Human small intestine. Glutaraldehyde and osmium fixation, uranyl and lead staining.

Magnification   15 000 ×

Refer to   Plates 3, 10a, 15, 19, 52, 63, 65, 67
             Section 7.2

Plate 68b *Plasma cell*

This cell represents the opposite functional pole from the lymphocyte shown above. In this case, a former lymphocyte recognised the antigen for which it was made and proliferated to form a clone, or family, of directly descended cells, most of which became plasma cells like this. While the job of the lymphocyte is to remember only one antigen, the job of the plasma cell is to manufacture only one antibody, a particular specific immunoglobulin molecule. For this purpose, the plasma cell is equipped with an elaborate protein synthetic apparatus.

The plasma cell nucleus (N) has a characteristic chromatin pattern, consisting of large dense peripheral blocks of heterochromatin, broken by 'spokes' of euchromatin, thus producing a 'clock-face' or 'cartwheel' pattern. The bulky cytoplasm is occupied almost entirely by dilated cisternae of the granular endoplasmic reticulum (R), except a compact central zone adjacent to the nucleus, which contains the elaborate Golgi apparatus (G). Notice the finely granular content of the cisternae of endoplasmic reticulum. This consists of immunoglobulin newly synthesised by the cell.

The surrounding connective tissue space (T) contains numerous collagen fibrils.

G   Golgi apparatus of plasma cell
N   Nucleus of plasma cell, showing typical
          chromatin pattern
R   Granular endoplasmic reticulum
T   Connective tissue containing collagen fibrils

Tissue Human small intestine. Glutaraldehyde and osmium fixation, uranyl and lead staining.

Magnification   10 000 ×

Refer to   Plates 3, 15, 19, 52, 63, 64, 65
             Section 7.2

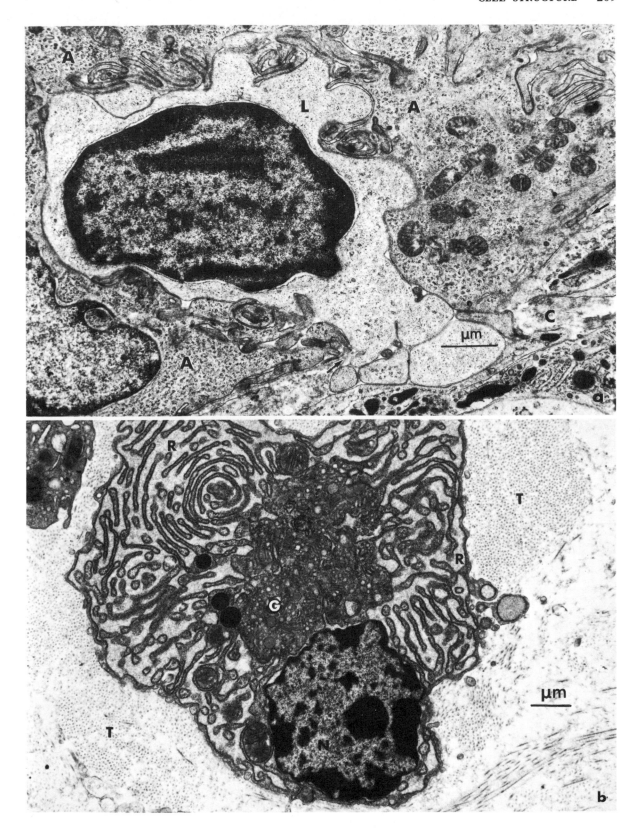

STORAGE AND PROTECTION

Plate 69a *Vessel containing red corpuscles*

This is a scanning electron micrograph of a cross-section of a blood vessel running through the connective tissue (C) of dermis in human skin. The adventitial connective tissue layer of the vessel wall can be seen (A), but the lumen is so crowded with red blood cells (R) that surface details of the endothelial layer cannot be made out. Numerous tangled collagen and elastic fibres make up the structure of the surrounding connective tissue.

A   Adventitial layer of blood vessel
C   Collagen fibres in connective tissue of dermis
R   Red blood cells

Tissue Human skin, fixed in formol saline, embedded in wax. A section was cut and dewaxed prior to air drying and carbon/platinum coating.

Magnification 900 ×

Refer to Plates 9, 79, 80
            Section 8.1, 9.1

Plate 69b *Blood film, acanthocytosis*

Most of the red blood corpuscles in this field are of abnormal shape, although a few nearly normal disc-shaped cells (D) are seen. The others are markedly 'spiky' in appearance, an abnormality described as acanthocytosis. This is one feature of a systemic disease characterised by congenital absence of the beta lipoproteins of the plasma. Scanning electron microscopy is a useful technique to demonstrate these and other abnormalities of red cell shape.

D   Nearly normal disc-shaped cell, surrounded by
             acanthocytes

Tissue Human blood film from a case of acanthocytosis (abetalipoproteinaemia).

Magnification 1500 ×

Refer to Plates 45, 55b, 56b, 58, 60
            Section 8.1

Blood sample by courtesy of C. Wardrope

Plate 69c *Red blood corpuscles*

The red corpuscle is a deceptively simple structure, but on section the varying planes through the cells present outlines of varying contour. The popular biconcave disc image of the red corpuscle in its resting state takes no account of its dynamic behaviour. When pushed by the force of the circulation, red corpuscles twist and undulate through the capillaries, bending and distorting with enormous resilience. Thus sections taken through fixed cells rarely reflect the symmetrical disc configuration of the idealised red corpuscle.

This micrograph shows blood cells within a small venule, lined by endothelial cells. The sections of these red corpuscles (R) reveal no internal cytoplasmic detail. When seen at much higher magnification, the membrane of the corpuscle has the conventional trilaminar pattern of membranes elsewhere, but there are no formed organelles. The marked cytoplasmic density reflects the high concentration of haemoglobin, which is not only osmiophilic but also contains substantial amounts of iron. Other components of the corpuscle include actin, although not in the form of identifiable microfilaments, and a membrane-associated protein called spectrin, which plays a part in the maintenance of red cell shape. The red corpuscle is a masterpiece of simplification of structure in the interests of the efficiency of its functions of storage of haemoglobin and carriage of blood gases.

Other features of the micrograph include a polymorphonuclear leucocyte (P) and a further red corpuscle (R₁) lying outside this venule in the connective tissue space. The leucocyte is a normal inhabitant of the intestinal lamina propria, but the red cell probably reached that location through minor trauma associated with the procedure for taking the biopsy. The elongated cell (M) is an extremely attenuated smooth muscle cell. The base of the epithelium contains a basal granulated intestinal endocrine cell (E). Finally, within the vascular lumen, notice two small non-nucleated structures (T) near a blood lymphocyte (L). These small cells are platelets, or thrombocytes, an essential part of the protective mechanisms of the vascular system.

E   Basal granulated intestinal endocrine cell
L   Blood lymphocyte within vascular lumen
M   Attenuated smooth muscle cell
P   Polymorphonuclear leucocyte in intestinal lamina
             propria
R   Red blood corpuscles in vessel
R₁   Red blood corpuscle, extravasated into connective
             tissue
T   Thrombocytes or platelets

Tissue Human small intestine. Glutaraldehyde and osmium fixation, uranyl and lead staining.

Magnification 5000 ×

Refer to Plates 45, 55b, 56b, 58, 60, 78a, 80a, 89
            Sections 8.1, 9.1

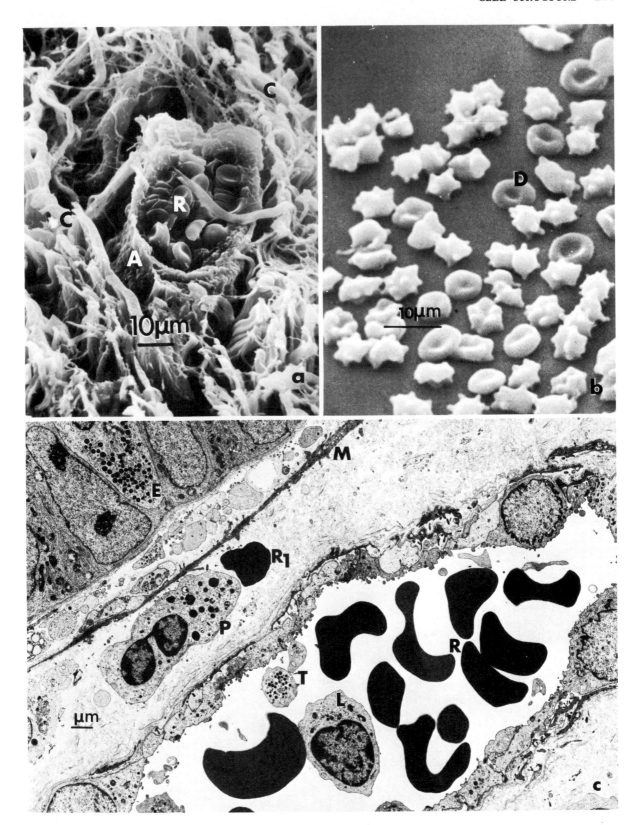

Plate 70a *Adipose tissue*

This scanning electron micrograph shows a view of a section of fatty tissue, lying deep to the dermis (D) in which cross-sectioned collagen fibres are seen. The outlines of the fat cells can be identified by the residual rim of cytoplasm (R) left round a space (S), where the globule of fat has been dissolved out during preparation.

The scored background is the metal stub, or specimen holder, normally hidden by the specimen but here visible through the spaces left by the dissolved fat.

D	Dermal connective tissue showing collagen fibres
R	Thin rim of cell membrane and cytoplasm left at periphery of fat cell
S	Space where fat globule has been dissolved away, showing the scored surface of the metal specimen carrier

Tissue Human skin, fixed in formol saline, embedded in wax, sections cut and then dewaxed, air dried, carbon/platinum coated.

Magnification 3500 ×

Refer to Plate 71
Sections 2.2, 8.2, 9.1

Plate 70b *Yellow fat cell*

This micrograph is not easy to interpret at first sight. It shows a small part of the cytoplasmic rim of a typical fat cell, the primary function of which is the storage of an energy reserve for the body. On one side there is an external lamina (L), which provides an outer envelope to fat cells. Scattered collagen fibrils (C) reinforce this lamina. Immediately internal to the lamina lies the cell membrane of the fat cell (arrows). The thin rim of cytoplasm contains only one recognisable structure, a mitochondrion (M). Beyond this cytoplasmic rim lies the stored fat (F), forming a droplet which is not demarcated from the cytoplasm by any limiting membrane. The fat has been dissolved during processing, leaving this empty outline of the space it formerly occupied.

C	Scattered collagen fibrils
F	Fat droplet, without limiting membrane
L	External lamina surrounding fat cell
M	Mitochondrion within thin rim of cytoplasm of fat cell

Tissue Rat parotid gland. Glutaraldehyde and osmium fixation, uranyl and lead staining.

Magnification 42 000 ×

Refer to Plate 71
Sections 2.2, 8.2, 9.1

Plate 70c *Brown fat cells*

This specialised form of adipose tissue differs in its function, its metabolism and its structure from the more conventional yellow depot fat illustrated above. In this low magnification view, there are several brown fat cells (B) around a small blood vessel containing a red corpuscle (R). This high degree of vascularity is a characteristic of brown fat.

The cells differ from yellow fat in that the stored lipid is in the form of multiple droplets (F) in each cell, whereas in yellow fat the lipid forms a single droplet. A second feature of note is the presence of large mitochondria (M) which occupy much of the cytoplasm remaining between the fat droplets. At high magnification these mitochondria are found to have closely packed, highly organised cristae, indicating some active role in cellular metabolism. A small unmyelinated nerve (N) lies between cells, reflecting their direct autonomic control.

In cells in general, a little less than half of the energy of oxidation is converted by the mitochondria to molecules of ATP, the remainder producing the heat of metabolism. Brown fat mitochondria are designed to be less efficient than other mitochondria in this respect, so that most of the energy of oxidation is dissipated as heat and very little goes into ATP production. This uncoupling of electron transport from ATP generation is of importance in providing heat in newborn animals and during the process of awakening from hibernation. There is now a suggestion that the purposeful burning of brown fat may play a part in the dissipation of surplus dietary calories which would otherwise be laid down as yellow fat.

B	Brown fat cells
F	Droplets of lipid in brown fat cells
M	Mitochondria of brown fat cell
N	Unmyelinated nerve terminal between fat cells
R	Red corpuscle in capillary vessel

Tissue Interscapular brown fat, mouse. Glutaraldehyde and osmium fixation, uranyl and lead staining.

Magnification 11 000 ×

Refer to Plates 40, 71, 89, 102
Section 8.2

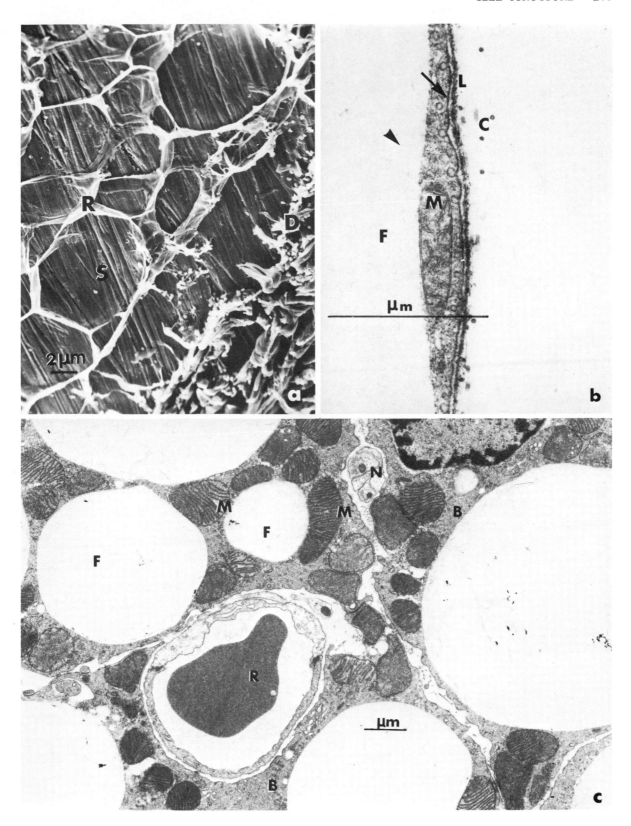

Plate 71 *Fat storing cell, liver*

Lipid is stored not only in adipose tissues; various other cell types can store fat in small amounts, whether for fuel, as in myocardium, or for the synthesis of steroid hormones, as in adrenal cortical cells. Damaged cells may also accumulate fat as is the case with hepatocytes after certain poisons or in response to anoxia.

In this micrograph showing human liver, however, the stored fat is not within a hepatocyte but is found in an interstitial fibroblast-like cell which lies beside the hepatocytes, just outside the cells lining the sinusoid. This is known as a fat-storing cell, or Ito cell, after its discoverer. The lipid stored here is particularly rich in vitamin A. This storage function, however, is perhaps not the primary role of the Ito cell, since fibroblasts elsewhere can also store vitamin A if it is present in excess. It is not clear how this storage function is related to the cell's role as a fibroblast. The fine collagen fibril framework of the liver lobule is probably produced and maintained by these cells, and when liver injury occurs, the Ito cells are reactivated to repair the damage by forming new fibrous tissue.

Notice the basic fibroblast-like features of the fat-storing cell, including a moderate-sized Golgi apparatus (G) and endoplasmic reticulum (R) with plentiful ribosomes. The lipid droplets appear empty, since their content has been dissolved away. Small bundles of collagen fibrils lie nearby (C), representing the reticulin framework of the liver lobule. The adjacent hepatocytes contain the usual organelles, including abundant particulate glycogen in the alpha configuration.

C   Collagen fibrils forming framework of liver lobule

G   Golgi apparatus of fat storing cell

R   Granular endoplasmic reticulum

Tissue Human liver. Glutaraldehyde and osmium fixation, uranyl and lead staining.

Magnification 18 000 ×

Refer to Plates 9, 45, 46, 70, 80
         Sections 5.4, 8.2, 9.1

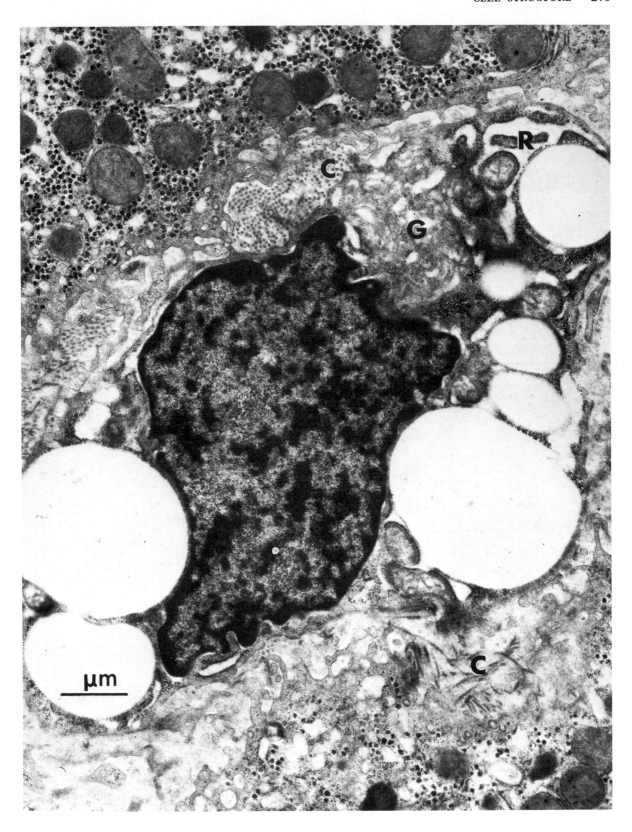

Plate 72a, 72b *Ectocervix, sdquamous cells*

These two scanning electron micrographs show features of the surface layer of the stratified, non-keratinising epithelium of the human cervix. The lower power micrograph in Plate 72a shows the pavement-like contours of the cells, with prominent linear cell junctions (J). At some points (*) a cell appears partly detached from the surface, prior to shedding. The higher power view of plate 72b shows details of the network of surface microridges (R) which cover such cells. In transmission electron microscopy, such profiles appear similar to short microvilli, but SEM allows recognition of the correct three dimensional configuration of these surface features.

J   Linear cell junctions marking cell boundaries
R   Microridges of squamous cell surface
*   Indicates partial detachment of cell from epithelial surface

Tissue Human cervix. Formalin fixed, prepared for SEM after dewaxing from paraffin wax block. Critical point dried, gold coated.

Magnification Plate 72a 1600 ×
                 Plate 72b 3000 ×

Refer to Plates 1, 5, 33b, 47, 48, 73, 76b, 79
            Sections 8.3, 9.1

Plate 72c, 72d *Epidermis, squamous cells*

These two scanning electron micrographs show the surface of a section of human skin, with its surface layer of keratinising squamous epithelium. In Plate 72c the image is similar to a light micrograph of skin, except that surface details of the layers can be seen, instead of thinly sectioned structures visualised by differential staining. Epidermis (E) and dermis (D) with its collagen bundles can be identified. Some of the epidermal layers can be made out, including the stratum basale (B), stratum spinosum (S), and stratum corneum (C), seen in more detail in Plate 72d. This shows the thin laminated plates of keratin which is all that remains of the keratinocytes, when they reach the stratum corneum.

B   Stratum basale or basal layer of the epidermis
C   Stratum corneum or keratinised layer of the epidermis, consisting of thin flattened plates
D   Dermis with cross-sectioned collagen bundles
E   Epidermis
S   Stratum spinosum or prickle cell layer

Tissue Human skin, fixed in formol saline, embedded in wax, sections cut and then dewaxed. Gold coated.

Magnification Plate 72c 190 ×
                 Plate 72d 2700 ×

Refer to Plates 1, 5, 33b, 47, 48, 73, 76b, 79
            Sections 8.3, 9.1

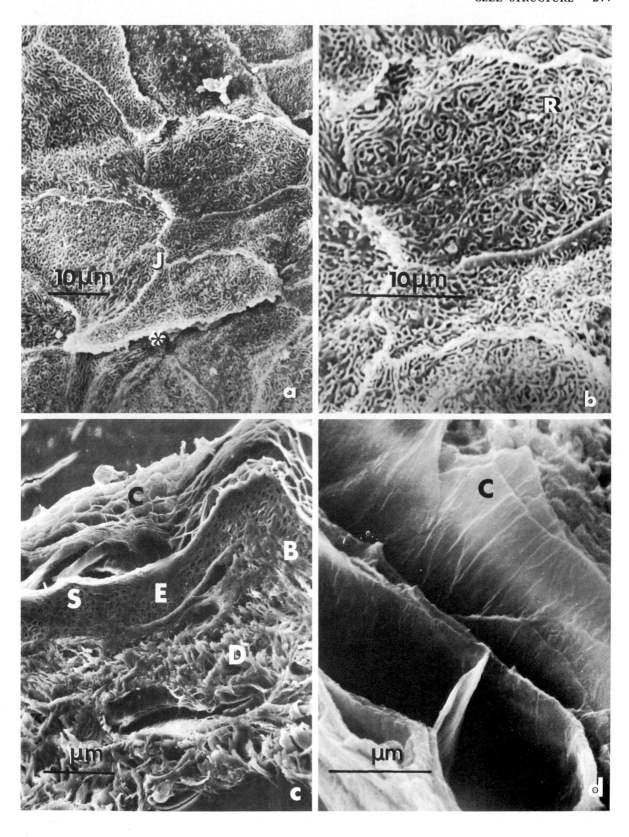

Plate 73 *Squamous epithelium, keratinocytes*

Keratinising squamous epithelium is the natural protective surface for areas exposed to wear and tear. The epithelial cells, as they mature and migrate towards the surface, become progressively filled with keratin filaments. The surface layer is simply a mat of dense keratin within the dead husks of the topmost cells of the epithelium. This layer is resistant to water and to abrasion. Throughout the thickness of the epithelium, the individual cells are held together by numerous desmosomes, to which the intracellular meshwork of filaments is connected. In this way, mechanical stresses at one point are spread throughout the epithelium, with resulting mechanical strength and resilience.

This micrograph shows parts of several cells in a squamous epithelium, all of which have fairly large nuclei (N). A nucleolus is seen in one of these (*). The intercellular spaces (S) are crossed by interlocking cytoplasmic processes, some of which are joined by desmosomes (D), although details of these are not well seen at this low magnification. Their position, however, even when cut obliquely, is marked by the insertion of the dense tonofilament bundles (B) which form a cytoskeletal network woven through the cytoplasm. It is the strength and resilience of this desmosome-filament network that confers on squamous epithelium its essential mechanical properties.

The primary metabolic function of these cells, therefore, is the manufacture of the keratin tonofilaments. Since this is protein for internal use, an elaborate granular endoplasmic reticulum is not called for. Instead, notice that the cytoplasm contains numerous polyribosomes, diffusely distributed throughout the cells. Mitochondria (M) are not particularly abundant.

B	Tonofilament bundles woven around the nucleus and inserted into desmosomes
D	Desmosomes, with attached tonofilaments
M	Mitochondria
N	Nucleus
S	Intercellular spaces crossed by cytoplasmic processes
*	Shows nucleolus of keratinocyte

Tissue Rat tongue. Glutaraldehyde and osmium fixation, uranyl and lead staining.

Magnification 14 000 ×

Refer to Plates 1, 5, 6, 27b, 33b, 72, 74, 75, 76b
Sections 4.8, 8.3

Micrograph by courtesy of H. S. Johnston

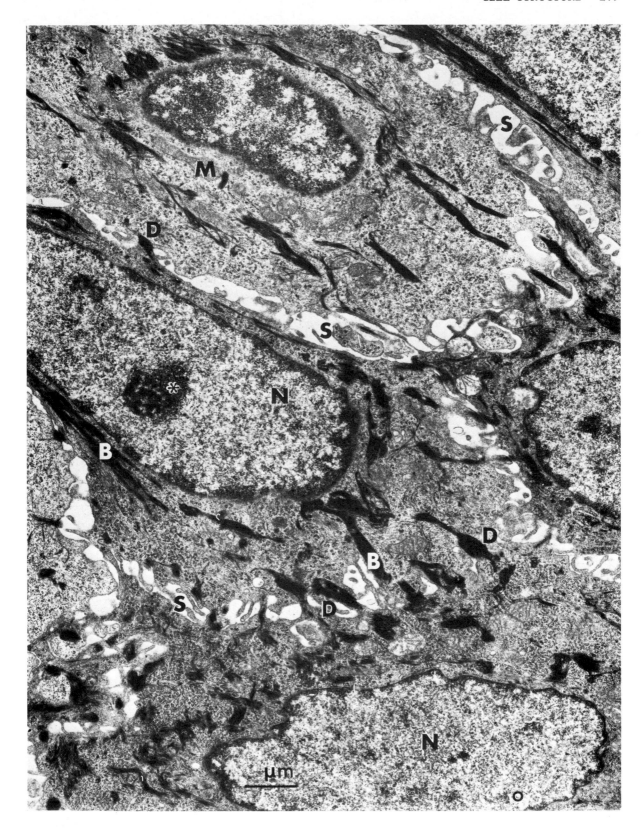

Plate 74a *Melanocyte in epidermis*

Protective mechanisms take many forms. The skin, exposed to potentially harmful ultraviolet radiation from the sun, has its own protective system based on the pigment melanin. This is manufactured by melanocytes in the basal layer of the epidermis and is passed to the keratinocytes. Its production is stimulated by sun exposure, as most holidaymakers know. The pigment forms a barrier to ultraviolet rays, preventing more serious damage to the tissues of the skin.

This micrograph shows the basal part of the epidermis to the left, with its underlying basal lamina (arrows) and connective tissue (C). Most of the cells are keratinocytes (K), with dense tonofilament bundles (T). One cell however is of different appearance, with slightly paler cytoplasm, no filament bundles and a nucleus (N) with a different chromatin pattern from that of the surrounding keratinocyte nuclei.

The most distinctive features of the melanocyte are its granules. Small dense granules are present throughout the cytoplasm, with no particular localisation in any one area of the cell. These represent the various stages of melanin granule formation, but the characteristic internal details of the granules are not seen at this low magnification. Next, notice that similarly dense, but rather large inclusions are also seen in surrounding keratinocytes. Often these appear to be compound granules, made up of two or more of the small melanin granules packed together (*). This is typical of melanin inclusions in the basal epidermal keratinocytes.

Melanin granules are actively transferred from elongated melanocyte processes, which appear to 'inject' their granules into the keratinocyte cytoplasm. The result is that the entire basal layer of the epidermis acquires a protective pigmentation through the activity of these scattered dendritic melanocytes.

C	Connective tissue underlying the epidermis
K	Keratinocytes, showing tonofilament bundles around their nuclei
N	Nucleus of melanocyte, without surrounding cytoplasmic tonofilaments
T	Tonofilament bundles in keratinocyte
*	Indicate compound melanin-containing granules in basal keratinocytes
Arrows	Indicate epidermal basal lamina

Tissue Human skin. Glutaraldehyde and osmium fixation, uranyl and lead staining.

Magnification 9000 ×

Refer to Plates 5, 8a, 72, 73, 75, 107
Sections 2.2, 4.8, 8.3, 8.4, 9.1

Micrograph by courtesy of C. Skerrow

Plate 74b *Melanocyte in epidermis*

The basal lamina (L) again separates the epidermal cells on the left from the connective tissue on the right (C) with its collagen fibrils.

Note the fine fibrils immediately below the basal lamina (arrows). These seem to form a fine fringe, running down from the lamina into the connective tissue space. These are anchoring fibrils, which strengthen the attachment of the basal lamina to the underlying tissue.

In this micrograph, parts of three keratinocytes are seen (K), but the main cell illustrated is a melanocyte (M). Its nucleus (N) lies towards the top of the field. The cytoplasm contains numerous melanin granules of varying density, reflecting different degrees of melanisation. In one of the less dense granules, a fine internal structure can be made out (*), representing the framework upon which melanin is synthesised and deposited as the granule matures. Each granule is limited by a membrane, which in most cases encloses only one clearly defined elliptical or rod-shaped granule. Note that this cell has rather little endoplasmic reticulum, but has a well developed Golgi apparatus (G).

The earliest form of melanin granule, often known as a premelanosome, is packaged in the Golgi apparatus and undergoes subsequent maturation through melanin synthesis after its release from the apparatus. As this proceeds, the delicate structural framework is progressively obscured by the accumulation of densely-staining melanin pigment.

C	Connective tissue containing collagen fibrils
G	Elaborate Golgi apparatus of melanocyte
K	Part of keratinocyte
L	Basal lamina
M	Melanocyte cytoplasm containing melanosomes
N	Nucleus of melanocyte
*	Indicates melanin granule with still visible internal substructure
Arrows	Indicate fringe of anchoring fibrils, linking the basal lamina to the underlying collagenous tissue

Tissue Human skin. Glutaraldehyde and osmium fixation, uranyl and lead staining.

Magnification 22 000 ×

Refer to Plates 5, 8a, 72, 73, 75, 107
Sections 2.2, 4.8, 8.3, 8.4, 9.1

Micrograph by courtesy of C. Skerrow

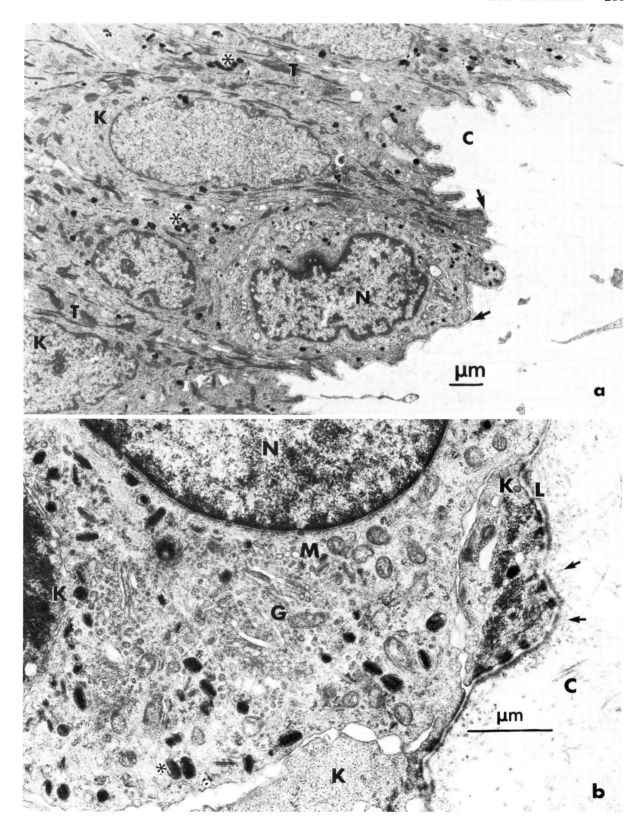

Plate 80a *Loose connective tissue*

Blood vessels run in loose connective tissue. This micrograph shows a small vein with surrounding collagen fibrils, a typical finding in tissue of this kind. The endothelial cells (E) surround a collapsed slit-like lumen. Smooth muscle (M) lies outside this endothelial tube, providing the necessary vascular contractility. Note the lamina densa (*) beneath the endothelium and around the muscle cells. In the connective tissue, cells are not present in this field, but bundles of collagen are seen cut at various angles. The interstitial space, although appearing empty, actually contains the carbohydrate-rich connective tissue ground substance, which is of fundamental importance for the various properties of these tissues.

E   Endothelial cells
M   Smooth muscle cells
*   Indicates lamina densa around components of the vessel

Tissue   Mouse testis, glutaraldehyde and osmium fixation, uranyl and lead staining.

Magnification   8000 ×

Refer to   Plates 9, 50c, 69a, 69c, 70, 71, 79, 81, 82, 83 Section 9.1

Plate 80b *Fibroblast, active*

The cells of connective tissue are rarely prominent in the resting condition, since their role is restricted to the slow process of tissue turnover. Resting fibrocytes are small inconspicuous cells, with little organised cyto-plasm. By contrast, the cells of growing connective tissue have all the features of synthetic activity that might be expected from the study of other cell types. These active cells, engaged in the formation of new connective tissue, are called fibroblasts.

This micrograph shows a fibroblast from an area of healing, where connective tissue is being actively formed. The cell is large and plump, with abundant cytoplasm showing elaborate membranes of granular endoplasmic reticulum with dilated cisternae (arrows) and a central Golgi apparatus (G). The nucleus (N) has a prominent nucleolus. Notice that there are fine bundles of filaments just within the cell surface (*), best seen where the surface is tangentially sectioned. These subsurface filaments are probably actin. The surrounding intercellular space shows masses of collagen fibrils (C), along with other amorphous background components of connective tissue.

C   Collagen fibrils in connective tissue space
G   Golgi apparatus of fibroblast
N   Nucleus of fibroblast with prominent nucleolus
*   Indicates subsurface cytoplasmic actin-like filaments
Arrows   Indicate dilated cisternae of granular endoplasmic reticulum

Tissue   Human small intestine, area of healing. Glutaraldehyde and osmium fixation, uranyl and lead staining.

Magnification   7000 ×

Refer to   Plates 9, 50c, 69a, 69c, 70, 71, 79, 81, 82, 83 Section 9.1

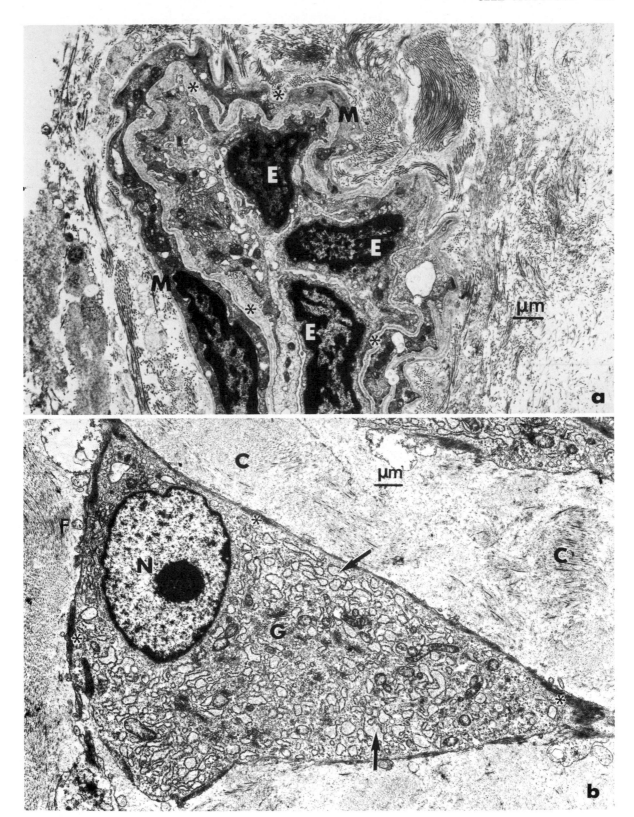

Plate 81a *Elastic tissue*

Elastic fibres have a variable appearance on electron microscopy. In this micrograph of a small arteriole, the elastic tissue (E) lies between the endothelium and the underlying smooth muscle (S). The elastica is pale and unstained, with virtually no internal structure, sandwiched between the lamina densa of the muscle cell and of the endothelium. Other features include micropinocytotic vesicles (V) at the surfaces of the endothelium, and cytoplasmic filaments (F) and mitochondria (M) in the muscle cell, close to its nucleus (N).

E	Elastic tissue
F	Cytoplasmic filaments of smooth muscle cell
M	Mitochondria of smooth muscle cell
N	Nucleus of smooth muscle cell
S	Smooth muscle cell
V	Micropinocytotic vesicles of endothelium

Tissue   Arteriole wall. Glutaraldehyde and osmium fixation, uranyl and lead staining.

Magnification   36 000 ×

Refer to   Plates 9, 69, 79, 80, 88
Section 9.1

Plate 81b *Elastic tissue*

Elastic tissue can also appear dense on electron micrographs. This micrograph shows an irregular dense mass of elastic tissue (E) with some internal irregular detail. The edges of the elastic fibre are irregular and fine filaments are gathered around the margin (*). Elsewhere, note the occasional isolated collagen fibrils (arrows) and the small bundles of long spacing collagen (C).

C	Long spacing collagen fibril bundles
E	Elastic fibre mass
*	Indicate fine fibrils at margin of elastic fibre mass
Arrows	Indicate long spacing collagen

Tissue   Human elastic tissue. Glutaraldehyde and osmium fixation, uranyl and lead staining

Magnification   14 000 ×

Refer to   Plates 9, 69, 79, 80, 88
Section 9.1

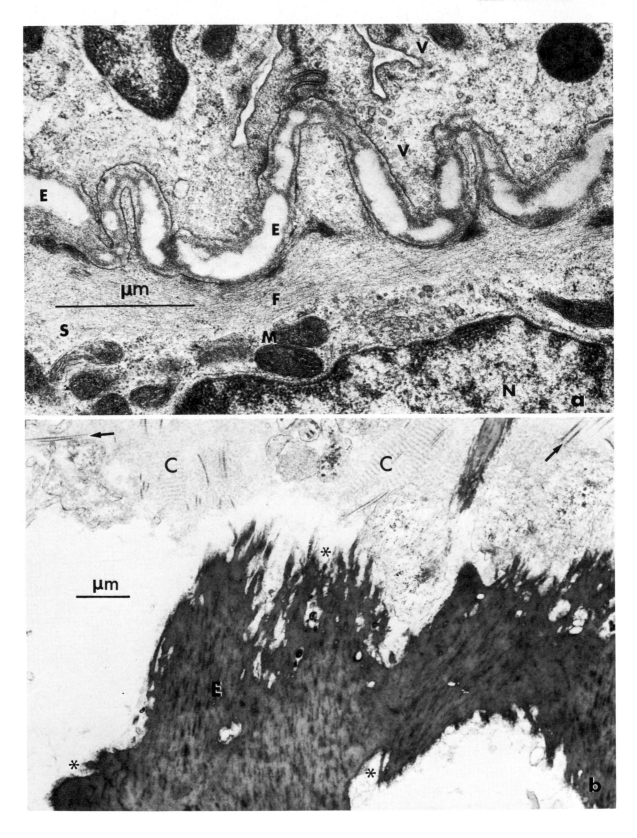

Plate 82 *Chondrocytes*

Cartilage is a form of firm but resilient connective tissue, not calcified like bone, but of more defined shape and solidity than fibrous tissue. The cells and the matrix are both distinctive in structural and biochemical terms, representing a modification of the basic pattern and role of the fibroblast.

In this micrograph, two pairs of chondrocytes are seen in a small group. Note the area of rarefied matrix around and between each pair (*), corresponding to the lacuna in which chondrocytes are seen to lie, under light microscopy. The rest of the matrix contains randomly orientated fibrils with the abundant ground substance which gives cartilage its mechanical properties. These fibrils are type II collagen, which differs from the type I collagen found in ordinary connective tissue.

The individual chondrocytes have the features of cells engaged in active synthesis. Note the elaborate granular endoplasmic reticulum, the cisternae of which (C) are typically filled with material of moderate density. This, coupled with the density of the ribosome-rich intercisternal cytoplasm, tends to render the chondrocyte rather non-photogenic. These cells also contain an elaborate Golgi apparatus (G), which participates in the synthesis of the carbohydrate-rich ground substance of cartilage. Other features of chondrocytes, not always seen, include glycogen aggregates and lipid droplets. These vary according to the metabolic state of the cells.

C   Dilated cisternae of granular endoplasmic reticulum

G   Elaborate Golgi apparatus of chondrocyte

★   Rarefied matrix adjacent to chondrocytes, corresponding to lacuna

Tissue Newborn rat femur. Glutaraldehyde and osmium fixation, uranyl and lead staining.

Magnification 18 000 ×

Refer to Plates 3, 15, 79, 80, 81, 83
          Section 9.2

Micrograph by courtesy of H. S. Johnston

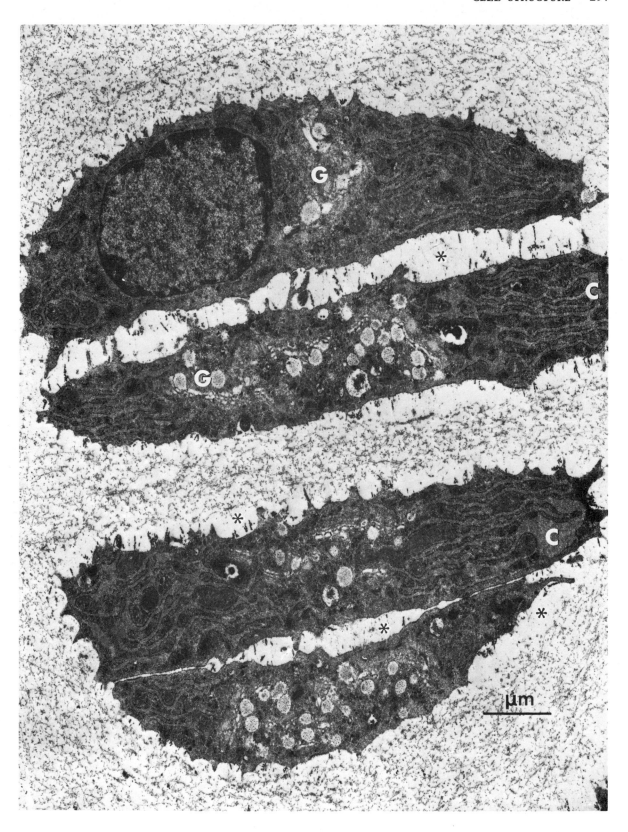

Plate 83a *Osteoblast*

The bone-forming cell is another variant on the basic fibroblast theme. The distinctive mechanical properties of bone are the result of the deposition of crystals of calcium phosphate in a matrix of collagen fibrils.

In this micrograph the osteoblast shows the usual structural features of the active protein-synthesising cell. Note the cisternae of granular reticulum filling most of the cytoplasm (R), and the presence of the Golgi apparatus (G), not particularly clearly seen in this micrograph. Several unremarkable mitochondria (M) are also seen.

Note that there is a mass of dense material on the left representing the calcified matrix of the bone. At the edges, the individual tiny needle-shaped crystals can just be made out. Between the calcified margin and the nearby osteoblast there lies a zone of uncalcified matrix, containing obliquely sectioned collagen fibrils (C) in a carbohydrate-rich ground substance. The process of calcification of the matrix involves the nucleation of calcium phosphate crystals in relation to specific points on the collagen fibrils, following which the crystals grow until they overshadow the fibrillar component of the matrix. The collagen remains, however, as an essential part of the calcified bone matrix, adding resilience to the otherwise brittle rigidity imposed by the inorganic constituents. In this area, calcification of the newly formed bone matrix has not yet begun. Uncalcified bone matrix is known as osteoid.

C   Collagen fibrils in uncalcified matrix around osteoblast
G   Ill-defined Golgi apparatus
M   Mitochondrion
R   Granular endoplasmic reticulum

Tissue Mouse femur. Glutaraldehyde and osmium fixation, uranyl and lead staining.

Magnification 21 000 ×

Refer to Plates 79, 80, 81, 82
Section 9.2

Micrograph by courtesy of H. S. Johnston

Plate 83b *Osteoclast*

The removal of bone is as essential as its formation in the remodelling processes involved in bone growth and in the normal turnover of bone in the adult skeleton. The osteoclast is the principal cell involved in bone resorption.

This micrograph shows an elongated osteoclast stretched over an area of resorbing bone matrix. The matrix is of variable density, allowing the fibrillar nature of the background to be recognised in places (F), while elsewhere the fibrils are obscured by the overall density of the tissue (X). The osteoclast generally has multiple nuclei, but only one is shown in the area illustrated in this case (N). The osteoclast cytoplasm contains vacuoles (V), which reflect the intracellular mechanisms for the breakdown of fragments of the bone matrix taken up into the osteoclast by phagocytosis. The configuration of the osteoclast surface in contact with the frayed surface of the bone matrix undergoing resorption is strikingly elaborate, with elongated cytoplasmic folds producing a greatly expanded cell surface at the area of maximum physiological activity (*).

F   Fibrillar area of matrix, showing lower density than surrounding areas
N   Nucleus of osteoclast
V   Cytoplasmic vacuoles in osteoclast
X   Denser area of bone matrix
*   Indicate complex infoldings of osteoclast surface at area of active resorption

Tissue Mouse femur, glutaraldehyde and osmium fixation. Uranyl and lead staining

Magnification 10 000 ×

Refer to   Plates 79, 80, 81, 82
Section 9.2

Micrograph by courtesy of H. S. Johnston

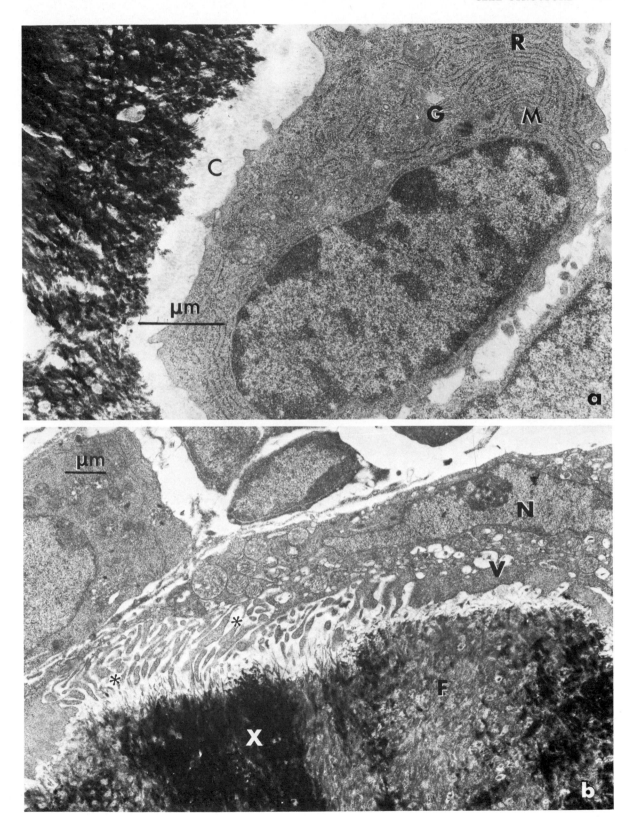

CONTRACTION

Plate 84a *Skeletal muscle, longitudinal section*

This low-magnification view of skeletal muscle shows parts of two cells, with an adjacent capillary vessel (V) lying in the surrounding connective tissue (T). At this magnification, the finer details of muscle structure are not clearly seen, but the basic cross-striation pattern is easily recognised, with prominent dense A bands (A) and pale I bands, the latter bisected by the Z lines (Z). Mitochondria are just visible between the myofibrils, but the components of the sarcoplasmic reticulum cannot be clearly distinguished. Just beneath the muscle surface, a tiny portion of a tangentially sectioned muscle nucleus can be seen in the peripheral rim of cytoplasm (N).

A	A band of muscle
N	Portion of muscle cell nucleus
T	Connective tissue around muscle cells
V	Capillary vessel
Z	Z line of muscle, bisecting the pale I band

Tissue Baboon muscle. Glutaraldehyde and osmium fixation, uranyl and lead staining.

Magnification   5000 ×

Refer to   Plates 33a, 53b, 85, 86, 87, 88, 89, 90
Sections 4.8, 10.1

Plate 84b *Skeletal muscle, transverse section*

Parts of seven muscle cells are seen in this micrograph, with surrounding connective tissue (T) in which vascular structures (V) can be seen. The internal layout of the cross-sectioned muscle cells can be recognised. Small dense mitochondria are scattered at random amongst the myofibril components. The plane of section is not exactly transverse, because some areas of cross-sectioned myofibrils show a dense A band configuration (A), while others show the pale I band pattern (I), where only thin actin filaments are present. Details of the sarcoplasmic reticulum are not clearly seen, but thin dense T tubules can just be picked out, forming a network between myofibrils at the junction between the A band and the I band areas (arrows).

A	A band area in cross-section
I	I band area in cross-section
T	T tubules
V	Capillary vessel
Arrows	Indicate thin dense T tubules

Tissue Baboon muscle. Glutaraldehyde and osmium fixation, uranyl and lead staining.

Magnification   4000 ×

Refer to   Plates 33a, 53b, 85, 86, 87, 88, 89, 90
Sections 4.8, 10.1

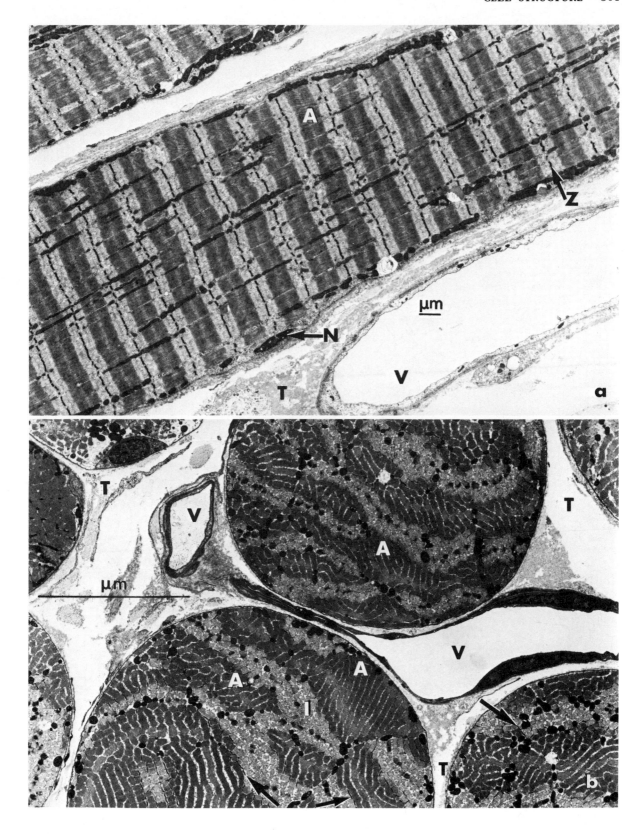

Plate 85a *Skeletal muscle, sarcoplasm*

In this longitudinal section of a skeletal muscle cell parts of several myofibrils can be seen. The Z line of one and two M lines of another are indicated. The plane of section has, however, displayed the sarcoplasmic components around one of these myofibrils to particular advantage. The segmental arrangement of mitochondria can be seen, one long mitochondrion lying parallel to the Z line (Mi). A single T tubule (T) can also be followed for some distance through the sarcoplasm, making contact at one point with the sarcoplasmic reticulum to form a triad (SR). Numerous glycogen particles are also found in the sarcoplasm (G).

G   Glycogen
M   M line
Mi   Mitochondrion
SR   Sarcoplasmic reticulum forming a triad
T   T tubule in longitudinal section
Z   Z line

Tissue   Human skeletal muscle. Glutaraldehyde and osmium fixation, uranium and lead staining.

Magnification   84 000 ×

Refer to   Plates 33a, 53b, 84, 86, 87, 88, 89, 90
          Sections 4.8, 10.1

Plate 85b *Insect flight muscle, filaments*

This cross-section of an insect flight muscle shows the geometrical array of thick and thin myofilaments which is characteristic of skeletal or striated muscle. The insert shows, at higher magnification, the hexagonal pattern formed by the thick myosin and thin actin filaments. Two other features of this muscle are the large mitochondria (M) with complex cristae and the open communication shown between the T tubules and the extracellular space (arrows). This relationship is not clearly demonstrable in most mammalian skeletal muscles. The two muscle cells which appear in this plate are separated by a connective tissue space, but each cell has an external or basal lamina closely applied to its surface (BL). In this specialised muscle, the T tubules and the sarcoplasmic reticulum form 'dyads' instead of triads as in the other muscle types illustrated.

BL   Basal or external lamina
M   Mitochondrion
Arrows   Indicate the point at which T tubules open to the intercellular space

Tissue   Grasshopper flight muscle. Glutaraldehyde and osmium fixation, uranium and lead staining.

Magnification   28 000 ×

Refer to   Plates 33a, 53b, 84, 86, 87, 88, 89, 90
          Sections 4.8, 10.1

Micrograph by courtesy of H. Elder

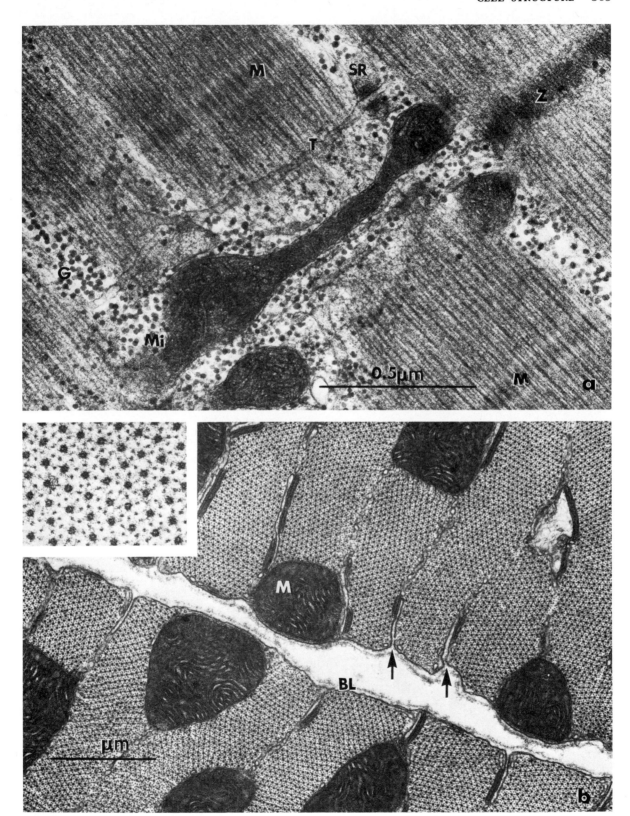

Plate 86 *Skeletal muscle, sarcomere*

This is a high-magnification micrograph of part of a skeletal muscle cell showing several parallel myofibrils crossing the plate vertically. They are separated by sarcoplasm. Part of a single sarcomere is seen. This plate should be compared directly with Plate 84a which shows the same features at lower magnification. Notice that the direction of the myofibrils in this plate is at right angles to that seen in Plate 84b.

The Z line bisects the pale I band which contains thin actin filaments. The thick filaments occupy the A band. At the point in the A band marked by A, the overlap between thick and thin filaments is seen. In the centre of the A band lies the H zone, in which only thick filaments are present. The beadings at the midpoints of the thick filaments line up to form the M line.

The five myofibrils which cross the field almost vertically are separated by sarcoplasm in which the components of the sarcoplasmic reticulum are seen (SR). The triads, with central T tubule and flanking foot processes, are present consistently at the region of the A-I junction, being particularly well seen at the points arrowed. The sarcomere pattern established by the arrangement of thick and thin filaments is followed by the sarcoplasmic reticulum and the mitochondria (Mi).

A	A band
H	H zone
I	I band
M	M line
Mi	Mitochondrion
SR	Sarcoplasmic reticulum
T	T Tubule with flanking foot processes
Z	Z line
Arrows	Indicate the triads located close to the A-I junction

Tissue   Rat skeletal muscle. Osmium fixation, phosphotungstic acid staining.

Magnification   88 000 ×

Refer to   Plates 33a, 53b, 84, 85, 87, 88, 89, 90
Sections 4.2, 4.8, 10.1

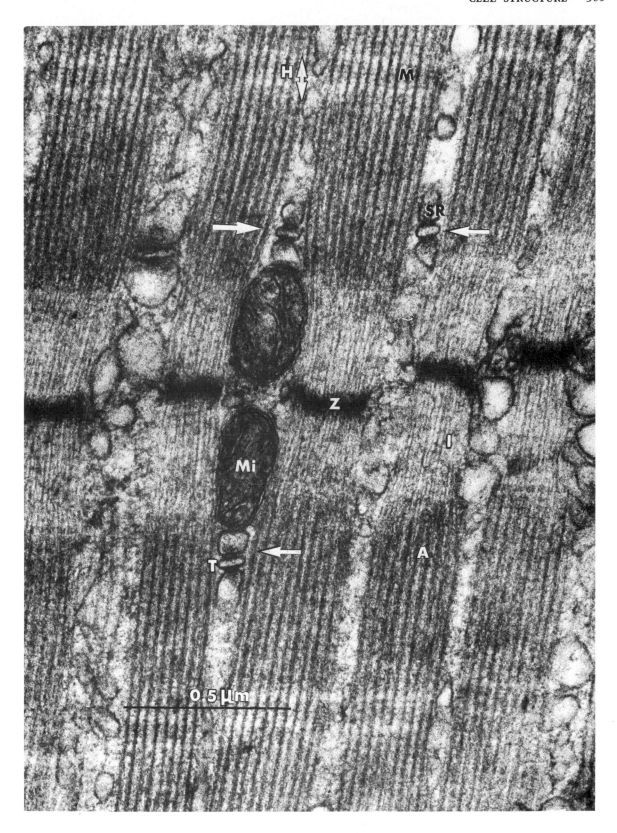

Plate 87a *Skeletal muscle, surface region*

This micrograph shows parts of two adjacent skeletal muscle cells separated by a connective tissue space (CT) in which collagen fibrils are seen. The sarcomere pattern with Z and M lines bisecting I and A bands respectively is essentially similar to that shown in the preceding plate. Immediately external to each cell lies a diffuse external lamina or basal lamina, representing the interface between muscle and connective tissue. In the lower muscle cell, a peripherally situated nucleus (N) can be seen lying between the myofibrils and the cell membrane. Each muscle cell has many such nuclei. At the cell surface (arrows) a few micropinocytotic vesicles or caveolae can be found. In this case the triads, when seen, are placed at the A-I junction (X). Single glycogen particles are quite plentiful in the sarcoplasm, a common finding in muscle cells.

CT	Connective tissue space
N	Nucleus of muscle cell
M	M Line
X	Triad situated at level of A-I junction
Arrows	Indicate surface caveolae or micropinocytotic vesicles

Tissue Human skeletal muscle. Glutaraldehyde and osmium fixation, uranium and lead staining.

Magnification 37 000 ×

Refer to Plates 33a, 53b, 84, 85, 86, 88, 89, 90
Sections 4.8, 10.1

Plate 87b *Skeletal muscle, transverse section*

This micrograph shows an area of cross-sectioned skeletal muscle in which the plane of section passes through different parts of adjacent myofibrils. In one area, the myofibrils have been sectioned through the A band (A), where the overlapping thick and thin filaments are both present. Nearby, the plane of section passes through the I band (I), where only cross-sections of thin filaments appear. Profiles of several mitochondria (M) lie between adjacent myofibrils, which are otherwise not clearly demarcated from each other. A complex dense pattern present in patches in the I band area corresponds to the Z line seen in face view (Z). Here, the thin filaments are linked in a square lattice pattern by the dense Z line material, which contains the muscle protein alpha actinin.

A	A band in cross-section showing thick and thin filaments
I	I band in cross-section showing thin filaments only
M	Mitochondria between myofibrils
Z	Z line in face view, showing square lattice pattern

Tissue Baboon muscle. Glutaraldehyde and osmium fixation, uranyl and lead staining.

Magnification 50 000 ×

Refer to   Plates 33a, 53b, 84, 85, 86, 88, 89, 90
Sections 4.8, 10.1

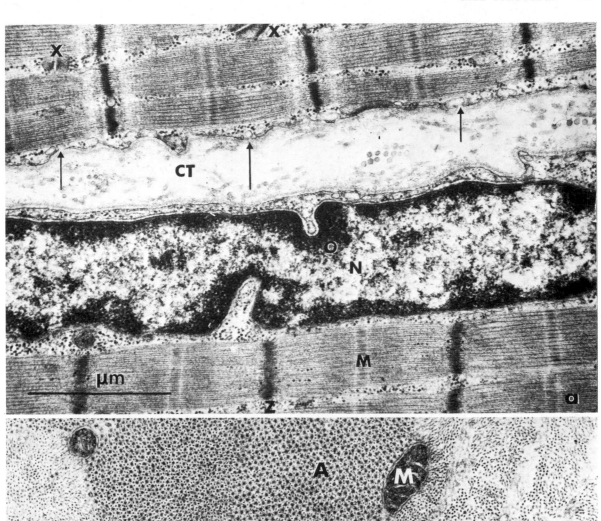

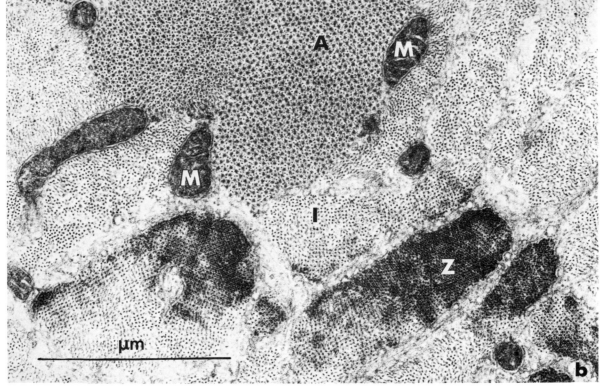

Plate 88 *Smooth muscle cells*

This is a low-magnification micrograph of several smooth muscle cells. Adjacent cells are separated by a connective tissue space (C) of variable width, in which sparse collagen fibrils can be seen at a number of points reinforcing the delicate basal lamina or lamina densa which surrounds each cell individually. There is one gap junction present, at which there is closer contact between the membranes of adjacent cells, providing a point where surface depolarisation may spread from cell to cell (GJ).

Most of the cytoplasm is occupied by filaments (F) with a predominantly longitudinal orientation without evidence of a sarcomere pattern of repeating units. Interspersed between the fine filaments are dense bodies of uncertain length which often appear cigar-shaped in section (D). Thickenings (T) on the inner aspect of the cell membrane of the muscle cell at different points form attachments for the myofilaments. Micropinocytotic vesicles appear at the cell surface at several points (P).

Few formed organelles are present in the cell. The mitochondria are relatively poorly developed (M) and form only a small proportion of the cytoplasmic volume. Ribosomes (R) and a small Golgi apparatus are sometimes present close to the nucleus in smooth muscle. The nucleus of one cell appears in two portions in this plate. The nucleus in smooth muscle (N) is commonly twisted or indented in configuration and a thin section which fails to pass through its centre may give a false impression that more than one nucleus is present in the cell. The

two nuclear profiles seen in the upper right hand corner of this plate are probably the result of such a plane of section effect.

The complete contrast between the fine structure of smooth and striated muscle underlines their known differences of function.

C	Connective tissue space with collagen fibrils, reinforcing external lamina
D	Dense bodies with cigar-shaped profile
F	Myofilaments
GJ	Gap junction
M	Mitochondrion
N	Nucleus
P	Micropinocytotic vesicles
R	Area of cytoplasm close to nucleus, containing ribosomes
T	Thickening on the inner surface of the cell membrane representing attachment area for myofilaments

Tissue   Chicken gizzard, smooth muscle. Osmium fixation, lead staining.

Magnification   14 000 ×

Refer to   Plates 84, 85, 86, 87, 89, 90
               Sections 4.8, 10.1

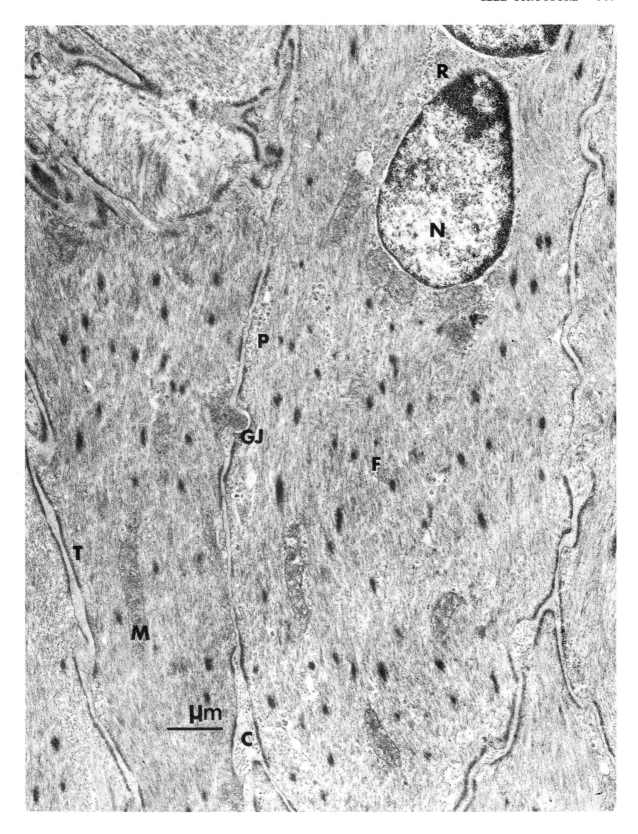

Plates 90a, 90b *Cardiac muscle, intercalated disc*

This micrograph shows part of the intercalated disc at high magnification. Two Z lines of adjacent myofibrils can be made out, with the other components of the sarcomere pattern. The transverse component of the intercalated disc, consisting of adhesion specialisations of zonula adhaerens type (A), lies at the level of the Z line of the sarcomere. The dense area on either side of this specialisation serves as the attachment for the actin filaments of the muscle. A desmosome or macula adhaerens is also seen (D), but is obliquely sectioned. The other component of the disc, the region of gap junction (G) specialisation, runs parallel to the myofibrils. This represents an area of close membrane contact believed to have increased permeability to ions, allowing the wave of depolarisation of the muscle cell surface, which initiates contraction, to pass from cell to cell, leading to the characteristic spreading activity of cardiac muscle. Oblique sectioning effects are responsible for the apparent blurring of the membranes seen at points along the junction (O). Part of a mitochondrion appears in one corner of the plate (M). The insert shows part of the disc at higher magnification. The close contact at the gap junction can be made out. The adhaerens area is partially obscured by oblique section.

A	Adhaerens area of disc
D	Adhesion portion of intercalated disc corresponding to desmosome
G	Gap junction area
M	Part of a mitochondrion
O	Oblique section of membranes
Z	The Z line of the sarcomere

Tissue   Rat heart. Glutaraldehyde and osmium fixation, uranium and lead staining.

Magnification   Plate 90a 92 000 ×
                  Plate 90b 110 000 ×

Refer to   Plates 4, 6, 89
             Sections 2.2, 4.8, 10.1

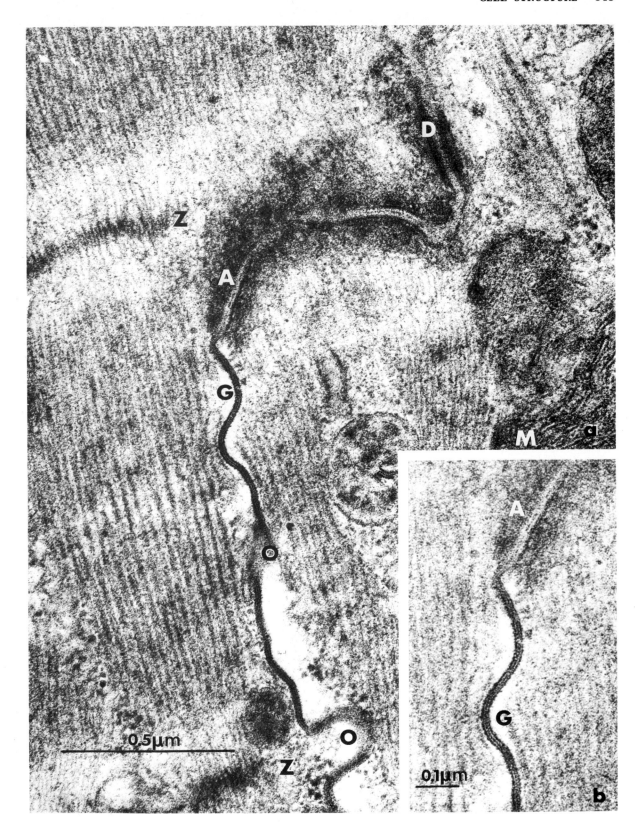

MOTILITY

Plate 91a *Ciliated epithelium, trachea*

This scanning electron micrograph shows a surface view of respiratory mucosa. Most of the cells are ciliated (C), providing a thick mat of motile processes which play an important role in the protective mechanisms of the respiratory tract. Mucus from tracheobronchial glands pours out onto this surface, where some strands remain after processing (*). The dome-shaped apices of the non-ciliated cells of the mucosa have scanty microvilli, which are tiny by comparison with the surrounding cilia.

C   Cilia, forming a surface mat
*   Indicate strands of mucus

Tissue   Rat trachea. Glutaraldehyde and osmium fixation, critical point drying, gold coated.

Magnification   3500 ×

Refer to   Plates 34c, 92, 93, 94
             Section 10.2

Plate 91b *Ciliated, epithelium, endometrium*

This scanning electron micrograph shows ciliated cells on the surface of human endometrium. A ciliated region takes up most of the picture, the long sinuous cilia being clearly seen. The other surface cells are covered by stubby microvilli (M). Note the difference between the length and diameter of cilia and of microvilli.

M   Microvilli on non-ciliated cells

Tissue   Human endometrium, fixed in formalin and embedded in paraffin wax. The block was then dewaxed and critical point dried before being coated with gold.

Magnification   11 000 ×

Refer to   Plates 34c, 92, 93, 94
           Section 10.2

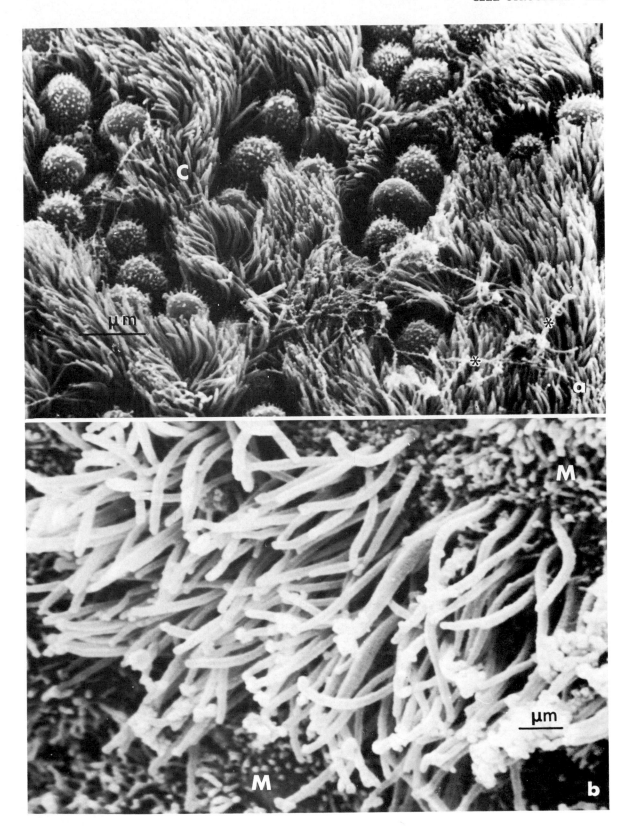

Plate 92 *Ciliated edometrial gland cells*

This micrograph shows part of an endometrial gland lumen (L) surrounded by columnar cells, two of which are ciliated. The gland is separated by a thin basal lamina (*) from the underlying stromal cells (S). As is usual in glandular epithelium, the columnar cells are joined together at their apices by junctional complexes (J). Note the presence of parts of two pale cells (P) which probably represent migrating lymphoid cells.

The ciliated columnar cells also have short microvilli between the cilia. The cilia themselves extend into the lumen and are cut across in oblique section, owing to their sinuous configuration. Note that each cilium originates in a basal body (B) in the apical cytoplasm. These structures, similar on cross-section to centrioles, are produced, as the cell matures, by the repeated replication of the centrioles of the cell. These then migrate to the surface and produce cilia. Details of the microtubules which form the axial filament complex, or axoneme, are not clearly seen at this low magnification.

Other cellular features seen here are the modest supranuclear Golgi systems of the columnar cells (G) and their mitochondria (M).

B	Basal body, serving as root of cilium
G	Golgi apparatus
J	Junctional complexes
L	Lumen of endometrial gland
M	Mitochondria
P	Processes of pale cells, probably lymphoid
S	Stromal cells surrounding endometrial gland
*	Indicate basal lamina

Tissue   Human endometrium. Glutaraldehyde and osmium fixation, uranyl and lead staining.

Magnification   7000 ×

Refer to   Plates 34c, 91, 93, 94
Section 10.2

Micrograph by courtesy of I. A. R. More

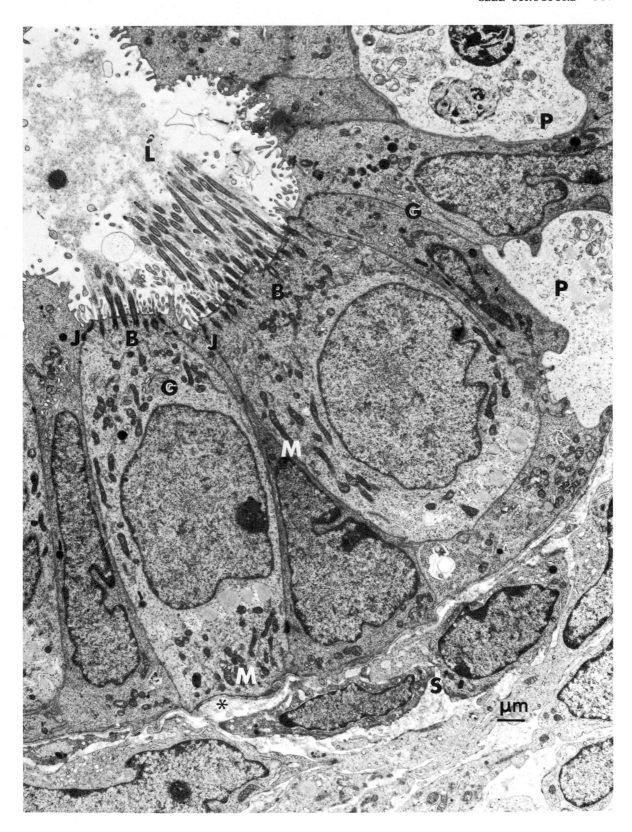

Plate 93 *Cilia, tracheal epithelium*

This is a micrograph, at moderate magnification, of the surface of a tracheal epithelial cell showing the cilia (C) with their internal axial filament complexes ending in basal bodies (B). Short rootlets anchor the cilia in the apical cytoplasm (R). In the upper part of the micrograph the cilia have been cut obliquely and at times nearly transversely, showing to some extent their typical 9+2 pattern of organisation. In the lower parts of the cilia the axial components are sectioned longitudinally, showing their tubular nature. Notice the dimensions of the cilia, compared with the size of the few microvilli (MV) which are also seen at the cell surface.

The surface membrane of the tracheal epithelial cell covers the entire cilium as well as the adjacent microvilli. A junctional complex (JC) marks the contact point between adjacent cells at the left of the field, but obliquity of section obscures its morphology. The arrows in the cytoplasm indicate a microtubule which lies in the plane of the section. Its diameter is comparable with that of the axial components of the cilia.

B	Basal body
C	Cilia
JC	Junctional complex
L	Lumen of trachea
MV	Microvillus
R	Rootlet of cilium
Arrows	Indicate a microtubule in the apex of the tracheal epithelial cell

Tissue Rat trachea. Glutaraldehyde and osmium fixation, uranium staining.

Magnification   36 000 ×

Refer to   Plates 10b, 34c, 91, 92, 94, 104c
             Sections 4.9, 4.10, 10.2

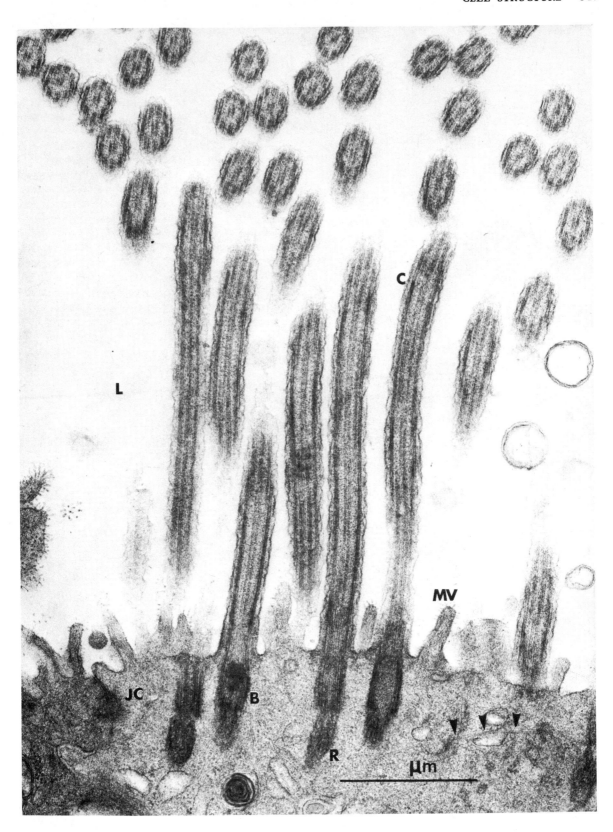

Plate 94 *Cilia, transverse section*

This micrograph shows details of the internal structure of cilia cut in cross-section. The components of the axoneme, or axial filament complex, are seen to be microtubular in nature. Two central tubules are linked by diffuse 'spokes' to the nine peripheral doublets, arranged on the circumference of the cilium, surrounded by the cell membrane. Notice that most of the central pairs are aligned in a roughly parallel fashion. The arrows, at right angles to the plane of these central tubules, show the direction of ciliary beat.

The individual doublets of the nine peripheral axonemal subunits are made up of two fused tubules, identified as subunits A and B. Two ill-defined arms project from each subunit A, extending towards the subunit B of the adjacent doublet. These are the dynein arms of the cilium, forming the basis of the ratchet mechanism which produces the bending upon which ciliary movement depends. Notice that several cross-sectioned microvilli are also seen between the cilia (V).

A	subunit A of axonemal peripheral doublet
B	Subunit B of axonemal peripheral doublet
V	Microvilli between cilia
Arrows	Indicate direction of beat, at right angles to the plane of the central doublet

Tissue Hen oviduct (magnum). Glutaraldehyde and osmium fixation. Uranyl and lead staining.

Magnification   150 000 ×

Refer to   Plates 10b, 34c, 91, 92, 93, 104c
Sections 4.9, 4.10, 10.2

Micrograph by courtesy of H. S. Johnston

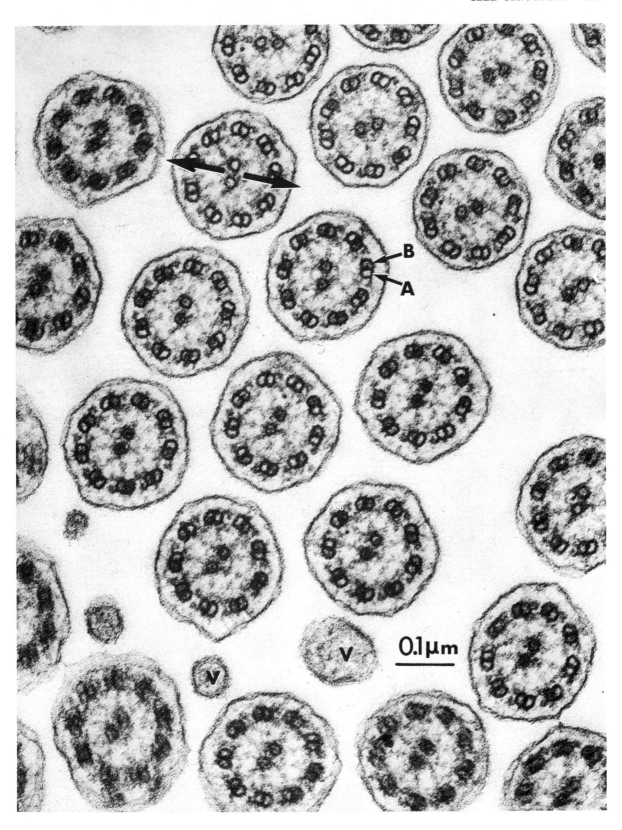

Plate 95a *Spermatozoa*

This scanning electron micrograph shows several bull spermatozoa lying on a metal stub, or specimen holder (S). The outline of the head (H) of the spermatozoon can be seen, with the tail (T) attached to it at the neck. The midpiece of the tail contains a tube of spirally orientated mitochondria. The surface impression of the mitochondrial spiral can be seen at some places on the tails (arrows).

H	Head of spermatozoon
S	Surface of the specimen stub
T	Tail of spermatozoon
Arrows	Indicate spiral pattern of mitochondrial sheath in midpiece

Tissue Bull spermatozoa. Glutaraldehyde and osmium fixation. Air dried. Coated with carbon/platinum.

Magnification 6000 ×

Refer to Plates 18a, 26a, 28a, 93, 94, 96
Sections 4.5, 4.6, 4.9, 10.2

Specimen by courtesy of E. Aughey and J. P. Renton

Plate 95b *Spermatozoon*

This micrograph shows the head of a developing spermatozoon in the testicular tubule. The distinctive granulated pattern of the nuclear material (N) is characteristic of the human spermatozoon. The dense structure applied closely to the outer aspect of the nucleus is the acrosomal cap (A), a flattened sac containing lysosomal enzymes used in the process of fertilisation of the ovum. The head of the developing sperm is buried in the cytoplasm of a surrounding Sertoli cell (S), which supports the maturation of the germ cell. Careful examination shows that both the germ cell and the supporting cell have intact cell membranes, so that their relationship is one of close contact, but not cytoplasmic fusion.

A	Acrosomal cap
N	Nucleus of spermatozoon
S	Sertoli cell cytoplasm

Tissue Human testis. Glutaraldehyde and osmium fixation, uranyl and lead staining.

Magnification 30 000 ×

Refer to Plates 18a, 26a, 28a, 93, 94, 96
Sections 4.5, 4.6, 4.9, 10.2

Plate 95c *Spermatozoon*

This micrograph shows some other features of the developing male germ cell and its relationships with the surrounding Sertoli cell. In this case the sperm head has been tangentially sectioned, showing the extreme density of the nuclear material (N). The acrosome again envelopes the sperm head (A). The cytoplasm of the germ cell has become elongated and orientated around the axonemal structures of the developing propulsion unit (X), also obliquely sectioned. Around the axonemal components are grouped the mitochondria (M) of the germ cell, which are beginning to form themselves into the midpiece mitochondrial spiral. There is an obvious functional significance in this close relationship between the motile structure of the flagellum and the source of cellular ATP, the fuel for movement. Notice the distinctive configuration of the germ cell mitochondria, both around the axoneme of this nearly mature cell, and in the peripheral cytoplasm of a neighbouring less mature germ cell (M_1), which also displays the Golgi apparatus (G) to advantage, in the active phase of acrosome formation.

Finally, note the distinctive features of the adjacent Sertoli cell (S). This cell has smooth endoplasmic reticulum (R) and occasional tubular mitochondrial cristae, typical features of a cell involved in steroid metabolism. Bundles of parallel microtubules can also just be made out (arrows).

A	Acrosome
G	Golgi apparatus of immature germ cell
M	Mitochondria forming a spiral around the axoneme of the developing spermatozoon
M_1	Similar mitochondria in immature germ cell
N	Nucleus of spermatozoon
R	Smooth endoplasmic reticulum
S	Sertoli cell cytoplasm
X	Axoneme of spermatozoon, obliquely sectioned
Arrows	Indicate microtobules in Sertoli cell cytoplasm

Tissue Rat testis. Glutaraldehyde and osmium fixation. Uranyl and lead staining.

Magnification 29 000 ×

Refer to Plates 18a, 26a, 28a, 93, 94, 96
Sections 4.5, 4.6, 4.9, 10.2

Micrograph by courtesy of H. S. Johnston

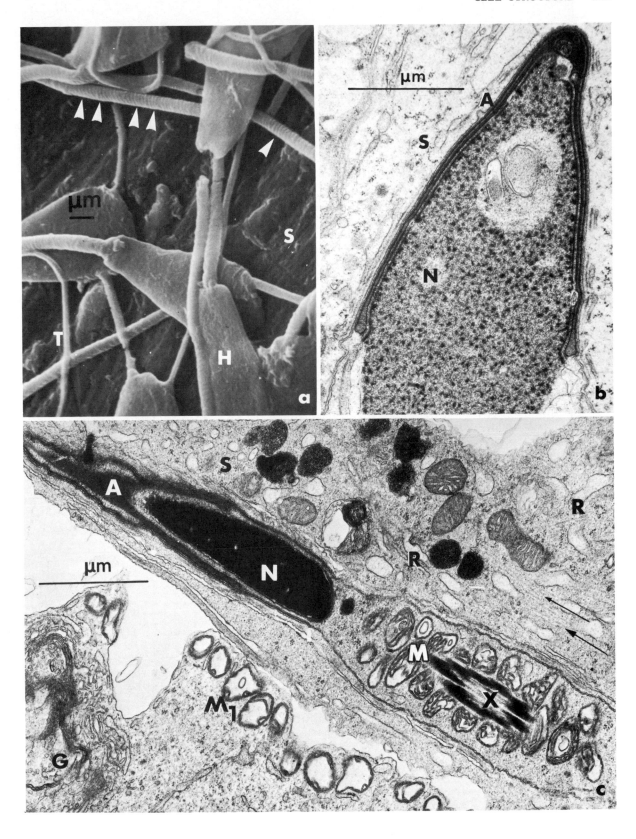

Plate 96a *Sperm tails, transverse section*

This micrograph shows a transverse section through several sperm tails. The different cross-sectional appearances are due to the plane of section passing through different parts of these several cells. In four cells, the mitochondrial spiral of the midpiece (M) is surrounded by varying amounts of cytoplasm. The central microtubular components of the axoneme (*) resemble closely the structures of the ciliary axoneme shown in Plate 94. In addition, however, there are nine broad dense additional components (D), each of which is associated with one of the peripheral doublets of the axoneme. As can be seen, each of these individual dense components has a distinctive structure, precisely related to its position around the axonemal circumference. The functional significance of these specialisations remains obscure. The remaining three identifiable sperm tails (T) have been cut across close to their tips. The cytoplasmic structures of the midpiece and the additional dense components have disappeared at this level, leaving only the axoneme and surrounding cell membrane. At this level, the sperm tail is indistinguishable from a cilium.

D   Dense additional components of sperm tail
M   Mitochondrial spiral of midpiece
T   Sperm tails sectioned close to the tip
*   Axonemal microtubules

Tissue   Rat testis. Glutaraldehyde and osmium fixation. Uranyl and lead staining.

Magnification   26 000 ×

Refer to   Plates 26a, 28a, 93, 94, 95
             Sections 4.5, 4.6, 4.9, 10.2

Micrograph courtesy of H. S. Johnston

Plate 96b *Sperm tail, longitudinal section*

This is a high-magnification micrograph of a sperm tail cut in longitudinal section, surrounded by the limiting membrane of the sperm. The 'fibrous' sheath lies within the limiting membrane. The central axial filament complex consists of microtubular components similar to those seen in the cilium. The dense additional components peripheral to the axial filament complex are seen in longitudinal section. They display no evident substructure. A fine periodicity is, however, seen along the central pair of the axial filament complex.

A   Axial filament complex
D   Dense additional component of sperm tail
F   Fibrous sheath

Tissue   Spermatozoon from rat epididymis. Glutaraldehyde and osmium fixation, uranium staining.

Magnification   47 000 ×

Refer to   Plates 26a, 28a, 93, 94, 95
             Sections 4.5, 4.6, 4.9, 10.2

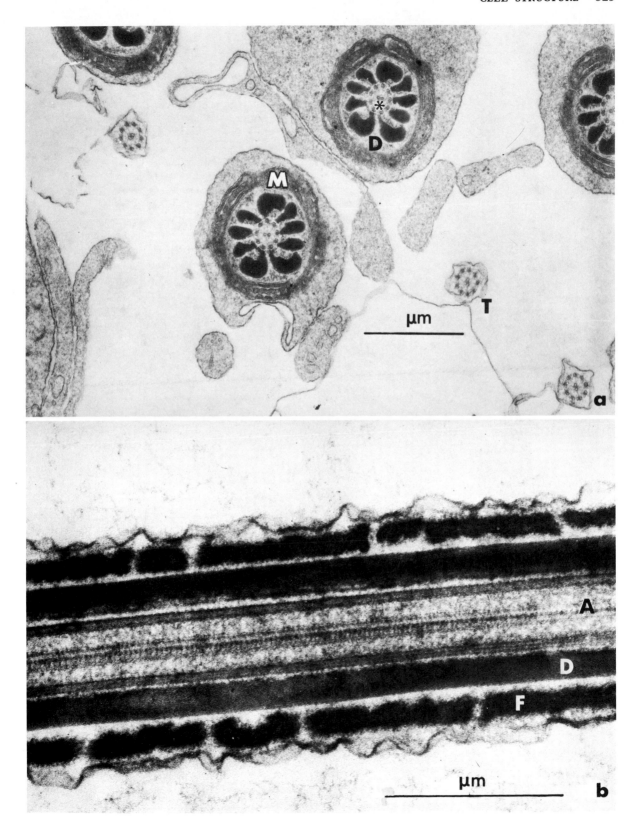

COMMUNICATION

Plate 97 *Neurone, central nervous system*

This micrograph gives some impression of the complexity of organisation of the nervous tissues. The cell body of a neurone lies in the centre of the field, with a large pale nucleus (N). To assist interpretation, the main cell processes have been outlined (*). The cytoplasm of the cell body, or perikaryon, contains aggregates of granular endoplasmic reticulum (R), mitochondria, and dense residual bodies or telolysosomes, while the neuronal processes have parallel microtubules and filaments, although these details are not clearly visualised at this low magnification. Note the complexity of the surrounding neuropil. Large and small myelinated axons (A) are seen cut at various angles. Numerous unmyelinated axons and glial cell processes fill the remaining space. The areas of pale cytoplasm (P) are astrocyte processes, which act as packing and insulation for many of the small nerve processes. Note the absence of a true connective tissue interstitial space in the substance of the nervous tissue.

A	Myelinated axons
N	Pale nucleus of neurone
P	Astrocyte processes serving as support and packing for neural structures
R	Granular endoplasmic reticulum
****	Indicate the outlines of the neurone and its processes

Tissue Adult mouse spinal cord, cervical level. Glutaraldehyde and osmium fixation. Uranyl and lead staining.

Magnification 9000 ×

Refer to Plates 98, 99
Section 11.1

Micrograph by courtesy of H. S. Johnston

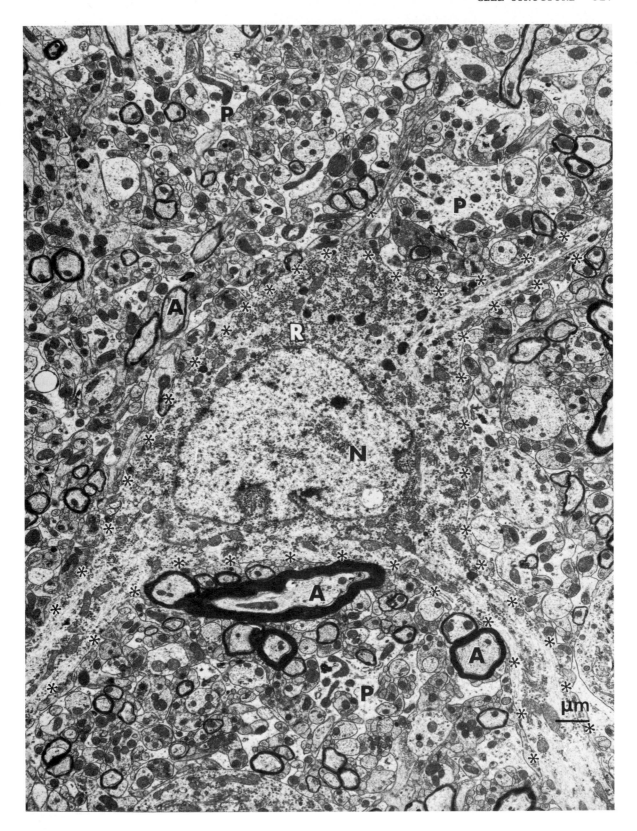

Plate 98a *Nerve cell, Nissl substance*

This area of neuronal cytoplasm shows several loose packets of aggregated membranes of the granular endoplasmic reticulum (R). These packets are recognised by their basophilic reaction in light microscopic sections, where they are identified as Nissl substance. Note the abundant polyribosomes between the roughly parallel membrane cisternae. Also seen are parts of the Golgi apparatus (G), several mitochondria (M) and occasional lysosomes (L). The nucleus of the nerve cell (N) is pale, with particularly diffuse chromatin.

G   Golgi apparatus
L   Lysosome-like dense bodies
M   Mitochondria
N   Nucleus of nerve cell
R   Aggregate of granular endoplasmic reticulum

Tissue   Adult mouse spinal cord, cervical level. Glutaraldehyde and osmium fixation. Uranyl and lead staining.

Magnification   18 000 ×

Refer to   Plates 97, 99, 100, 101, 102, 103
             Sections 4.2, 4.9, 11.1

Micrograph courtesy of H. S. Johnston

Plate 98b *Nerve cell process*

This micrograph shows the dendrite of a Purkinje cell of the cerebellum, containing numerous parallel microtubules (T) running horizontally across the field. In nerve cells, these are known as neurotubles. Surrounding this cell process, there are several other neuronal and glial processes cut at various angles. The pale cytoplasm (A) is part of an astrocyte, enveloping assorted nerve twigs, some of which contain numerous synaptic vesicles (S), associated in places with synaptic membrane specialisations (arrows).

Within the Purkinje cell, note the strikingly parallel orientation of all of the cytoplasmic structures. Smooth-surfaced membrane cisternae (*) and mitochondria (M) appear to be organised into this parallel configuration by the abundant surrounding microtubules.

Note the virtual absence of intercellular spaces in the brain, other than the narrow cleft which represents the standard minimum separation of two unspecialised contacting cell membranes. The role of the connective tissues elsewhere, as channels for the diffusion of tissue fluids, nutrients and metabolites, is taken on by the neuroglial cells of the brain, which envelope the neural processes and which form a network connected to the cerebral capillary vessels.

A   Astrocyte cytoplasm, enveloping nerve cell processes
M   Mitochondria of Purkinje cell
S   Synaptic vesicles
T   Parallel microtubules
*   Indicate smooth-surfaced membrane cisternae within the dendrite
Arrows   Indicate synaptic membrane specialisations

Tissue   Rat cerebellum, glutaraldehyde and osmium fixation, uranium staining.

Magnification   30 000 ×

Refer to   Plates 97, 99, 100, 101, 102, 103
           Sections 4.2, 4.9, 11.1

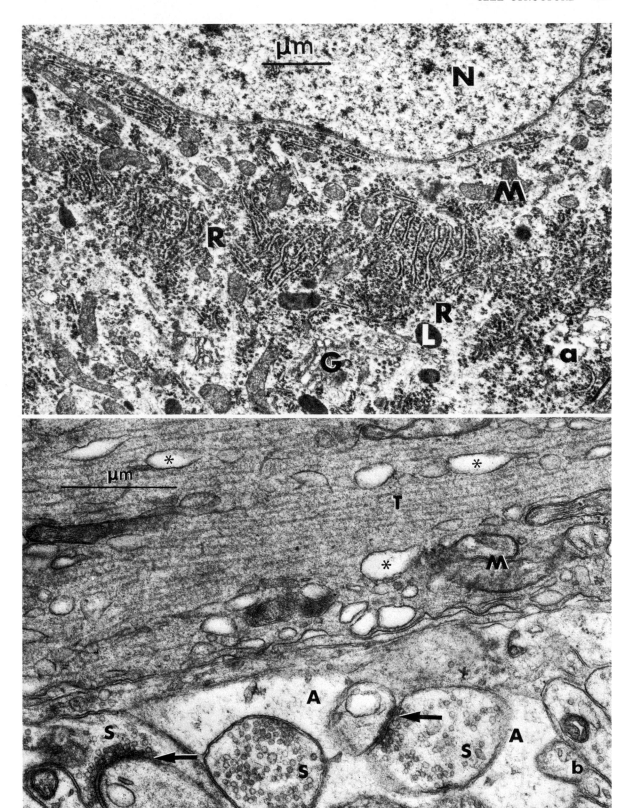

Plate 99 *Nerve tissue, capillary vessels*

This low-magnification survey view of spinal cord shows the typical dense packing of the nerve and glial cell processes which make up the substance of the nervous tissues. Note the many myelinated axons cut at various angles, as well as the much more numerous unmyelinated cell processes.

In this field, two capillary vessels are seen. The lumen (L) in each case is completely empty, and the vessel appears unusually rounded. This is explained by perfusion fixation. The tissue was fixed by introducing glutaraldehyde fixative under physiological pressure into the vascular system, to ensure optimum tissue preservation. Thus the usual plasma coagulum seen in vessels fixed by immersion is absent in this preparation.

Note that the endothelial lining of the capillaries is continuous, without fenestrations and with little evidence of micropinocytosis. A typical basal lamina surrounds the endothelium, but this is not underlain by a connective tissue space as in other tissues. Instead the neuroglial cells known as astrocytes have numerous foot processes applied to the outer aspect of the capillary, producing an almost complete glial sheath around the blood vessel (*). Other processes, of these and other astrocytes, form a complex network throughout the tissues, surrounding neural processes and synapses and, presumably, providing metabolic stability and physiological insulation as well as simple physical support.

L   Lumen of blood capillary
*   Indicates glial processes surrounding capillary vessel

Tissue   Adult mouse spinal cord, cervical level. Glutaraldehyde and osmium fixation. Uranyl and lead staining.

Magnification   7000 ×

Refer to   Plates 52, 53, 54, 97, 98, 100, 101, 102, 103
Sections 6.4, 11.1

Micrograph by courtesy of H. S. Johnston

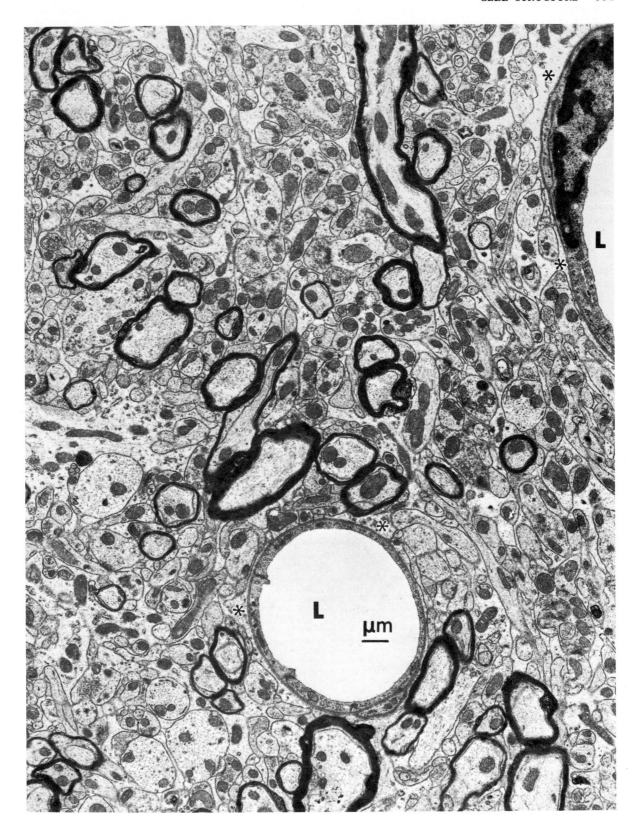

Plate 100a *Unmyelinated nerve*

This is a cross-section of a peripheral nerve, showing parts of several Schwann cell tubes separated by cross-sectioned collagen fibrils (C). A thin process, probably of a fibroblast, lies in the interstitium (F). One myelinated nerve lies in the corner, but the centre of the field is occupied by three unmyelinated axons (A) embedded within the cytoplasm of a single Schwann cell unit, the nucleus of which occupies much of the area (X). In the axons, cross-sectioned neurotubules are present, along with thinner neurofilaments. Each axon is suspended in its position in the Schwann cell by a mesaxon (*) formed by the invaginated Schwann cell surface membrane. Each Schwann cell unit is surrounded by a typical closely fitting basal or external lamina (L).

A   Unmyelinated axons
C   Cross-sectioned collagen fibrils
F   Fibroblast process amongst collagen fibrils
L   Basal lamina of Schwann cell
X   Nucleus of Schwann cell

Tissue   Human peripheral nerve. Glutaraldehyde and osmium fixation, uranyl and lead staining.

Magnification   11 000 ×

Refer to   Plates 28b, 97, 98, 99, 101, 102, 103
             Sections 4.8, 4.9, 11.1

Plate 100b *Myelinated nerve*

This micrograph shows parts of several myelinated nerve axons surrounded by collagen fibrils (C). As above, a single Schwann cell unit fills the main part of the field. In this case the nucleus (X) is also seen, but the cell contains only a single myelinated axon (A). Neurotubules in cross-section (arrows) alternate with filament bundles and there are three mitochondria within the axoplasm. Details of the mesaxon are not clearly visible, but the points of origin of the outer and inner mesaxons have been marked (arrows). Note also an area where the outer layers of the mesaxon are not completely fused with the myelin layer (*). This emphasises the fact that the myelin sheath is a derivative of the cell membrane, formed by the packing and fusion of multiple layers of an elongated mesaxon. A well-defined basal lamina (L) surrounds the Schwann cell, as in the picture above.

A   Axon of myelinated nerve
C   Collagen fibrils in connective tissue
L   Basal lamina of Schwann cell
X   Nucleus of Schwann cell
*   Indicates area of loose membrane packing at edge of myelin sheath
Arrows   Indicate points of origin of outer and inner mesaxons

Tissue   Human peripheral nerve. Glutaraldehyde and osmium fixation, uranyl and lead staining.

Magnification   10 000 ×

Refer to   Plates 28b, 97, 98, 99, 101, 102, 103
           Sections 4.8, 4.9, 11.1

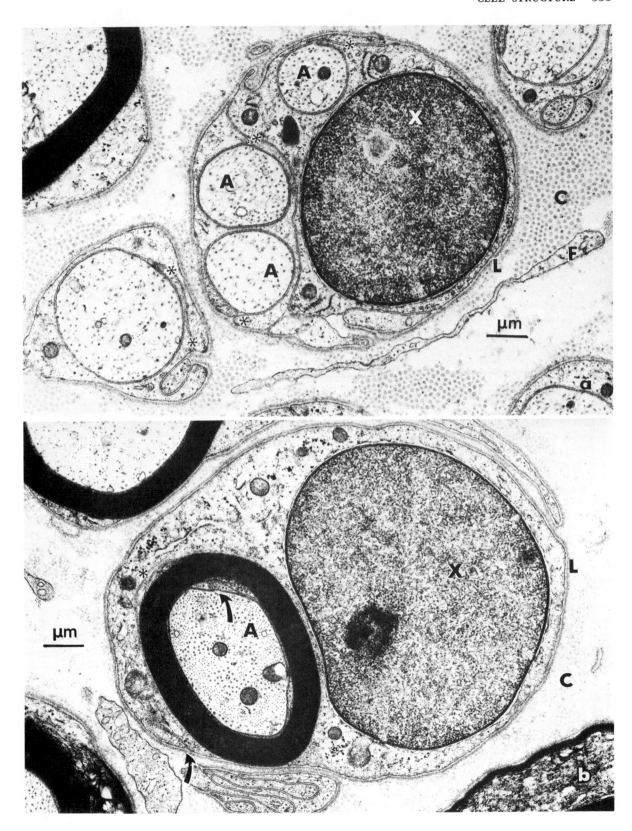

Plate 101a *Myelin sheath*

Myelin is produced by the compact packing of layer upon layer of Schwann cell membrane. This micrograph shows the rim of a myelinated axon (A), in which neurotubules in cross-section are clearly seen. The surrounding Schwann cell cytoplasm (S) contains some particulate glycogen (G). A basal lamina (L) separates the nerve from the surrounding collagen fibrils (C) and adjacent unmyelinated nerves (X). The origin of the mesaxon is clearly seen on the left of the micrograph (*) and the mode of formation of the myelin by close packing of this double layer of membrane material is well shown. From this it is clear that the myelin sheath is a membrane derivative, and is part of the Schwann cell rather than part of the nerve axon itself.

A   Axon of myelinated nerve
C   Collagen fibrils in transverse section
G   Glycogen particles in Schwann cell cytoplasm
L   Basal lamina surrounding Schwann cell
S   Schwann cell cytoplasm
X   Unmyelinated axon
*   Indicate the origin of the mesaxon and its fusion with the outer layer of the myelin sheath

Tissue Human peripheral nerve. Glutaraldehyde and osmium fixation, uranyl and lead staining.

Magnification 37 000 ×

Refer to Plates 22a, 28b, 85a, 97, 98, 99, 100, 102, 103
         Sections 4.11, 11.1

Plate 101b *Myelin sheath*

This higher magnification view of a segment of the myelin sheath is bounded by collagen fibrils (C) on the right, representing the connective tissue, and by part of the nerve axon on the left (A) in which cross-sectioned neurotubules can be seen (T). The basal lamina (L) separates the Schwann cell from the collagen. A rim of Schwann cell cytoplasm (S) contains some glycogen particles (G) and a mitochondrion. Note the fusion of the outer mesaxon with the outer layer of the myelin sheath (*). The myelin has a regular major and minor periodicity. The innermost membrane (arrow) is the limiting membrane of the nerve axon.

A   Axon of myelinated nerve
C   Collagen fibrils in cross-section
G   Glycogen particles
L   Basal lamina of Schwann cell
S   Cytoplasm of Schwann cell
T   Neurotubules in cross-section
*   Indicates fusion of outer mesaxon with myelin sheath

Tissue Human peripheral nerve. Glutaraldehyde and osmium fixation, uranyl and lead staining.

Magnification 54 000 ×

Refer to Plates 22a, 28b, 85a, 97, 98, 99, 100, 102, 103
         Sections 4.11, 11.1

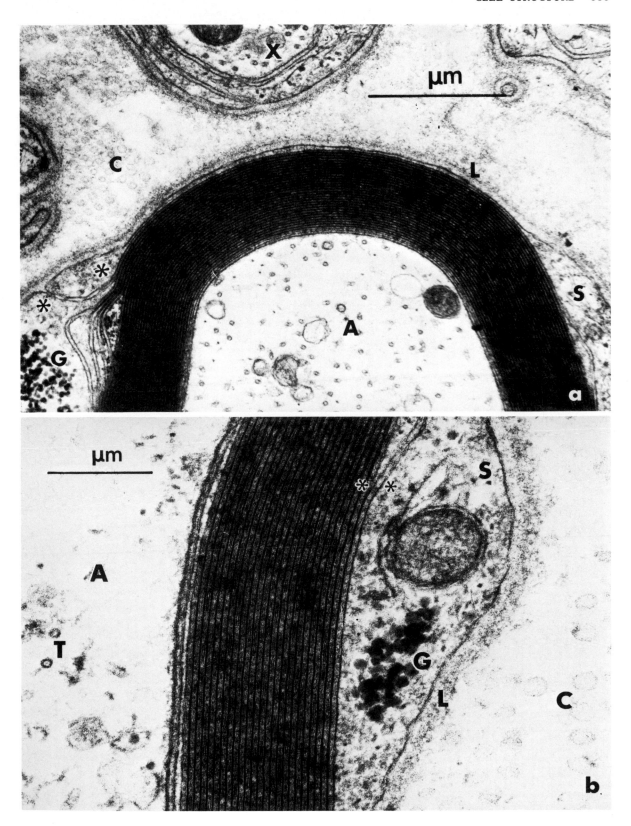

Plate 102a *Nerve ending, salivary gland*

This shows the basal portion of a salivary gland acinar cell (A), with plentiful ribosomes and strands of granular endoplasmic reticulum (R). A basal lamina (L) separates the cell from the underlying connective tissue (C). Part of a nearby capillary vessel (V) displays micropinocytotic activity (arrows). The nuclei (N) of the endothelial cell and the acinar cell are identified. There is a round cytoplasmic process (P) containing numerous small vesicles embedded in a groove at the base of the acinar cell. This is a small unmyelinated nerve terminal, with synaptic vesicles, representing the neuro-effector junction through which nervous control of salivary secretion is achieved.

A	Acinar cell cytoplasm
C	Connective tissue space
L	Basal lamina
P	Cytoplasmic process of nerve terminal
R	Granular endoplasmic reticulum
V	Capillary endothelial cell
Arrows	Indicate micropinocytotic vesicles

Tissue   Rat parotid gland. Glutaraldehyde and osmium fixation, uranyl and lead staining.

Magnification   52 000 ×

Refer to   Plates 70c, 98, 103
Section 11.1

Plate 102b *Nerve ending, salivary gland*

Between two zymogenic cells lies a pale cytoplasmic process, which again has the features of a neuro-effector junction. The nerve process contains numerous empty cytoplasmic vesicles (V) as well as two dense-core vesicles, presumably representing the presence of two different transmitter substances. Several mitochondria (M) are also present. Within the zymogenic cells, much of the granular endoplasmic reticulum is obliquely sectioned (X), causing blurring of the cisternal outlines. Similarly, the obliquely sectioned perinuclear cisterna (C) is not clearly visualised.

C	Obliquely sectioned perinuclear cisterna, imperfectly visualised
M	Mitochondria aggregated within nerve terminal
V	Synaptic vesicles
X	Obliquely sectioned cisternae of granular endoplasmic reticulum

Tissue   Rat parotid gland. Glutaraldehyde and osmium fixation, uranyl and lead staining.

Magnification   37 000 ×

Refer to   Plates 70c, 98, 103
Section 11.1

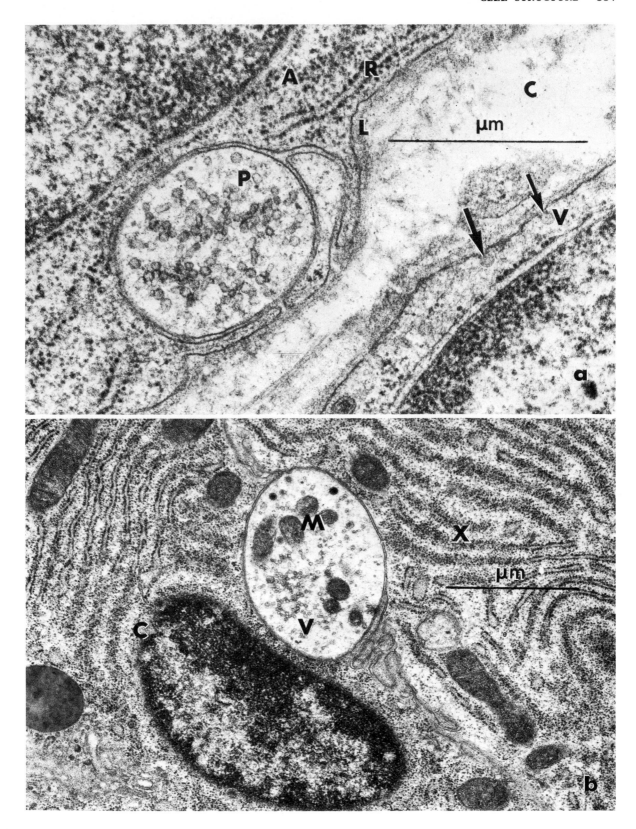

Plate 103 *Motor end-plate, skeletal muscle*

This is a low-magnification micrograph showing a motor nerve terminal entering into contact with a skeletal muscle cell at the motor end-plate. The muscle cell occupies the main part of the field while the nerve terminal reaches it from the connective tissue space in the upper part of the micrograph.

The motor nerve (Ne) enters a depression on the surface of the muscle cell at the point marked X-X and branches to form terminals in close contact with the infolded muscle cell surface. The small branches of the nerve, two of which are seen in this part of the end-plate, are distinguished by the presence of numerous small vesicles (V), which contain the stored transmitter substance, acetylcholine, released from the nerve terminals on stimulation. The surface membrane of the nerve terminal is not folded or specialised in any obvious way.

The corresponding surface of the muscle cell is thrown into elaborate folds around each nerve terminal. The space between the nerve and the muscle cell is filled with material continuous with the external lamina or basal lamina of the muscle cell. This homogeneous material also fills the clefts formed by the infolding of the muscle surface, termed the subsynaptic gutter (SG). This specialisation of the muscle cell membrane presents an increased surface area at this important interface.

The myofibril component of the muscle cell is not seen in this plate, but one muscle cell nucleus is present in its typical position close to the surface of the cell (N). Adjacent to this are found many mitochondria (M) and other sarcoplasmic components.

CT	Connective tissue space
M	Mitochondrion
N	Nucleus of muscle cell
Ne	Axon of motor nerve
SG	Subsynaptic gutter
V	Vesicles in nerve terminals
X-X	Indicates the point at which the nerve enters into close contact with muscle

Tissue   Rat skeletal muscle. Osmium fixation, phosphotungstic acid staining.

Magnification   17 000 ×

Refer to   Plates 84, 85, 86, 87, 98, 102
Sections 10.1, 11.1

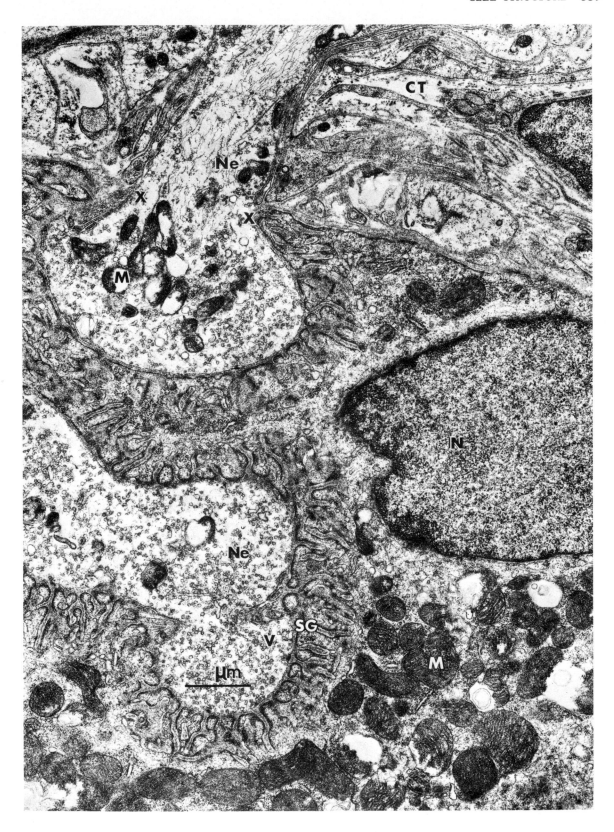

Plate 104a *Light-sensitive membranes, chloroplast*

This micrograph shows part of a plant cell. The cell wall (W), made of cellulose, forms a thick layer external to the cell membrane. The cytoplasm appears as a rim surrounding a large intracellular vacuole (V). The main structural feature of the cell is the chloroplast, in which the membrane-associated pigment chlorophyll is located. The chloroplast, like the mitochondrion, has two distinct membranes (M) and contains internal stacks of membrane lamellae. The grana of the chloroplast (G) are areas where these lamellae are closely packed. The background or stroma (S) of the chloroplast contains finely granular material and starch granules (arrow). Within the chloroplast the complex processes of photosynthesis take place under the stimulus of light.

G   Grana of chloroplast
M   Double limiting membrane of chloroplast
S   Stroma of chloroplast
V   Intracellular vacuole
W   Cell wall
Arrow   Indicates starch granule

Tissue Bean leaf. Glutaraldehyde and osmium fixation, uranium and lead staining.

Magnification 27 000 ×

Refer to Plates 7, 22, 23
          Sections 2.2, 4.5, 11.2

Plate 104b *Light-sensitive membranes, retinal rod*

Two parallel retinal rod outer segments are seen here, separated by a narrow space. The entire structure is filled with closely packed parallel membrane pairs or discs (arrows), with which the visual pigment is associated. Many of the membranes are cut slightly obliquely (O), blurring their profiles, but at some points the individual membrane pair forming a single disc can be clearly distinguished. Similarly the limiting membrane of the outer segment is clearly seen at some points but is inapparent owing to oblique cut elsewhere (X). This complex lamellar structure is designed to receive the stimulus of light and transform it into a neural signal which is sent to the brain.

O   Obliquely sectioned areas of membranes
X   Obliquely cut limiting membrane
Arrows   Indicate membrane pair forming a single disc

Tissue Rat retina. Glutaraldehyde and osmium fixation, uranium and lead staining.

Magnification 32 000 ×

Refer to Plates 93, 94
          Sections 4.10, 10.2, 11.2

Micrograph by courtesy of J. Shaw Dunn

Plate 104c *Retinal rod, outer and inner segments*

Parts of two outer segments are present, with their characteristic stacked lamellae. The connection (C) is seen between one of these outer segments (OS) and its inner segment (IS) which contains the other cytoplasmic organelles. This connection has some features of a cilium, including an organised central core. The component tubules of the core (T) originate in a centriole, which is not clearly sectioned in this case. The abundant mitochondria usually seen in the inner segment have not been included in this micrograph.

C   Connection between outer and inner segments
IS   Inner segment
OS   Outer segment
T   Tubular core similar to that of cilium

Tissue Rat retina. Glutaraldehyde and osmium fixation, uranium and lead staining.

Magnification 41 000 ×

Refer to Plates 93, 94
          Sections 4.10, 10.2, 11.2

Micrograph by courtesy of J. Shaw Dunn

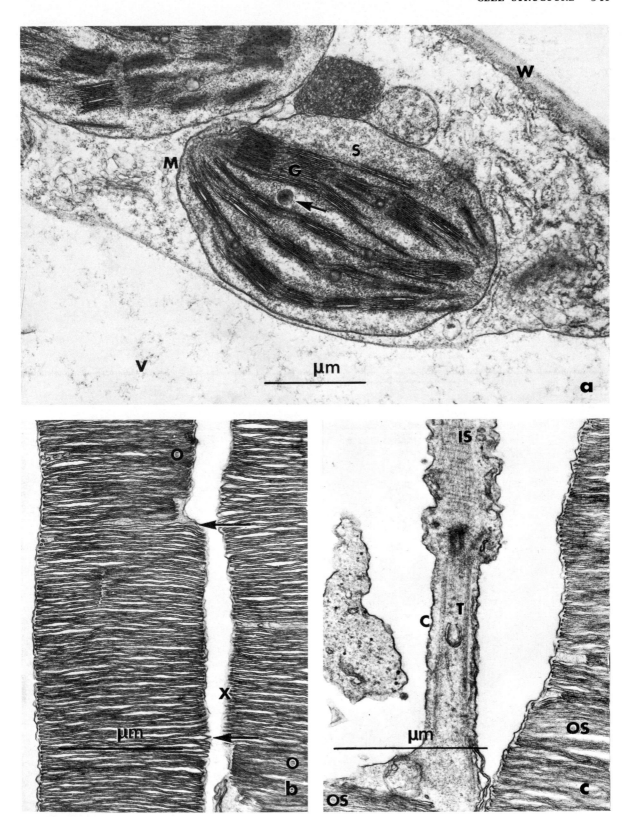

PROBLEM CELLS

Plate 105 *Langerhans cell*

This micrograph shows part of a Langerhans cell (L) between epidermal keratinocytes (K). The Langerhans cell is quite distinct from the keratinocytes surrounding it. The desmosomes seen here (D) are linking adjacent keratinocytes; the Langerhans cell itself contains no tonofilaments and is not attached by desmosomes to surrounding cells. These features suggest that this is in some way an 'interloper' cell type, in the same way that intra-epithelial lymphocytes are immigrants in many epithelial sheets. In fact, recent evidence suggests that the Langerhans cell is a type of intra-epidermal macrophage, the function of which may be related to the recognition of foreign antigens. Perhaps, for example, this cell recognises and picks up antigenic molecules which penetrate the skin. It may then migrate through the dermis and along lymphatics to the regional lymph nodes, where it could pass the antigen to lymphoid cells for defensive action. Although this is not yet certain, such a hypothesis would account for the observed distribution of Langerhans cells in the epidermis, dermis, lymphatic vessels and lymph nodes.

The distinctive ultrastructural feature of this cell is the Langerhans granule, many of which are seen here (G). Most of these are straight rod-shaped structures, with a characteristic central density which shows a periodic cross-striation. Only one granule, the 'handle' of which is obliquely sectioned and thus blurred, shows the terminal dilatation (*) which has given rise to the term 'tennis racquet' granules. The part played by these granules in cell function remains obscure. It is not even certain how or where the granules are formed, although they can be observed in continuity with the cell membrane on occasion.

This distinctive granule, however, provides pathologists with a valuable marker for cells of this family in a poorly understood group of diseases known as histiocytosis X, characterised by cellular infiltrates with macrophage-like characteristics, in which Langerhans granules are typically found.

D   Desmosomes linking adjacent keratinocytes
G   Langerhans granules
K   Keratinocyte cytoplasm
L   Langerhans cell cytoplasm
*   Indicates a 'tennis racquet' granule, with a terminal dilatation. The 'handle' is blurred through an oblique sectioning effect

Tissue   Human epidermis. Glutaraldehyde and osmium fixation, uranyl and lead staining.

Magnification   41 000 ×

Refer to   Plates 58a, 73, 74, 75
         Sections 7.1, 8.3, 12.1

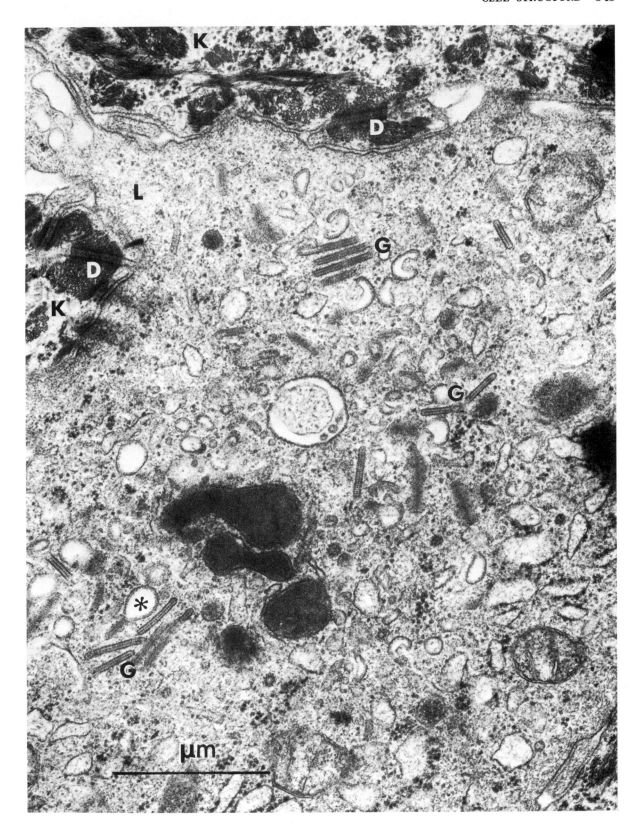

ULTRASTRUCTURAL PATHOLOGY

Plate 106a *Intracytoplasmic virus particles*

This is a cell from a mouse salivary gland tumour, showing numerous cytoplasmic virus particles lying close to the nucleus (N), which appears unaffected. The virus particles within the cytoplasm (V_1) are gathered around several large intracytoplasmic vacuoles, into which they seem to be pushing. The particles which lie inside the vacuoles (V_2) have acquired an extra envelope, presumably derived from the limiting membrane of the vacuole. Similar virus particles are found in various animal tumours. The electron microscope plays a valuable part in virus research since it is the only instrument which is capable of resolving and directly imaging the structural details of particles of viral dimensions.

N   Nucleus of tumour cell
V_1 Intracytoplasmic virus particles
V_2 Virus particles within cytoplasmic vacuole

Tissue Mouse salivary tumour. Glutaraldehyde and osmium fixation, uranyl and lead staining.

Magnification   50 000 ×

Refer to   Plate 102b
             Sections 13.1, 13.2

Plate 106b *Intranuclear virus particles*

Viruses of the herpes family are typically found within the nucleus, as well as in the cytoplasm of infected cells. Often the viruses gather in such numbers that they form an intranuclear inclusion body, which can be clearly seen by light microscopy. This micrograph shows part of such an intranuclear inclusion from a pancreatic acinar cell in a case of cytomegalic inclusion body disease in a baby. The virus responsible, known as CMV for short, produces large cells with dilated nuclei, in which prominent owl's eye inclusion bodies are found. Electron microscopy shows numerous circular virus particles of uniform shape and size, often containing a central dense nucleoid. These features are typical of a virus of the herpes family. The dense clumps of material around which the particles are gathered are the residual chromatin of the nucleus. Electron microscopy can be of value to the pathologist in the accurate diagnosis of human disease, whether by confirming the presence of virus particles in cases of suspected infection, or by identifying the type of virus by particular structural features, such as its size and shape.

Tissue Human pancreas. Formaldehyde and osmium fixation, uranyl and lead staining.

Magnification   70 000 ×

Refer to   Plates 10a, 13
             Sections 13.1, 13.2

Micrograph by courtesy of A. A. M. Gibson

Plate 106c *Negatively stained virus*

This preparation was made by mixing a drop of fluid scraped from a skin lesion, with a drop of heavy metal stain. The mixture was then spread on a thin film and examined under the electron microscope, without embedding or sectioning. Viruses present in the fluid are outlined by the surrounding dense stain, which penetrates cracks between fine surface details and reveals the minute structure of the particle. This simple and rapid technique is known as negative staining. It is an excellent method for identifying viruses if they are present in sufficient numbers.

In this case, the skin lesion was from the hand of a sheep farmer. A viral skin infection known as orf can be caught from sheep, but it is difficult to distinguish from other types of skin infection. Electron microscopy, as described above, can give a quick and certain answer to such a question, since the orf virus, a member of the pox virus family, has a highly characteristic structure. The virus is elongated, with rounded ends, and has a criss-cross basket-weave surface pattern. In the days before smallpox was eradicated, this technique of negative staining was used on occasion to distinguish between smallpox, caused by a pox virus, and chickenpox, caused by a virus of the herpes family. The rapid answer provided by electron microscopy was of importance in deciding on the necessary public health measures, should a case of smallpox be discovered.

Tissue Negatively stained specimen of scrapings from a skin lesion in a case of orf.

Magnification   70 000 ×

Refer to   Plates 116d, 116e
             Sections 13.1, 13.2

Plate 106d *Giardia lamblia*

This micrograph shows a view of the surface of small intestine, with longitudinally sectioned microvilli (M). The intestinal 'fuzzy coat' or glycocalyx forms a dense thin layer immediately above the tips of the microvilli (G). Within the lumen (L) there lies a large cell with a highly characteristic internal structure. Several centrioles are seen (C), although the flagella to which they are connected are not present in this plane of section. Other features include vesicles, ribosomes and surface specialisations. This is an intestinal parasite known as Giardia lamblia, which inhabits the lumen and adheres to the villous surface, sometimes causing mild intestinal upsets and even malabsorption.

C   Centrioles
G   Glycocalyx, or cell coat
L   Lumen of intestine
M   Microvilli of absorptive cell

Tissue Human small intestinal biopsy in a case of Giardia lamblia infection. Glutaraldehyde and osmium fixation, uranyl and lead staining.

Magnification   26 000 ×

Refer to   Plates 7, 14, 30, 31, 67b, 76b, 113
             Sections 13.1, 13.2

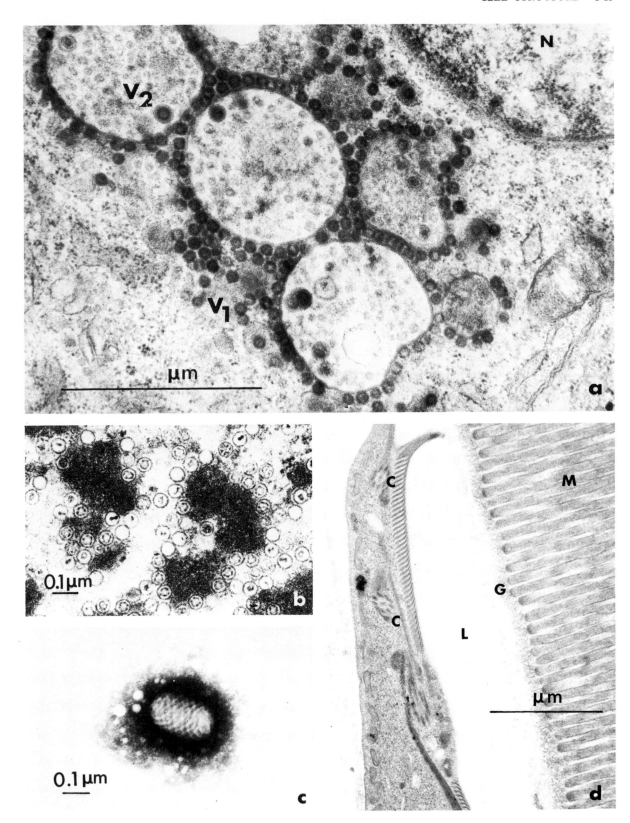

Plates 107a, 107b *Malignant melanoma*

Electron microscopy is of practical value in the investigation of various types of human tumour. In this case, the type of tumour could not be identified with confidence by the pathologist, using only light microscopy. The tumour cells were elongated and closely packed, without distinguishing features. In Plate 107a, the low magnification electron micrograph shows parts of several cells. The nucleus (N) of one has a large nucleolus. The cytoplasm is relatively unremarkable, apart from the presence of a few small granules (arrows). At higher magnification, in Plate 107b, these granules are seen to have the highly distinctive structural features of premelanosomes, or unmelanised melanin granules. The recognition of this pattern of subcellular differentiation allows the pathologist to reach the correct diagnosis of non-pigmented or amelanotic melanoma.

N   Nucleus of tumour cell, containing large nucleolus
Arrows   Indicate small cytoplasmic granules which prove to be premelanosomes at higher magnification

Tissue Biopsy of tumour in lymph nodes in occipital region. Glutaraldehyde and osmium fixation, uranyl and lead staining.

Magnification   Plate 107a 13 000 ×
                  Plate 107b 94 000 ×

Refer to   Plates 8a, 74, 75b
             Sections 8.4, 13.1, 13.2

Plate 107c *Squamous carcinoma*

The histopathologist, using only light microscopy, can sometimes find it difficult to distinguish between certain poorly differentiated tumours. In some cases the tumour may have arisen from melanocytes, while in other cases it may be of squamous cell origin. Such tumours can, on occasion, look alike under the light microscope.

This micrograph is from a deposit of secondary tumour in a case where the primary source of the tumour was not known. Histological examination of this poorly differentiated tumour failed to reveal its origins, with some pathologists identifying it as a melanoma and others as a squamous cell cancer. This micrograph was taken from remnants of the tumour which had been stored in formalin fixative. The preservation of cellular detail is not very good, but it is sufficient to allow the recognition of desmosomes (D) and tonofilament bundles (T), which provide excellent evidence that this tumour is a squamous cell cancer. There were no melanosomes to be seen under the electron microscope.

D   Desmosomes
T   Tonofilament bundles

Tissue Secondary tumour in skin of chest wall. Formalin and osmium fixation, uranyl and lead stain.

Magnification   37 000 ×

Refer to   Plates 1, 5, 6, 73, 75a
             Sections 2.2, 4.8, 13.1, 13.2

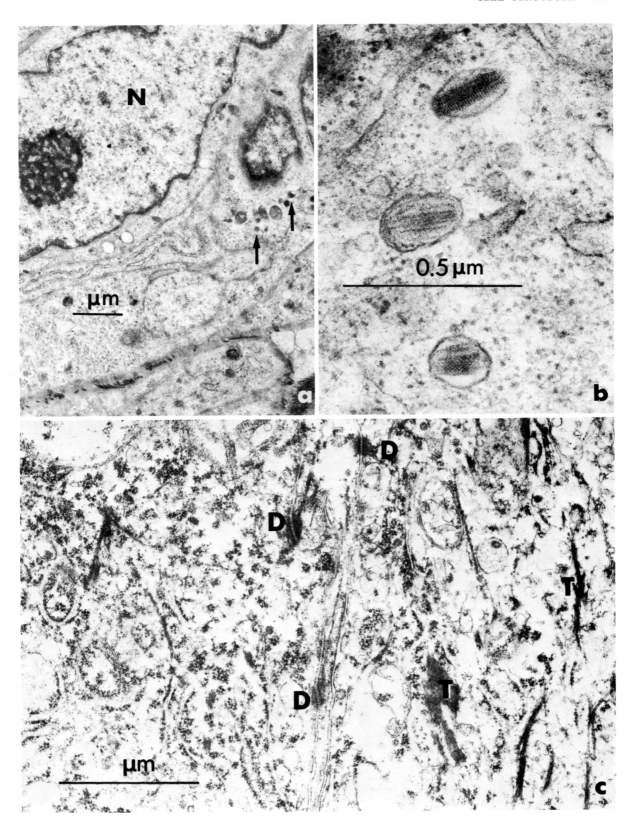

Plate 111 *Glycogen storage disease*

This micrograph shows liver cells around a sinusoid. Two liver cell nuclei are seen (N) and a Kupffer cell (K) can be identified in the sinusoid, the lumen of which is plugged by a small aggregate of platelets (P), forming a thrombus. The cytoplasm of the liver cells is largely filled by masses of pale finely granular material (G). This is an abnormal accumulation of cytoplasmic glycogen, which comes about because of a lack of an essential enzyme in the metabolic pathway of glycogen. This causes liver enlargement and abnormality of function. The enzyme defect is due to a genetic abnormality, which affects every cell of the body, but the liver is particularly involved because of its central role in glycogen metabolism.

G   Masses of stored cytoplasmic glycogen in hepatocytes
K   Kupffer cell in sinusoid
N   Nucleus of hepatocyte
P   Platelet aggregate in sinusoid lumen

Tissue   Human infant liver, glycogen storage disease. Glutaraldehyde and osmium fixation, uranyl and lead staining.

Magnification   4000 ×

Refer to   Plates 22a, 33, 67c, 78, 85a, 101b
Sections 4.11, 8.7, 13.1, 13.2

Micrograph by courtesy of A. A. M. Gibson

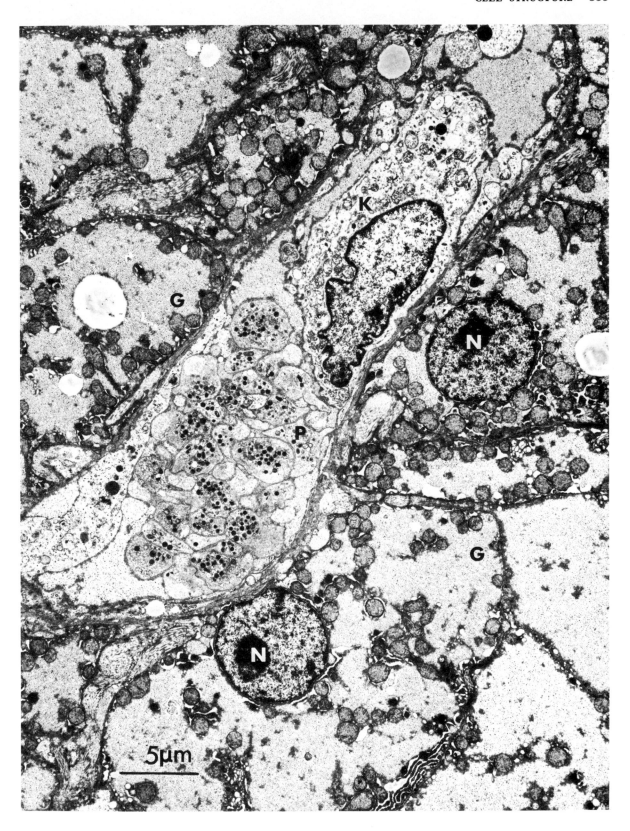

5μm

Plate 112a *Lipid storage disease*

In this type of storage disease, the enzyme lacking from the cells of the body is a gangliosidase, required to catabolise complex lipids found predominantly in the brain. For this reason, the brain is the organ most affected by the metabolic disruption caused by the accumulation of lipid. This micrograph of part of a brain cell shows how the cytoplasm is filled with large complex laminated structures. The distinctive pattern is due to the tendency of complex lipids to form lamellar membrane-like structures. These are essentially residual bodies, containing lipid which cannot be broken down further, for want of the necessary gangliosidase. The children affected by this disease become blind and mentally defective in infancy or early childhood, and eventually die.

Tissue   Human brain biopsy, Tay-Sachs disease. Glutaraldehyde and osmium fixation, uranyl and lead staining.

Magnification   70 000 ×

Refer to   Plates 97, 98, 99
Sections 13.1, 13.2

Plate 112b *Amyloid disease*

In a disease known as amyloidosis, a distinctive waxy deposit of amorphous material called amyloid is laid down in connective tissues, often in response to some chronic over-stimulation of the immune system. It occurs, for example, as a complication of chronic infections such as tuberculosis, and in the chronic inflammatory autoimmune disease, rheumatoid arthritis. Amyloidosis can affect many organs and tissues, but its most serious effects are often on the kidneys, where it can cause the nephrotic syndrome, through severe loss of serum proteins in the urine.

The histological indentification of amyloid depends on various semi-specific staining reactions with a range of dyestuffs, the results of which can on occasion be capricious. When problems are encountered, the electron microscope can help by showing the characteristic ultrastructural features of amyloid, which exists as a dense mat of fibrillar material quite unlike any normal connective tissue component. This micrograph shows

the dense bundles of amyloid fibrils (F) beside a cell process.

F   Amyloid fibrils

Tissue   Human amyloidosis.

Magnification   24 000 ×

Refer to   Plates 79, 80
Sections 13.1, 13.2

Plate 112c *Whipple's disease*

This condition is a rare and formerly mysterious disease, predominantly affecting the small intestine, causing malabsorption. It is characterised by accumulations of large abnormal macrophages containing distinctive PAS positive granules, throughout the lamina propria of the gut.

Early views of this disease assumed that it was some type of storage disorder, or perhaps a primary malabsorptive disease. Electron microscopy, however, has shown that the lamina propria in Whipple's disease contains numerous small rod-shaped bacilli, which are taken up by macrophages and only partially destroyed. The macrophage inclusions are, therefore, phagolysosomes, or secondary lysosomes, containing masses of residual bacterial debris. The organism has not yet been positively identified, and the possible contribution of a host factor, such as an immunological weakness or a macrophage defect, has not been fully evaluated. Whipple's disease, however, has been brought a little closer to a solution by the electron microscopic identification of a probable bacterial cause. Antibiotic treatment nowadays usually brings about a cure. This micrograph shows an area of macrophage cytoplasm containing two large membrane-limited inclusions in which the remains of numerous rod-shaped bacteria (B) are accumulated.

B   Bacterial remnants in macrophage

Tissue   Human small intestinal biopsy, Whipple's disease. Osmium fixation, uranyl and lead staining.

Magnification   33 000 ×

Refer to   Plates 24, 25, 63, 67, 76b
Sections 7.1, 13.1, 13.2

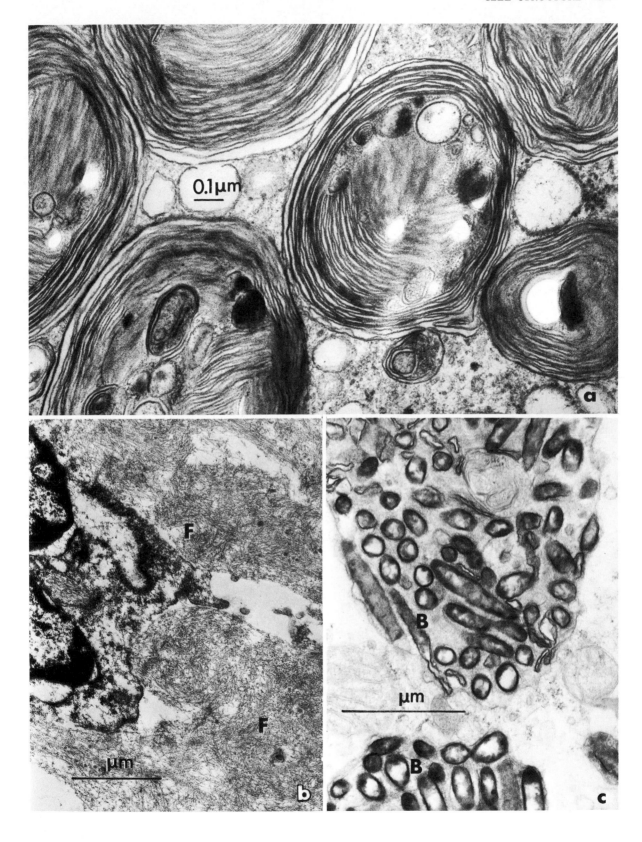

Plates 113a, 113b *Absorption and malabsorption*

These two plates contrast the healthy enterocytes in Plate 113a, with abnormal enterocytes in Plate 113b. The abnormality present in the lower micrograph is a consequence of a probable allergic response to the protein found in wheat, known as gluten. Patients with this abnormal responsiveness to gluten suffer from malabsorption and diarrhoea, which can be cured by the removal of gluten from the diet. This disease is often known as gluten-enteropathy, or coeliac disease. The intestinal epithelium is severely affected in coeliac disease, with loss of villi and the formation of an abnormal 'flat' mucosa. The individual enterocytes show various abnormalities including cytoplasmic vacuolation (V), the presence of abnormal numbers of lysosome-like dense bodies (D) and marked shortening of the microvilli which form the striated border (B). In addition, there is a relative increase in the numbers of intraepithelial lymphocytes (L), suggesting an abnormality of immunological mechanisms. By comparison, the enterocytes above show orderly cytoplasmic organelles, normal numbers of lysosomes and a healthy microvillous border.

B   Striated border composed of microvilli
D   Dense bodies of lysosome type
L   Intraepithelial lymphocyte
V   Cytoplasmic vacuolation

Tissue Human small intestine. Glutaraldehyde and osmium fixation, uranyl and lead staining.

Magnification   Plate 113a 9000 ×
                   Plate 113b 7000 ×

Refer to   Plates 7, 14, 27a, 47, 48
              Sections 6.1, 13.1, 13.2

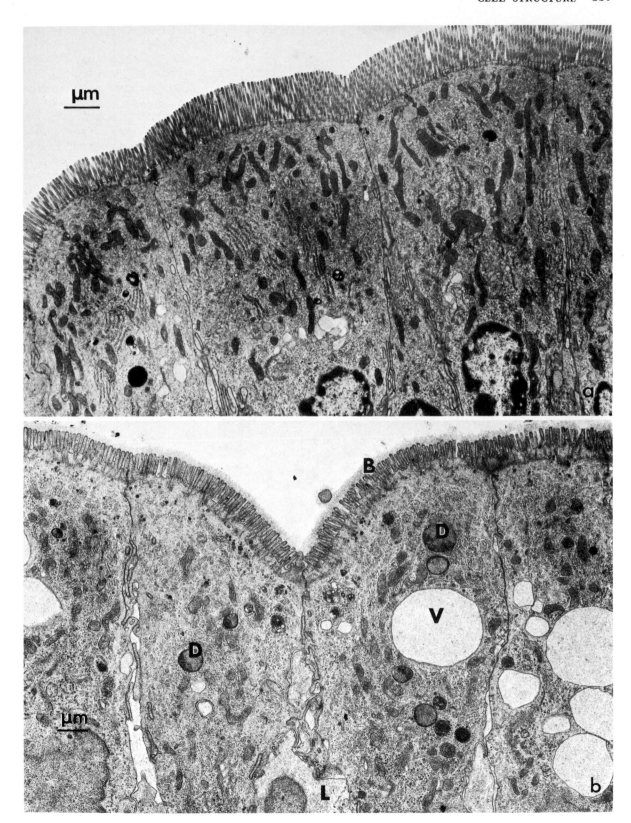

Plates 114a, 114b, 114c, 114d *Colonic mucosa in health and disease*

These four plates provide an indication of how mucosal surface morphology may vary in different diseases. They are all scanning electron micrographs of human large bowel mucosa.

Plate 114a shows the pattern of normal mucosa at moderate magnification, with several orifices of crypts of Lieberkühn (C) surrounded by collars of cells which form crypt units. The distinctive pattern of the normal mucosa can also be seen at lower magnification in Plate 49. The very different pattern of the small intestine can be seen, for comparison, in Plates 47 and 48.

Plate 114b shows the irregular folded surface of a simple mucosal tumour. The abnormal epithelial surface is thrown into irregular ridges which lack the orderly pattern of the normal mucosa.

Plate 114c shows an area of intact mucosa in ulcerative colitis. Strands of abnormal surface exudate, consisting of fibrin, polymorphs and red cells, cover part of the mucosa (E) but elsewhere the epithelial cell surfaces can be seen, with an opening of a crypt (C). Again, the normal crypt unit pattern has been lost in this abnormally flattened mucosa.

Plate 114d, finally, shows a grossly irregular mucosal surface, which was not expected from histological examination. This was a case of mild inflammation of the bowel. The patient, however, had received, several years previously, a course of radiotherapy for pelvic cancer. Some damage to the bowel mucosa is inevitable in such cases. It remains uncertain whether the mucosal changes seen here are simply due to inflammation, or perhaps reflect the damage caused to the bowel by previous radiation.

These micrographs show that scanning electron microscopy is able to demonstrate striking mucosal changes in various diseases of the large bowel. It is too early yet to say whether any of these images will prove to be of practical value to the diagnostic pathologist, but there can be no doubt that they help to provide a fuller and more graphic presentation of the morphology of disease than can be obtained by histological means alone.

C   Opening of crypt of Lieberkühn
E   Exudate covering part of mucosa in ulcerative
           colitis

Tissue   a. Normal human rectum
           b. Human neoplastic polyp
           c. Human ulcerative colitis
           d. Human rectal mucosa following radio-
              therapy

Glutaraldehyde and osmium fixation, tissue dried by critical point method. Gold coated for SEM.

Magnification   Plate 14a 2200 ×
                  Plate 114b 160 ×
                  Plate 114c 130 ×
                  Plate 114d 1300 ×

Biopsies by courtesy of A. Wong

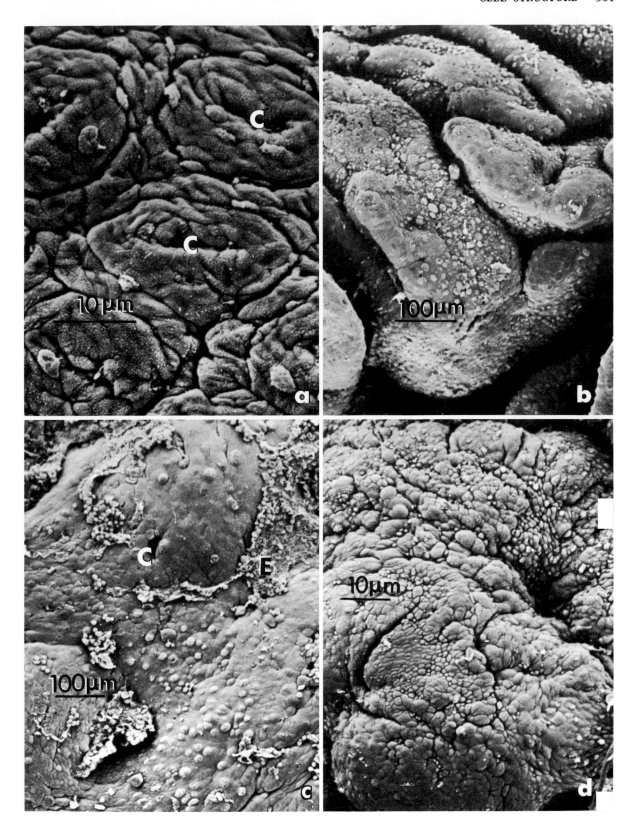

TECHNIQUES AND APPLICATIONS

Plate 115a *Tracer technique, phagocytosis*

This is a low-magnification micrograph of a cell suspension obtained from the peritoneal cavity of a mouse. The cells, two of which are macrophages, have been incubated for one hour in a suspension of colloidal thorium (Thorotrast). Aggregates of finely divided metallic tracer (T) are seen in places between the cells. At several points tracer particles are seen close to the surface of one macrophage (X). Most of the tracer has, however, already been ingested by these cells and appears in large phagocytic vacuoles in their cytoplasm (V). These vacuoles represent an attempt by the cells to digest the foreign material. They can be shown to contain hydrolytic enzymes including acid phosphatase and are thus identified as secondary lysosomes. The primary lysosomes of the macrophages are too small to be seen clearly at this magnification. The Golgi apparatus (G) and granular endoplasmic reticulum (ER) of the macrophage can be distinguished. The cell surface is irregular in outline. The nuclei of these cells (N) show the pattern of euchromatin and heterochromatin particularly clearly. As well as the two macrophages, parts of several lymphocytes (L) and a single degenerate polymorph (P) can be seen in this field. These cells have not ingested the tracer.

ER	Granular endoplasmic reticulum
G	Golgi apparatus
L	Lymphocyte
N	Nucleus of macrophage
P	Degenerate polymorph
T	Particulate tracer between cells
V	Phagocytic vacuole
X	Tracer at macrophage surface

Tissue Mouse peritoneal cells. Glutaraldehyde and osmium fixation, uranium and lead staining.

Magnification 20 000 ×

Refer to Plates 7, 24, 25, 33b, 65, 67
Sections 2.2, 4.6, 7.1, 14.6

Plate 115b *Ruthenium red stain, cell coat*

This micrograph shows parts of two lymphoid cells from mouse peritoneal cavity. The nuclei (N) of the cells can be seen, along with the Golgi apparatus (G) of one cell. During processing the cells were exposed to ruthenium red, a dye which becomes attached to the surface coat or glycocalyx and which takes up osmium preferentially during post-fixation, thus marking the position of the cell coat by a dense staining reaction. A thin layer of increased density appears on the exposed surfaces of the cells (arrows). Where the cells lie in contact the staining reaction forms a particularly prominent layer between them.

G	Golgi apparatus
N	Nucleus
Arrows	Indicate thin cell coat stained by ruthenium red at the exposed cell surface

Tissue Mouse peritoneal cells. Glutaraldehyde and osmium fixation, uranium and lead staining following ruthenium red.

Magnification 31 000 ×

Refer to Plates 7, 24, 25, 33b, 65, 67
Sections 2.2, 4.6, 7.1, 14.6

Plate 115c *Ruthenium red stain, cell coat*

This plate shows a high-magnification view of the surface of a macrophage seen in a 'glancing' section, which emphasises the surface irregularity of the cell. The cell was exposed to ruthenium red in order to stain the surface coat or glycocalyx. The densely stained material, patchily distributed, lies external to the outer lamina of the trilaminar cell membrane, details of which can be made out at the point arrowed.

Arrow	Indicates glycocalyx external to trilaminar cell membrane

Tissue Mouse peritoneal macrophage. Glutaraldehyde fixation, uranium and lead staining following ruthenium red.

Magnification 59 000 ×

Refer to Plates 7, 24, 25, 33b, 65, 67
Sections 2.2, 4.6, 7.1, 14.6

Micrographs by courtesy of I. A. Carr

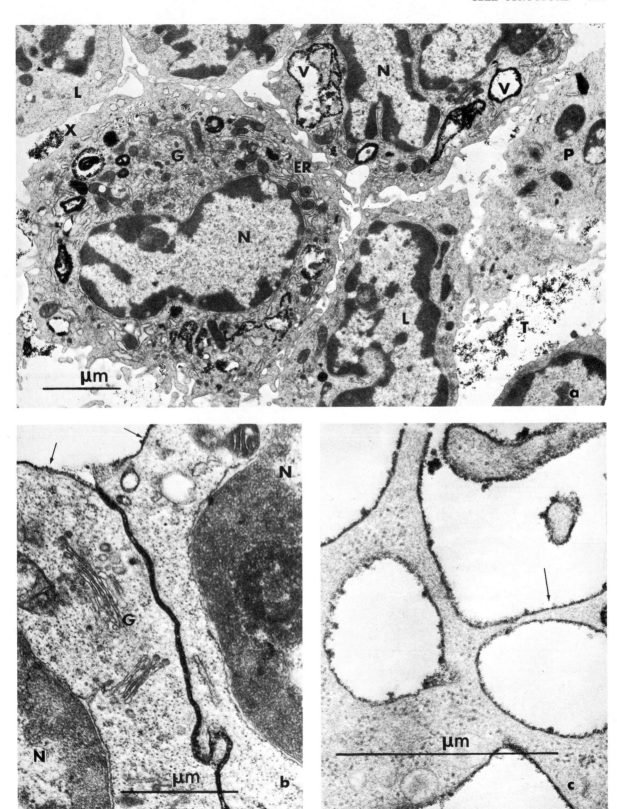

Plate 116a *Autoradiography, RNA synthesis*

This micrograph shows the nucleus of a fibroblast grown in culture and labelled for one hour in a medium containing tritium-labelled uridine. This metabolite is incorporated in newly synthesized RNA. By autoradiography the site of synthesis can be located by the position of silver grains (G) produced by the radioactive label. Most of the grains in this case are seen over areas of euchromatin, in keeping with the view that these are the regions involved in the active transcription of genetic information. The techniques of autoradiography involve spreading a thin layer of photographic emulsion over the section and leaving it in contact for up to 3 months. The silver grains, produced by the foci of radioactivity, are made visible by conventional photographic development and fixation. The gelatine is then removed from the emulsion and the section examined by electron microscopy. The silver grains are seen as coiled threads of high density. This technique allows the subcellular localisation of radioactive tracers to be achieved. The positions of the silver grains mark the location of the label in relation to the cell components as seen by the electron microscope.

G   Silver grains

Tissue   Fibroblast from cell culture labelled with tritiated uridine. Glutaraldehyde fixation, uranium and lead staining.

Magnification   29 000 ×

Refer to   Plate 10
             Sections 3.1, 14.6

Micrograph by courtesy of I. More

Plate 116b *Shadowing, DNA molecule*

This dense thread is a single molecule of DNA isolated from an adenovirus and spread on a carbon film. The DNA was prepared for examination using the Kleinschmidt protein monolayer technique. The contrast seen in this micrograph was introduced by shadowing the specimen with a fine layer of carbon and platinum, evaporated at high temperature under vacuum. The specimen was rotated during shadowing to build up an even layer of the coating. By using variations of this type of technique, much information can be obtained, concerning the nature of macromolecules.

Specimen   Adenovirus DNA. Carbon platinum shadowing.

Magnification   36 000 ×

Refer to   Plate 10
             Sections 3.1, 14.6

Micrograph by courtesy of E. A. C. Follet

Plate 116c *Special techniques: shadowing*

In this preparation, coarse aggregates of chromatin appear as thick threads spread on a thin carbon film. The dense circular

profiles are spheres of latex rubber used as a marker to determine the direction of the shadowing. In this case shadowing was carried out from two different directions. The latex spheres show this double shadowing effect clearly. From the length of the shadow and the diameter of the latex sphere it is possible to calculate the precise angle of shadowing. Conversely given a known angle of shadowing, the height of any detail of the specimen can be calculated by measurement of the shadow which it casts.

Arrows   Indicate the two directions of shadowing

Specimen   Isolated chromatin and latex spheres. Double shadowed.

Magnification   70 000 ×

Refer to   Plate 10
             Sections 3.1, 14.6

Plate 116d *Negative staining, viruses*

This preparation shows a number of human wart virus particles spread on a carbon film. They are rounded structures, 55 nm in diameter, with surface subunits which are demonstrated by the negative staining technique. The layer of stain forms the dense background of the micrograph and the stain penetrates between and outlines the surface details of the viruses. The virus particles themselves are not stained and thus appear as 'empty' areas in the otherwise dense preparation.

Specimen   Human wart virus. Negatively stained with phosphotungstic acid.

Magnification   150 000 ×

Refer to   Plate 10
             Sections 3.1, 14.6

Micrograph by courtesy of E. A. C. Follett

Plate 116e *Negative staining, viruses*

This preparation shows several of the characteristic rod-shaped units of tobacco mosaic virus. A fine periodicity can be made out and there is a central channel in each virus rod, along which the stain has penetrated. The external boundaries of the virus are also outlined. These rods are sometimes broken into smaller fragments, as can be seen here. One such fragment apparently lies vertical to the carbon film on which the preparation has been spread. It appears as a round structure with a central hole.

Arrow   Indicates 'end on' virus subunit

Specimen   Tobacco mosaic virus. Negatively stained with uranyl acetate.

Magnification   140 000 ×

Micrograph by courtesy of E. A. C. Follett

Refer to   Plates 13, 106
             Sections 13.1, 13.2, 14.6

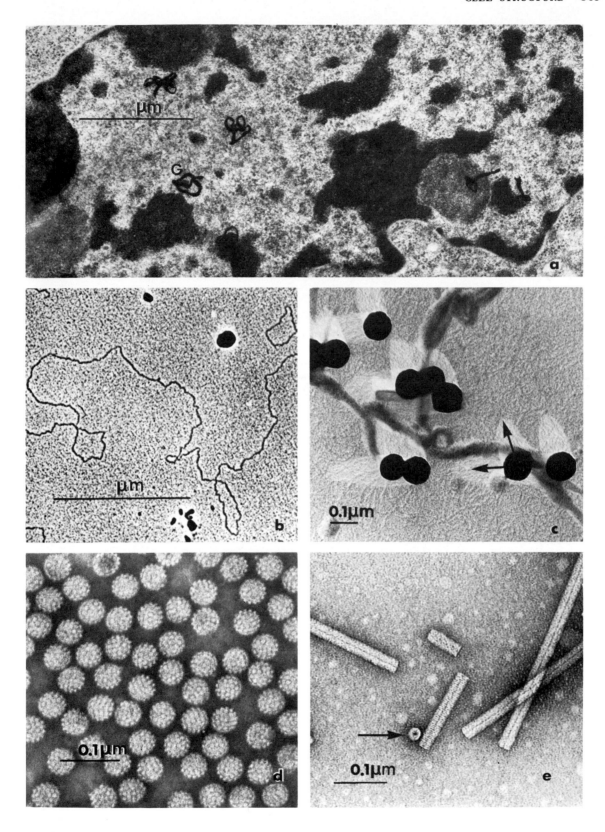

Further reading

SCIENTIFIC JOURNALS

Almost all journals in the biomedical field now regularly contain articles which refer to the findings of electron microscopical investigations. There are, however, various specialist journals, especially in the fields of cell biology, anatomy and pathology, which lay particular emphasis on ultrastructural studies. These include the following:

American Journal of Anatomy
American Journal of Pathology
Anatomical Record
Cell & Tissue Research
International Review of Cytology
Journal of Cell Biology
Journal of Cell Science
Journal de Microscopie
Journal of Microscopy
Journal of Submicroscopic Cytology
Journal of Ultrastructure Research
Micron
Tissue & Cell
Ultrastructural Pathology
Virchows Archiv B Cell Pathology

There are also regular international congresses of electron microscopy, the proceedings of which are published. In particular, in the field of scanning electron microscopy, a useful publication known as Scanning Electron Microscopy is published yearly by SEM Inc., under the editorship of Dr Om Johari, detailing the proceedings of the annual SEM meetings held in the U.S.A.

CHAPTER 1

The study of biological structure

Beck, F. & Lloyd, J. B. (1974) *The Cell in Medical Science.* London: Academic Press.
Bloom, W. & Fawcett, D. W. (1975) *A Textbook of Histology.* Philadelphia: Saunders.
Brinkley, B. R. & Porter, K. R. (1977) *International Cell Biology.* New York: Rockefeller University Press.
Butler, P. J. G. & Klug, A. (1978) The assembly of a virus. *Sci. Am.*, **239**, 52.
Constantinides, P. (1974) *Functional Electronic Histology.* Amsterdam: Elsevier.
David, H. (1978) Cellular Pathobiology. In: Johannessen, J. V. (ed.) *Electron Microscopy in Human Medicine.* Volume 2, pp. 1–148. New York: McGraw-Hill.
De Robertis, E. D. P., Nowinski, W. W. & Saez, F. A. (1975) *Cell Biology.* Philadelphia: Saunders.
Ghadially, F. N. (1981) *Ultrastructural Pathology of the Cell and Matrix.* London: Butterworths.
Goldstein, L. & Prescott, D. M. (1980) *Cell Biology. A Comprehensive Treatise.* New York: Academic Press.
Gunning, B. & Steer, M. W. (1975) *Plant Cell Biology. An Ultrastructural Approach.* New York: Crane, Russak.
Ham, A. W. (1974) Histology. Philadelphia: Lippincott.
Hodges, G. M. & Hallowes, R. C. (1979, 1980) *Biomedical Research Applications of Scanning Electron Microscopy.* Vols 1, 2. New York: Academic Press.
Horne, R. W. (1974) *Virus Structure.* New York: Academic Press.
Johannessen, J. V. (ed.) *Electron Microscopy in Human Medicine.* (Eleven volumes) New York: McGraw Hill.
Kessel, R. G. & Kardon, R. H. (1978) *Tissues and Organs. A Text Atlas of Scanning Electron Microscopy.* San Francisco: Freeman.
MacClean, N. (1977) *The Differentiation of Cells.* London: Edward Arnold.
Malkinson, A. M. (1975) *Hormone Action.* London: Chapman & Hall.
Motta, P., Andrews, P. M. & Porter, K. R. (1977) *Microanatomy of Cell and Tissue Surfaces. An Atlas of Scanning Electron Microscopy.* Philadelphia: Lea & Febiger.
Pfeiffer, C. J., Rowden, G. & Weibel, J. (1974) *Gastrointestinal Ultrastructure.* New York: Academic Press.
Rhodin, J. A. G. (1974) *Histology. A Text and Atlas.* London: Oxford University Press.
Sandborn, E. B. (1972) *Light and Electron Microscopy of Cells and Tissues.* New York: Academic Press.
Simons, K., Garoff, H. & Helenius, A. (1982) How an animal virus gets into and out of its host cell. *Sci. Am.*, **246**, 46.

Toner, P. G., Carr, K. E. & Wyburn, G. M. (1971) *The Digestive System. An Ultrastructural Atlas and Review.* London: Butterworths.

Tribe, M. A., Eraut, M. R. & Snook, R. K. (1975) *Electron Microscopy and Cell Structure.* Cambridge: Cambridge University Press.

Weiss, L. & Greep, O. R. (1977) *Histology.* New York: McGraw-Hill.

CHAPTER 2

Biological membranes and the cell surface

Membrane structure and function

Abercrombie, M. (1980) Contact inhibition and malignancy. *Nature*, **281**, 259.

Crowther, R. A., Finch, J. T. & Pearse, B. M. F. (1976) On the structure of coated vesicles. *J. Mol. Biol.*, **103**, 785.

Davson, H. & Danielli, J. F. (1952) *The Permeability of Natural Membranes.* Cambridge: Cambridge University Press.

Fox, C. F. (1972) The structure of cell membranes. *Sci. Am.*, **226**, 31.

Hull, B. E. & Staehelin, L. A. (1979) The terminal web. A re-evaluation of its structure and function. *J. Cell Biol.*, **81**, 67.

Kaplan, J. (1981) Polypeptide binding membrane receptors. Analysis and classification. *Science*, **212**, 14.

Knutton, S. & Robertson, J. D. (1976) Regular structures in membranes: The luminal plasma membrane of the cow urinary bladder. *J. Cell Sci.*, **22**, 355.

Lodish, H. F. & Rothman, J. E. (1979) The assembly of cell membranes. *Sci. Am.*, **240**, 38.

Luftig, R., Wehrli, E. & McMillan, P. (1977) The unit membrane image: A re-evaluation. *Life Sci.*, **21**, 285.

Mooseker, M. S. & Tilney, L. G. (1975) Organisation of an actin filament-membrane complex: Filament polarity and membrane attachment in the microvilli of intestinal epithelial cells. *J. Cell Biol.*, **67**, 725.

Nicolson, G. L. (1976) Trans-membrane control of the receptors on normal and tumor cells. 1. Cytoplasmic influence over cell surface components. *Biochim. Biophys. Acta*, **457**, 57.

Poste, G. & Nicolson, G. (1977) *Dynamic Aspects of Cell Surface Organisation.* (Cell surface reviews, volume 3.) New York: Elsevier/North Holland.

Quinn, P. J. (1976) *The Molecular Biology of Cell Membranes.* London: Macmillan.

Roth, J. (1978) The lectins. Molecular probes in cell biology and membrane research. *Exp. Pathol. Suppl.*, **3**, 7.

Singer, S. J. (1975) The molecular organisation of membranes. *Ann. Rev. Biochem.*, **43**, 805.

Singer, J. S. & Nicolson, G. L. (1972) The fluid mosaic model of the structure of cell membranes. *Science*, **175**, 720.

Vehen, A. & Mosher, D. F. (1978) High molecular weight cell surface associated glycoprotein lost in malignant transformation. *Biochim. Biophys. Acta*, **516**, 1.

Willingham, M. C. (1976) Cyclic AMP and cell behaviour in cultured cells. *Int. Rev. Cytol.*, **44**, 319.

Cell adhesion specialisations

Allen, T. D. & Potten, C. S. (1975) Desmosomal form fate and function in mammalian epidermis. *J. Ultrastruct. Res.*, **51**, 94.

Bennett, M. V. L. (1973) Function of electrotonic junctions in embryonic and adult tissues. *Fed. Proc.*, **32**, 65.

Caputo, R. & Peluchetti, D. (1977) The junctions of normal human epidermis. A freeze-fracture study. *J. Ultrastruct. Res.*, **61**, 44.

Claude, P. & Goodenough, D. A. (1973) Fracture faces of zonulae occludentes from 'tight' and 'leaky' epithelia. *J. Cell Biol.*, **58**, 390.

Cobb, J. S. L. & Bennett, T. (1969) A study of nexuses in visceral smooth muscle. *J. Cell Biol.*, **41**, 287.

Friend, D. S. & Gilula, N. B. (1972) Variation in tight and gap junctions in mammalian tissues. *J. Cell Biol.*, **53**, 758.

Goodenough, D. A. (1974) Bulk isolation of mouse hepatocyte gap junctions. Characterisation of the principal protein, connexin. *J. Cell Biol.*, **61**, 557.

Goodenough, D. A. (1975) The structure and permeability of isolated hepatic gap junctions. *J. Cell Biol.*, **45**, 272.

Kachar, B. & Reese, T. S. (1982) Evidence of lipidic nature of tight junction structure. *Nature*, **296**, 464.

Kelly, D. (1976) The hemidesmosome: New fine structural features revealed by freeze-fracture techniques. *Cell Tissue Res.*, **172**, 289.

Leloup, R., Laurent, L., Ronveaux, M. F., Drochmans, P. & Wanson, J. C. (1979) Desmosomes and desmogenesis in the epidermis of calf muzzle. *Biol. Cell.*, **34**, 137.

Loewenstein, W. R. (1970) Intercellular communication. *Sci. Am.*, **222**, 79.

McNutt, N. S. & Weinstein, R. S. (1970) The ultrastructure of the nexus. A correlated thin-section and freeze-cleave study. *J. Cell Biol.*, **47**, 666.

McNutt, N. S. & Weinstein, R. S. (1973) Membrane ultrastructure at mammalian intercellular junctions. *Progress in Biophysics and Molecular Biology*, **26**, 45.

Peracchia, C. (1977) Gap junction structure and function. *Trends Biochemical Sci.*, **2**, 26.

Revel, J. P., Yancey, S. B., Meyer, D. J. & Nicholson, B. (1980) Cell junctions and intercellular communication. *In Vitro*, **16**, 1010.

Staehelin, L. A. (1974) Structure and function of intercellular junctions. *Int. Rev. Cytol.*, **39**, 191.

Staehelin, L. A. & Hull, B. E. (1978) Junctions between living cells. *Sci. Am.*, **238**, 41.

Unwin, P. W. T. & Zampighis, G. (1980) Structure of the junction between communicating cells. *Nature*, **283**, 545.

Wade, J. B. (1974) The structure of the zonula occludens. *J. Cell Biol.*, **60**, 168.

Cell coat

Bennett, G. (1970) Migration of glycoprotein from Golgi apparatus to cell coat in the columnar cells of the duodenal epithelium. *J. Cell Biol.*, **45**, 668.

Bennett, G., Leblond, C. P. & Haddad, A. (1974) Migration of glycoproteins from the Golgi apparatus to the surface of various cell types as shown by radioautography after labelled fucose injection into rats. *J. Cell Biol.*, **60**, 258.

Ito, S. (1974) Form and function of the glycocalyx on free cell surfaces. *Phil. Trans. Roy. Soc. B.*, **268**, 55.

Luft, J. H. (1976) The structure and properties of the cell surface coat. *Int. Rev. Cytol.*, **45**, 291.

Rambourg, A. & Leblond, C. P. (1967) Electron microscope observations on the carbohydrate-rich cell coat present at the surface of cells in the rat. *J. Cell Biol.*, **32**, 27.

Basement membrane

Briggaman, R. A., Dalldorf, F. G. & Wheeler, C. E. (1971) Formation and origin of basal lamina and anchoring fibrils in adult human skin. *J. Cell Biol.*, **51**, 384.

Kefalides, N. A. (1971) Chemical properties of basement membranes. *Int. Rev. Exp. Pathol.*, **10**, 1.

Kefalides, N. A. (1973) Structure and biosynthesis of basement membranes. *Int. Rev. Connect. Tiss. Res.*, **6**, 63.

Kefalides, N. A. (1978) *Biology and Chemistry of Basement Membranes*. New York: Academic Press.

Pierce, G. B. & Nakane, P. K. (1969) Basement membranes; synthesis and deposition in response to cellular injury. *Lab. Invest.*, **21**, 27.

Vracko, R. & Benditt, E. P. (1972) Basal lamina: The scaffold for orderly cell replacement. Observations on regeneration of injured skeletal muscle fibers and capillaries. *J. Cell Biol.*, **55**, 406.

Walker, F. (1973) The origin, turnover and removal of glomerular basement membrane. *J. Pathol.*, **110**, 233.

CHAPTER 3

Structure and function of the nuclear components

Nucleus and nucleolus

Back, F. (1976) The variable condition of euchromatin and heterochromatin. *Int. Rev. Cytol.*, **45**, 25.

Baserga, R. (1981) The cell cycle. *New Engl. J. Med.*, **304**, 453.

Bouteille, M., Kalifat, S. R. & Delarue, J. (1967) Ultrastructural variations of nuclear bodies in human diseases. *J. Ultrastruct. Res.*, **19**, 474.

Brown, D. D. (1981) Gene expression in eukaryotes. *Science*, **211**, 667.

Busch, H. & Smetana, K. (1970) *The Nucleolus*. New York: Academic Press.

Ghosh, S. (1976) Nucleolar structure. *Int. Rev. Cytol.*, **44**, 1.

Gilbert, W. (1981) DNA sequencing and gene structure. *Science*, **214**, 1305.

Goodenough, U. (1978) *Genetics*. Philadelphia: Holt-Saunders.

Kornberg, R. D. & Klug, A. (1981) The nucleosome. *Sci. Am.*, **244**, 48.

Mazia, D. (1974) The cell cycle. *Sci. Am.*, **230**, 55.

Miller, O. L. & Beattie, B. R. (1969) Visualization of nucleolar genes. *Science*, **164**, 955.

Miller, O. L. (1973) The visualization of genes in action. *Sci. Am.*, **228**, 34.

More, I. A. R. & McSeveney, D. (1980) The three dimensional structure of the nucleolar channel system in the endometrial glandular cell: Serial sectioning and high voltage microscopic studies. *J. Anat.*, **130**, 673.

Nurse, P. (1980) Cell cycle control — both deterministic and probabilistic. *Nature*, **280**, 9.

Ris, H. & Korenberg, J. (1979) Chromosome structure and levels of chromosome organisation. In: Prescott, D. M. & Goldstein, L. (eds) *Cell Biology. A Comprehensive Treatise*. Volume 2, 267. London: Academic Press.

Robbins, E. & Gonatas, N. K. (1964) The ultrastructure of a mammalian cell during the mitotic cycle. *J. Cell Biol.*, **21**, 429.

Robbins, E. & Jentzsch, G. (1969) Ultrastructural changes in the mitotic apparatus at the metaphase–to–anaphase transition. *J. Cell Biol.*, **40**, 678.

Stubblefield, E. (1973) The structure of mammalian chromosomes. *Int. Rev. Cytol.*, **35**, 1.

Watson, J. (1975) *The Molecular Biology of the Gene*. New York: Addison-Wesley.

Nuclear envelope

Abelson, H. T. & Smith, G. H. (1970) Nuclear pores: The pore–annulus relationship in thin section. *J. Ultrastruct. Res.*, **30**, 558.

Faberge, A. C. (1973) Direct demonstration of eight-fold symmetry in nuclear pores. *Z. Zellforsch. mikrosk. Anat.*, **136**, 183.

Faberge, A. C. (1974) The nuclear pore complex: Its free existence and an hypothesis as to its origin. *Cell Tissue Res.*, **151**, 403.

Franke, W. W. (1970) On the universality of nuclear pore complex structure. *Z. Zellforsch. mikrosk. Anat.*, **105**, 405.

Franke, W. W. (1977) Structure and function of nuclear membranes. In: Garland, P. B. & Mathias, A. P. (eds) Biochemistry of the Cell Nucleus. *Biochem. Soc. Symp.*, **42**, 125–135. London: The Biochemical Society.

Severs, N. J. & Jordan, E. G. (1978) Nuclear pores. Can they expand and contract to regulate nucleocytoplasmic exchange? *Experientia*, **34**, 1007.

CHAPTER 4

Structure and function of the cytoplasmic components

Ribosomes and endoplasmic reticulum

Bolender, R. P. & Weibel, E. R. (1973) A morphometric study of the removal of phenobarbital-induced membranes from hepatocytes after cessation of treatment. *J. Cell Biol.*, **56**, 746.

Cardell, R. R. (1977) Smooth endoplasmic reticulum in rat hepatocytes during glycogen deposition and depletion. *Int. Rev. Cytol.*, **48**, 221.

Devine, C. E., Somlyo, A. V. & Somlyo, A. P. (1973) Sarcoplasmic reticulum and mitochondria as calcium accumulation sites in smooth muscle. *Phil. Trans. Roy. Soc. Lond. (B)*, **265**, 17.

Jamieson, J. D. & Palade, G. E. (1967) Intracellular transport of secretory proteins in the pancreatic exocrine cell. I. Role of the peripheral elements of the Golgi. *J. Cell Biol.*, **34**, 577.

Jamieson, J. D. & Palade, G. E. (1967) Intracellular transport of secretory proteins in the pancreatic exocrine cell. II. Transport to condensing vacuoles and zymogen granules. *J. Cell Biol.*, **34**, 597.

Jamieson, J. D. & Palade, G. E. (1968) Intracellular transport of secretory proteins in the pancreatic exocrine cell. III. Dissociation of intracellular transport from protein synthesis. *J. Cell Biol.*, **39**, 580.

Jamieson, J. D. & Palade, G. E. (1968) Intracellular transport of secretory proteins in the pancreatic exocrine cell. IV. Metabolic requirements. *J. Cell Biol.*, **39**, 589.

Jones, A. L. & Fawcett, D. W. (1966) Hypertrophy of the agranular endoplasmic reticulum in hamster liver induced by phenobarbital. *J. Histochem. Cytochem.*, **14**, 215.

Nanninga, N. (1973) Structural aspects of ribosomes. *Int. Rev. Cytol.*, **35**, 135.

Pfuderer, P. & Schwartzendruber, D. C. (1966) The configuration of isolated polysomes. *J. Cell Biol.*, **30**, 193.

Wool, I. G. (1979) The structure and function of eukaryotic ribosomes. *Ann. Rev. Biochem.*, **48**, 719.

Annulate lamellae

Bhawan, J., Ceccacci, L. & Cranford, J. (1978) Annulate lamellae in a malignant mesenchymal tumour. *Virchows Arch. B. Cell Pathol.*, **26**, 261.

Gulyas, B. J. (1971) The rabbit zygote: Formation of annulate lamellae. *J. Ultrastruct. Res.*, **35**, 112.

Gulyas, B. J. (1975) The dependence of annulate lamellae formation on the nucleus in parthenogenetic rabbit eggs. *Cell Tissue Res.*, **162**, 475.

Kessel, R. G. (1968) Annulate lamellae. *J. Ultrastruct. Res. Suppl.*, **10**.

Wischnitzer, S. (1970) The annulate lamellae. *Int. Rev. Cytol.*, **27**, 29.

Golgi apparatus

Beams, H. W. & Kessel, R. G. (1968) The Golgi apparatus: Structure and function. *Int. Rev. Cytol.*, **23**, 209.

Neutra, M. & Leblond, C. P. (1969) The Golgi apparatus. *Sci. Am.*, **220**, 100.

Novikoff, A. B. (1977) Cytochemical contribution to the differentiation of GERL from the Golgi apparatus. *Histochem. J.*, **9**, 525.

Novikoff, P. M., Novikoff, A. B., Quintana, N. & Hauw, J. J. (1971) Golgi apparatus, GERL, and lysosomes of neurons in rat dorsal root ganglia, studied by thick sections and thin section cytochemistry. *J. Cell Biol.*, **50**, 859.

Rambourg, A., Clermont, Y. & Marraud, A. (1974) Three-dimensional structure of the osmium-impregnated Golgi apparatus as seen in the high voltage electron microscope. *Amer. J. Anat.*, **140**, 27.

Whaley, W. G. (1975) *The Golgi Apparatus.* New York: Springer-Verlag.

Young, R. W. (1973) The role of the Golgi complex in sulfate metabolism. *J. Cell Biol.*, **57**, 175.

Mitochondria

Arntzen, C. J. & Armond, P. (1977) In: Sanadi, R. & Vernon, L. (eds) *Current Topics in Bioenergetics.* New York: Academic Press.

Barnardt, T. & Afzelius, B. A. (1972) The matrix granules of mitochondria. A review. *Sub-cell. Biochem.*, **1**, 375.

Bernstein, L. H. & Wollman, S. H. (1975) Association of mitochondria with desmosomes in the rat thyroid gland. *J. Ultrastruct. Res.*, **53**, 87–92.

Borograd, L. (1975) Evolution of organelles and eukaryotic genomes. *Science*, **188**, 891.

Fernandez-Moran, H., Oda, T., Blair, P. V. & Green, D. E. (1964) A macromolecular repeating unit of mitochondrial structure and function. *J. Cell Biol.*, **22**, 63.

Flavell, R. (1972) Mitochondria and chloroplasts as descendants of prokaryotes. *Biochem. Genet.*, **6**, 275.

Munn, E. A. (1974) *The Structure of Mitochondria.* London: Academic Press.

Roodyn, D. B. & Wilkie, D. (1968) *The Biogenesis of Mitochondria.* London: Methuen.

Sjostrand, F. S. (1978) The structure of mitochondrial membranes. A new concept. *J. Ultrastruct. Res.*, **64**, 217.

Lysosomes and microbodies

Afzelius, B. (1965) The occurrence and structure of microbodies. A comparative study. *J. Cell Biol.*, **26**, 835.

Arstila, A. U., Shelburn, J. D. & Trump, B. F. (1972) Studies on cellular autophagocytosis: A histochemical study on sequential alterations of mitochondria in the glucagon-induced autophagic vacuoles of rat liver. *Lab. Invest.*, **27**, 317.

Biempica, L. (1966) Human hepatic microbodies with crystalloid cores. *J. Cell Biol.*, **29**, 383.

Daems, W. T. (1966) The fine structure of mouse liver microbodies. *J. Microsc.*, **5**, 295.

De Duve, C. (1969) The peroxisome: A new cytoplasmic organelle. *Proc. Roy. Soc. Series B. Biol. Sci.*, **173**, 71.

De Duve, C. & Baudhuin, P. (1966). Peroxisomes. *Physiol. Rev.*, **46**, 323.

De Duve, C. & Wattiaux, R. (1966) Functions of lysosomes. *Ann. Rev. Physiol.*, **28**, 435.

Dingle, J. T. & Fell, H. B. (1969–76) Lysosomes in Biology and Pathology. Amsterdam: North Holland.

Hand, A. R. (1973) Morphologic and cytochemical identification of peroxisomes in the rat parotid and other exocrine glands. *J. Histochem. Cytochem.*, **21**, 131.

Holtzmann, E. (1975) *Lysosomes. A Survey.* New York: Springer-Verlag.

Hruban, Z. & Rechcigl, M. (1969) Microbodies and related particles. *Int. Rev. Cytol., Suppl.* 1.

Koobs, D. H., Schultz, R. L. & Jutzy, R. V. (1978) The origin of lipofuscin and possible consequences to the myocardium. *Arch. Path. Lab. Med.*, **102**, 66.

Masters, C. & Holmes, R. (1977) The metabolic roles of peroxisomes in mammalian tissues. *Int. J. Biochem.*, **8**, 549.

Masters, C. & Holmes, R. (1977) Peroxisomes: New aspects of cell physiology and biochemistry. *Physiol. Rev.*, **57**, 816.

Moody, D. E. & Reddy, J. K. (1976) Morphometric analysis of the ultrastructural changes in rat liver induced by the peroxisome proliferator Sa H 42–348. *J. Cell Biol.*, **71**, 768.

Novikoff, P. M. & Novikoff, A. B. (1972) Peroxosomes in absorptive cells of mammalian small intestine. *J. Cell Biol.*, **53**, 532.

Pfeifer, U. & Scheller, H. (1975) A morphometric study of cellular autophagy including diurnal variations in kidney tubules of normal rats. *J. Cell Biol.*, **64**, 608.

Pitt, D. (1975) *Lysosomes and Cell Function.* London: Longman.

Shnitka, T. K. (1966) Comparative ultrastructure of hepatic microbodies in some mammals and birds in relation to species differences in uricase activity. *J. Ultrastruct. Res.*, **16**, 598.

Sternlieb, I. & Quintana, N. (1977) The peroxisomes of human hepatocytes. *Lab. Invest.*, **36**, 140.

Filaments, tubules and centrioles

Allen, R. D. (1975) Evidence for firm linkages between microtubules and membrane-bounded vesicles. *J. Cell Biol.*, **64**, 497.

Barrett, L. A. & Dawson, R. B. (1972) Microtubules and erythroid cell shape. *Fed. Proc.*, **31**, 629.

Bauduin, H., Stock, C., Vincent, D. & Genier, J. F. (1975) Microfilamentous system and secretion of enzyme in the exocrine pancreas. Effect of Cytochalasin B. *J. Cell Biol.*, **66**, 165.

Behnke, O. (1970) Microtubules in disk-shaped blood cells. *Int. Rev. Exp. Pathol.*, **9**, 1.

Berns, M. W., Rattner, J. B., Brenner, S. & Meredith, S. (1977) The role of the centriolar region in animal cell mitosis. *J. Cell Biol.*, **72**, 351.

Brinkley, B. R., Miller, C. L., Fuseler, J. W., Pepper, D. A. & Wible, L. J. (1977) The cytoskeleton and cell transformation in malignancy. Microtubules, microfilaments, and growth properties in vitro. *Cancer Bull.*, **29**, 13.

Byers, H. R. & Porter, K. R. (1977) Transformations in the structure of the cytoplasmic ground substance in erythrophores during pigment aggregation and dispersion. I. A study using whole cell preparations in stereo high voltage electron microscopy. *J. Cell Biol.*, **75**, 541.

Craig, S. W. & Pardo, J. V. (1979) Alpha-actinin localization in the junctional complex of intestinal epithelial cells. *J. Cell Biol.*, **80**, 203.

De Rosier, D. J., Tilney, L. G. & Egelman, E. (1980) Actin in the inner ear. The remarkable structure of the stereocilium. *Nature*, **287**, 291.

Dirksen, E. R. (1971) Centriole morphogenesis in developing ciliated epithelium of the mouse oviduct. *J. Cell Biol.*, **51**, 286.

Dustin, P. (1978) *Microtubules*. Berlin: Springer-Verlag.

Dustin, P. (1980) Microtubules. *Sci. Am.*, **243**, 58.

Erickson, H. P. (1975) The structure and assembly of microtubules. *Ann. N.Y. Acad. Sci.*, **253**, 60.

Franke, W. W., Schmid, E., Osborn, M. & Weber, K. (1978) Different intermediate-sized filaments distinguished by immunofluorescence microscopy. *Proc. Nat. Acad. Sci. USA.*, **75**, 5034.

Franke, W. W., Schmid, E., Osborn, M. & Weber, K. (1979) Intermediate sized filaments of human endothelial cells. *J. Cell Biol.*, **81**, 570.

Goldman, J. E., Schaumburg, H. H. & Norton, W. T. (1978) Isolation and characterisation of glial filaments from human brain. *J. Cell Biol.*, **78**, 426.

Langford, G. M. (1980) Arrangement of subunits in microtubules with 14 protofilaments. *J. Cell Biol.*, **87**, 521.

Lazarides, E. (1980) Intermediate filaments as mechanical integrators of cellular space. *Nature*, **283**, 249.

Lazarides, E. & Weber, K. (1974) Actin antibody: The specific visualisation of actin filaments in non-muscle cells. *Proc. Nat. Acad. Sci. USA.*, **71**, 2268.

Lloyd, C. W., Smith, C. G., Woods, A. & Rees, D. A. (1977) Mechanisms of cellular adhesion. II. The interplay between adhesion, the cytoskeleton and morphology in substrate-attached cells. *Exp. Cell Res.*, **110**, 427.

Mooseker, M. S. (1976) Brush border motility. Microvillar contraction in triton-treated brush borders isolated from intestinal epithelium. *J. Cell Biol.*, **71**, 417.

Mooseker, M. K., Pollard, T. D. & Fujiwara, K. (1978) Characterisation and localization of myosin in the brush border of intestinal epithelial cells. *J. Cell Biol.*, **79**, 444.

Pepe, F. A. (1972) The myosin filament. *J. Cell Biol.*, **52**, 255.

Porter, K. R. & Tucker, J. B. (1981) The ground substance of the living cell. *Sci. Am.*, **244**, 40.

Reaven, E. P. & Reaven, G. M. (1977) Distribution and content of microtubules in relation to the transport of lipid. *J. Cell Biol.*, **75**, 559.

Rodewald, R., Newman, S. B. & Karnovsky, M. J. (1976) Contraction of isolated brush borders from the intestinal epithelium. *J. Cell Biol.*, **70**, 541.

Wilson, L. (1975) Action of drugs on microtubules. *Life Sci.*, **17**, 303.

Wolosewick, J. J. & Porter, K. R. (1979) Microtrabecular lattice of the cytoplasmic ground substance. Artifact or reality. *J. Cell Biol.*, **82**, 114.

Wuerker, R. B. & Kirkpatrick, J. B. (1972) Neuronal microtubules, neurofilaments and microfilaments. *Int. Rev. Cytol.*, **33**, 45.

Metabolites

Babcock, M. B. & Cardell, R. R. (1974) Hepatic glycogen patterns in fasted and fed rats. *Amer. J. Anat.*, **140**, 299.

Biava, C. (1963) Identification and structural forms of human particulate glycogen. *Lab. Invest.*, **12**, 1179.

Crichton, R. R. (1971) Ferritin: Structure, synthesis and function. *New Engl. J. Med.*, **284**, 1413.

De Bruijn, W. C. (1973) Glycogen, its chemistry and morphologic appearance in the electron microscope. *J. Ultrastruct. Res.*, **42**, 29.

Ghadially, F. N. (1979) Haemorrhage and haemosiderin. *J. Submicr. Cytol.*, **11**, 271.

Le Beux, Y. J. (1969) An unusual ultrastructural association of smooth membranes and glycogen particles. The glycogen body. *Z. Zellforsch. mikrosk. Anat.*, **101**, 433.

Munro, H. N. & Linder, M. C. (1978) Ferritin: Structure, biosynthesis, and role in iron metabolism. *Physiol. Rev.*, **58**, 317.

Revel, J. P. (1964) Electron microscopy of glycogen. *J. Histochem. Cytochem.*, **12**, 104.

Richter, G. W. (1978) The iron-loaded cell. The cytopathology of iron storage. *Amer. J. Pathol.*, **91**, 361.

CHAPTER 5

Secretion

Exocrine secretion

Brown, R. E. & Schazki, P. F. (1971) Intracisternal granules in the human pancreas. *Arch. Pathol.*, **91**, 351.

Bruni, C. & Porter, K. R. (1965) The fine structure of the parenchymal cell of the normal rat liver. I. General observations. *Amer. J. Pathol.*, **46**, 691.

Castle, J. D., Jamieson, J. D. & Palade, G. E. (1972) Radioautographic analysis of the secretory process in the parotid acinar cell of the rabbit. *J. Cell Biol.*, **53**, 290.

Davis, B. D. & Tai, P. C. (1980) The mechanism of protein secretion across membranes. *Nature*, **283**, 433.

Gillespie, E. (1975) Microtubules, cyclic AMP, calcium and secretion. *Ann. N.Y. Acad. Sci.*, **253**, 771.

Hand, A. R. & Oliver, C. (1977) Cytochemical studies of GERL and its role in secretory granule formation in exocrine cells. *Histochem. J.*, **9**, 375.

Ito, S. & Schofield, G. C. (1974) Studies on the depletion and accumulation of microvilli and changes in the tubulovesicular compartment of the mouse parietal cells in relation to gastric acid secretion. *J. Cell Biol.*, **63**, 364.

Ito, S. & Winchester, R. J. (1963) The fine structure of the gastric mucosa in the bat. *J. Cell Biol.*, **16**, 541.

Jamieson, J. D. & Palade, G. E. (1967) Intracellular transport of secretory proteins in the pancreatic exocrine cell. I. Role of the peripheral elements of the Golgi. *J. Cell Biol.*, **34**, 577.

Jamieson, J. D. & Palade, G. E. (1967) Intracellular transport of secretory proteins in the pancreatic exocrine cell. II. Transport to condensing vacuoles and zymogen granules. *J. Cell Biol.*, **34**, 597.

Jamieson, J. D. & Palade, G. E. (1968) Intracellular transport of secretory proteins in the pancreatic exocrine cell. III. Dissociation of intracellular transport from protein synthesis. *J. Cell Biol.*, **39**, 580.

Jamieson, J. D. & Palade, G. E. (1968) Intracellular transport of secretory proteins in the pancreatic exocrine cell. IV. Metabolic requirements. *J. Cell Biol.*, **39**, 589.

Jamieson, J. D. & Palade, G. E. (1971) Condensing vacuole conversion and zymogen granule discharge in pancreatic exocrine cells : Metabolic studies. *J. Cell Biol.*, **48**, 503.

Jezequel, A. M., Arakawa, K. & Steiner, J. W. (1965) The fine structure of the normal neonatal mouse liver. *Lab. Invest.*, **14**, 1894.

Jones, A. L. & Fawcett, D. W. (1966) Hypertrophy of the agranular endoplasmic reticulum in hamster liver induced by phenobarbital. *J. Histochem. Cytochem.*, **14**, 215.

Miyah, K., Abraham, J., Linthicum, S. & Wagner, R. (1976) Scanning electron microscopy of hepatic ultrastructure. *Lab. Invest.*, **35**, 369.

Motta, P. & Fumagalli, G. (1975) Structure of rat bile canaliculi as revealed by scanning electron microscopy. *Anat. Rec.*, **182**, 499.

Novikoff, A., Mori, M., Quintana, N. & Yam, A. (1977) Studies of the secretory process in the mammalian exocrine pancreas. I. The condensing vacuoles. *J. Cell Biol.*, **75**, 148.

Rubin, W., Ross, L. L., Sleisenger, M. H. & Jeffries, G. H. (1968) The normal human gastric epithelium. A fine structural study. *Lab. Invest.*, **19**, 598.

Endocrine secretion

Capella, C., Solcia, E., Frigerio, B., Buffa, R., Usellini, L. & Fontana, P. (1977) The endocrine cells of the pancreas and related tumours. *Virchows Arch. A. Path. Anat. Histol.*, **373**, 327.

Christensen, A. K. (1965) The fine structure of testicular interstitial cells in guinea pigs. *J. Cell Biol.*, **26**, 911.

De Kretser, D. M. (1967) The fine structure of the testicular interstitial cells in men of normal androgenic status. *Z. Zellforsch. mikrosk. Anat.*, **80**, 594.

Forssmann, W. G., Orci, L., Pictet, R., Renold, A. D. & Rouiller, C. (1969) The endocrine cells in the epithelium of the gastrointestinal mucosa of the rat. An electron microscope study. *J. Cell Biol.*, **40**, 692.

Fujita, H. (1975) Fine structure of the thyroid cell. *Int. Rev. Cytol.*, **40**, 197.

Fujita, T. & Kobayashi, S. (1977) Structure and function of gut endocrine cells. *Int. Rev. Cytol. (Suppl.)*, **187**.

Greider, M. H., Benscombe, S. A. & Lechago, J. (1970) The human pancreatic islet cells and their tumours. I. The normal pancreatic islets. *Lab. Invest.*, **22**, 344.

Idelman, S. (1970) Ultrastructure of the mammalian adrenal cortex. *Int. Rev. Cytol.*, **29**, 181.

Lacy, P. E. (1974) Structure and function of the endocrine cell types of the islets. *Adv. Metabol. Disord.*, **7**, 171.

Like, A. A. (1967) The ultrastructure of the secretory cells of the islets of Langerhans in man. *Lab. Invest.*, **16**, 937.

Nunez, E. A., Hedhammar, A., Wu, F. M., Whalen, J. P. & Krook, L. (1974) Ultrastructure of the parafollicular (C) cells and the parathyroid cell in growing dogs on a high calcium diet. *Lab. Invest.*, **31**, 96–108.

Orci, L., Perrelet, A. & Friend, D. S. (1977) Freeze-fracture of membrane fusions during exocytosis in pancreatic B-cells. *J. Cell Biol.*, **75**, 23.

Pantic, V. R. (1975) The specificity of pituitary cells and regulation of their activities. *Int. Rev. Cytol.*, **40**, 153.

Pearse, A. G. E., Polack, J. M. & Bloom, S. R. (1977) The newer gut hormones. Cellular sources, physiology, pathology and clinical aspects. *Gastroenterology*, **72**, 746.

Rhodin, J. A. G. (1971) The ultrastructure of the adrenal cortex of the rat under normal and experimental conditions. *J. Ultrastruct Res.*, **34**, 23.

Smith, R. E. & Farquhar, M. G. (1966) Lysosome function in the regulation of the secretory process in cells of the anterior pituitary gland. *J. Cell Biol.*, **31**, 319.

Symington, T. (1970) *Functional pathology of the Human Adrenal Gland.* Edinburgh : Livingstone.

CHAPTER 6

Absorption and permeability

Absorption

Anderson, J. H. & Taylor, A. B. (1973) Scanning and transmission electron microscopic studies of jejunal microvilli of the rat, hamster and dog. *J. Morphol.*, **141**, 281.

Andrews, P. M. (1975) Scanning electron microscopy of human and rhesus monkey kidneys. *Lab. Invest.*, **32**, 510.

Bulger, R. E., Siegel, L. F. & Pendergrass, R. (1974) Scanning and transmission electron microscopy of the rat kidney. *Amer. J. Anat.*, **139**, 438.

Friedman, H. I. & Cardell, R. R. (1977) Alterations in the endoplasmic reticulum and Golgi complex of intestinal epithelial cells during fat absorption and after termination of this process. *Anat. Rec.*, **188**, 77.

Graham, R. C. & Karnovsky, M. J. (1966) The early stages of absorption of injected horseradish peroxidase in the proximal tubules of mouse kidney. Ultrastructural cytochemistry by a new technique. *J. Histochem. Cytochem.*, **14**, 291.

Lipkin, M. (1973) Proliferation and differentiation of gastrointestinal cells. *Physiol. Rev.*, **53**, 981.

Mueller, J. C., Jones, A. L. & Long, J. A. (1972) Topographical and subcellular anatomy of the guinea pig gallbladder. *Gastroenterology*, **63**, 856.

Pfaller, W. & Klima, J. (1976) A critical re-evaluation of the structure of the rat uriniferous tubules as revealed by scanning electron microscopy. *Cell Tiss. Res.*, **166**, 91.

Potten, C. S. & Allen, T. D. (1977) Ultrastructure of cell loss in intestinal mucosa. *J. Ultrastruct. Res.*, **66**, 272.

Tischer, C. C., Bulger, R. E. & Trump, B. F. (1966) Human renal ultrastructure. I. Proximal tubule of healthy individuals. *Lab. Invest.*, **15**, 1357.

Toner, P. G. & Carr, K. E. (1969) The use of scanning electron microscopy in the study of the intestinal villi. *J. Path. Bact.*, **97**, 611.

Permeability

Anderson, A. D. & Anderson, N. J. (1975) Studies on the structure and permeability of the microvasculature in normal rat lymph nodes. *Amer. J. Pathol.*, **80**, 387.

Becker, R. P. & De Bruyn, P. P. (1976) The transmural passage of blood cells into myeloid sinusoids and the entry of platelets into the sinusoidal circulation: A scanning electron microscope investigation. *Amer. J. Anat.*, **145**, 183.

Bruns, R. R. & Palade, G. E. (1968) Studies on blood capillaries. I. General organization of blood capillaries in muscle. *J. Cell Biol.*, **37**, 244.

Clementi, F. & Palade, G. E. (1969) Intestinal capillaries. I. Permeability to peroxidase and ferritin. *J. Cell Biol.*, **41**, 33.

Dermer, G. B. (1970) The pulmonary surfactant content of the inclusion bodies found within type II alveolar cells. *J. Ultrastruct. Res.*, **33**, 306.

Friedrich, H. H. (1968) The tridimensional ultrastructure of fenestrated capillaries. *J. Ultrastruct. Res.*, **23**, 444.

Fujita, T., Tokunaga, J. & Edanaga, M. (1976) Scanning electron microscopy of the glomerular filtration membrane in the rat kidney. *Cell Tissue Res.*, **166**, 299.

Goerke, J. (1979) Lung surfactant. *Biochim. Biophys. Acta*, **344**, 241.

Groniowski, J., Walski, M. & Biczysko, W. (1972) Application of scanning electron microscopy for studies of the lung parenchyma. *J. Ultrastruct. Res.*, **38**, 473.

Karnovsky, M. J. (1968) The ultrastructural basis of transcapillary exchanges. *J. Gen. Physiol.*, **52**, 643.

Kikkawa, Y. (1970) Morphology of alveolar lining layer. *Anat. Rec.*, **167**, 389.

Ladman, A. J. & Finley, T. N. (1966) Electron microscopic observations of pulmonary surfactant and the cells which produce it. *Anat. Rec.*, **154**, 372.

Latta, H. (1970) The glomerular capillary wall. *J. Ultrastruct. Res.*, **32**, 526.

Maul, G. G. (1971) Structure and formation of pores in fenestrated capillaries. *J. Ultrastruct. Res.*, **36**, 768.

Motta, P. (1975) A scanning electron microscopic study of the rat liver sinusoid. *Cell Tiss. Res.*, **164**, 371.

Motta, P. & Porter, K. R. (1974) Structure of the rat liver sinusoids and associated tissue spaces as revealed by scanning electron microscopy. *Cell Tissue Res.*, **148**, 111.

Owen, R. L. (1977) Sequential uptake of horseradish peroxidase by lymphoid follicle epithelium of Peyer's patches in the normal unobstructed mouse intestine: An ultrastructural study. *Gastroenterology*, **72**, 440.

Rhodin, J. A. G. (1968) Ultrastructure of mammalian venous capillaries, venules and small collecting veins. *J. Ultrastruct. Res.*, **25**, 452.

Simionescu, M., Simionescu, N. & Palade, G. E. (1976) Segmental differentiations of cell junctions in vascular endothelium. Arteries and veins. *J. Cell Biol.*, **68**, 705.

Spinelli, F. (1974) Structure and development of the renal glomerulus as revealed by scanning electron microscopy. *Int. Rev. Cytol.*, **39**, 345.

Sueishi, K., Tanaka, T. & Oda, T. (1977) Immunoultrastructural study of surfactant system. *Lab. Invest.*, **37**, 136.

Weibel, E. R. & Palade, G. E. (1964) New cytoplasmic components in arterial endothelia. *J. Cell Biol.*, **23**, 101.

Williams, M. C. & Wissig, S. L. (1975) The permeability of muscle capillaries to horseradish peroxidase. *J. Cell Biol.*, **66**, 531.

Wisse, E. (1970) An electron microscopic study of the fenestrated endothelial lining of rat liver sinusoids. *J. Ultrastruct. Res.*, **31**, 125.

CHAPTER 7

Defence

Adamson, I. Y. R. & Bowden, D. H. (1978) Adaptive responses of the pulmonary macrophagic system to carbon. 2. Morphologic studies. *Lab. Invest.*, **38**, 430.

Allison, A. C. (1978) Macrophage activation and non-specific immunity. *Int. Rev. Exp. Pathol.*, **18**, 303.

Bainton, D. F. (1973) Sequential degranulation of the two types of polymorphonuclear leukocyte granules during phagocytosis of microorganisms. *J. Cell Biol.*, **58**, 249.

Bainton, D. F. & Farquhar, M. G. (1970) Segregation and packaging of granule enzymes in eosinophil leukocytes. *J. Cell Biol.*, **45**, 54.

Behnke, O. (1968) An electron microscope study of the megakaryocyte of the rat bone marrow. *J. Ultrastruct. Res.*, **24**, 412.

Carr, I. (1973) *The Macrophage: A Review of Ultrastructure and Function.* London: Academic Press.

Collet, A. J. (1970) Fine structure of the alveolar macrophage of the cat and modifications of its cytoplasmic components during phagocytosis. *Anat. Rec.*, **167**, 277.

De Duve, C. & Wattiaux, R. (1966) Functions of lysosomes. *Ann. Rev. Physiol.*, **28**, 435.

Goodall, R. J., Lai, Y. F. & Thompson, J. E. (1972) Turnover of plasma membrane during phagocytosis. *J. Cell Sci.*, **11**, 569.

Gordon, S. & Cohn, Z. A. (1973) The macrophage. *Int. Rev. Cytol.*, **36**, 171.

Klebanoff, S. J. & Clarck, R. A. (1978) *The Neutrophil. Function and Clinical Disorders.* Amsterdam: North Holland.

Luk, S. C., Nopajaroonsri, C. & Simon, G. T. (1973) The architecture of the normal lymph node and haemolymph node: A scanning and transmission electron microscope study. *Lab. Invest.*, **29**, 258.

McConnel, I., Munro, A. & Waldman, A. (1981) *The Immune System.* London: Blackwell Scientific.

Miller, F., De Harven, E. & Palade, G. E. (1966) The structure of eosinophil leukocyte granules in rodents and in man. *J. Cell Biol.*, **31**, 349.

Papadimitriou, J. M. (1978) Endocytosis and formation of macrophage polykarya: An ultrastructural study. *J. Pathol.*, **126**, 215.

Ryan, G. B. & Majno, G. (1977) Acute inflammation. A review. *Amer. J. Pathol.*, **86**, 183.

Scott, R. E. & Horn, R. G. (1970) Ultrastructural aspects of neutrophil granulocyte development in humans. *Lab. Invest.*, **23**, 202.

Simson, J. V. & Spicer, S. S. (1973) Activities of specific cell constituents in phagocytosis (endocytosis). *Int. Rev. Exp. Pathol.*, **12**, 79.

Spicer, S. S. & Hardin, J. H. (1969) Ultrastructure, cytochemistry and function of neutrophil leukocyte granules. A Review. *Lab. Invest.*, **20**, 488.

Steinman, R. M., Brodie, S. E. & Cohn, Z. A. (1976) Membrane flow during pinocytosis. A stereologic analysis. *J. Cell Biol.*, **68**, 665.

Stossel, T. P. (1974) Phagocytosis. *New Engl. J. Med.*, **290**, 717; 774; 833.

Van Der Rhee, H. J., Van Der Burgh-Dewinter, C. P. M. & Daems, W. Th. (1979) The differentiation of monocytes into macrophages, epithelioid cells and multinucleated giant cell in subcutaneous granulomas. *Cell Tiss. Res.*, **197**, 355.

Weiss, L. (1974) A scanning electron microscopic study of the spleen. *Blood*, **43**, 665.

Wisse, E. & Knook, D. L. (1977) *Kupffer Cells and Other Liver Sinusoidal Cells*. Amsterdam: Elsevier/North Holland.

CHAPTER 8

Storage and protection

Storage

Bessis, M. (1973) *Living Blood Cells and Their Ultrastructure*. New York: Springer-Verlag.

Cushman, S. W. (1976) Structure–function relationships in the adipose cell. *J. Cell Biol.*, **46**, 326.

Moe, D. & Surgenor, N. (1974) *The Red Blood Cell*. New York: Academic Press.

Napolitano, L. (1963) The differentiation of white adipose cells. An electron microscope study. *J. Cell Biol.*, **18**, 663.

Napolitano, L. (1965) The fine structure of adipose tissues. In: Reynold, A. E. & Cahill, G. F. (eds) *Handbook of Physiology*, V. Washington: American Physiological Society.

Seemayer, T. A., Knaack, J., Wang, N-S, & Ahmed, M. N. (1975) On the ultrastructure of hibernoma. *Cancer*, **36**, 1785.

Simpson, C. F. & Kling, J. M. (1967) The mechanism of denucleation in circulating erythroblasts. *J. Cell Biol.*, **35**, 237.

Slavin, B. G. (1972) The cytophysiology of mammalian adipose cells. *Int. Rev. Cytol.*, **33**, 297.

Smith, R. E. & Horwitz, B. A. (1969) Brown fat and thermiogenesis. *Physiol Rev.*, **49**, 330.

Protection

Andrews, P. M. & Porter, K. R. (1973) The ultrastructural morphology and possible functional significance of mesothelial microvilli. *Anat. Rec.*, **177**, 409.

Baradi, A. F. & Rao, S. N. (1976) A scanning electron microscope study of mouse peritoneal mesothelium. *Tissue Cell*, **8**, 159.

Breathnach, A. S. (1971) *An Atlas of the Ultrastructure of Human Skin*. London: Churchill.

Breathnach, A. S. (1975) Aspects of epidermal ultrastructure. *J. Invest. Dermatol.*, **65**, 2.

Briggaman, R. A. & Wheeler, C. E. (1975) The epidermal-dermal junction. *J. Invest. Dermatol.*, **65**, 71.

Elias, P. M. & Friend, D. S. (1975) The permeability barrier in mammalian epidermis. *J. Cell Biol.*, **65**, 180.

Fedorko, M. E. (1977) The functional capacity of guinea pig megakaryocytes. *Lab. Invest.*, **36**, 310.

Hashimoto, K. (1971) Cementosome: A new interpretation of the membrane-coating granule. *Arch. Dermatol.*, **240**, 349.

Hicks, R. M. (1975) The mammalian urinary bladder. An accommodating organ. *Biol. Rev.*, **50**, 215.

Ihzumi, T., Hattori, A., Sanada, M. & Muto, M. (1977) Megakaryocyte and platelet formation: A scanning electron microscope study in mouse spleen. *Arch. Histol. Jap.*, **40**, 305.

Koss, L. G. (1969) The asymmetric unit membranes of the epithelium of the urinary bladder of the rat. An electron microscopic study of a mechanism of epithelial maturation and function. *Lab. Invest.*, **21**, 154.

Lavker, R. M. (1976) Membrane coating granules: The fate of the discharged lamellae. *J. Ultrastruct. Res.*, **55**, 79.

Lavker, R. M. & Matoltsy, A. G. (1970) Formation of horny cells. *J. Cell Biol.*, **44**, 501.

Matoltsy, A. G. (1975) Desmosomes, filaments and keratohyalin granules: Their role in the stabilization and keratinization of the epidermis. *J. Invest. Dermatol.*, **65**, 127.

Matoltsy, A. G. (1976) Keratinization. *J. Invest. Derm.*, **65**, 20.

Minsky, B. D. & Chalpowski, F. J. (1978) Morphometric analysis of the translocation of luminal membrane between cytoplasm and cell surface of transitional epithelium during the expansion–contraction cycles of mammalian urinary bladder. *J. Cell Biol.*, **77**, 685.

Montagna, W. (1974) *The Structure and Function of Skin*. New York: Academic Press.

Parakkal, P. F. (1967) An electron microscopic study of esophageal epithelium in the newborn and adult mouse. *Amer. J. Anat.*, **121**, 175.

Parakkal, P. F. (1974) Cyclical changes in the vaginal epithelium of the rat seen by scanning electron microscopy. *Anat. Rec.*, **178**, 529.

Seiji, M., Fitzpatrick, T. B. & Birbeck, M. S. C. (1961) The melanosome: A distinctive subcellular particle of mammalian melanocytes and the site of melanogenesis. *J. Invest. Derm.*, **36**, 243.

Severs, N. J. & Warren, R. C. (1978) Analysis of membrane structure in the transitional epithelium of rat urinary bladder. I. The luminal membrane. *J. Ultrastruct. Res.*, **64**, 124–140.

Squier, C. A. & Rooney, L. (1976) The permeability of keratinized and nonkeratinized oral epithelium to lanthanum in vivo. *J. Ultrastruct. Res.*, **54**, 286.

Warren, R. C. & Hicks, R. M. (1978) Chemical dissection and negative staining of the bladder luminal membrane. *J. Ultrastruct. Res.*, **64**, 327–340.

Wolff, K. & Wolff-Schreiner, E. C. (1975) Trends in electron microscopy of skin. *J. Invest. Dermatol.*, **65**, 39.

Zelickson, A. S. (1967) *Ultrastructure of Normal and Abnormal Skin*. London: Kimpton.

CHAPTER 9

Mechanical support

Fibrous tissue

Balazs, E. A. (1970) *Chemistry and Molecular Biology of the Intercellular Matrix*. Volumes 1–3. New York: Academic Press.

Bornstein, P. (1974) The biosynthesis of collagen. *Ann. Rev. Biochem.*, **43**, 567.

Bradamante, Z, & Svajger, A. (1977) Pre-elastic (oxytalan) fibres in the developing elastic cartilage of the external ear of the rat. *J. Anat.*, **123**, 735.

Briggaman, R. A., Dalldorf, F. G. & Wheeler, C. E. (1971) Formation and origin of basal lamina and anchoring fibrils in adult human skin. *J. Cell Biol.*, **51**, 384.

Brinkmann, G. L. (1968) The mast cell in normal human bronchus and lung. *J. Ultrastruct. Res.*, **23**, 115.

Cotta-Pereira, G., Rodrigo, F. G. & David-Ferreira, J. F. (1976) Oxytalan, elaunin and elastic fibres in the human skin. *J. Invest. Derm.*, **66**, 143.

Cox, R. W., Grant, R. A. & Kent, C. M. (1972) The interpretation of electron micrographs of negatively stained native collagen. *J. Cell Sci.*, **10**, 547.

Glauert, A. M. & Mayo, C. R. (1973) The study of the three-dimensional structural relationships in connective tissues by high voltage electron microscopy. *J. Microsc.*, **97**, 83.

Gotte, L., Giro, M. G., Volpin, D. & Horne, R. W. (1974) The ultrastructural organisation of elastin. *J. Ultrastruct. Res.*, **46**, 23.

Greenlee, T. K., Ross, R. & Hartmann, J. L. (1966) The fine structure of elastic fibers. *J. Cell Biol.*, **30**, 59.

Kessler, S. & Kuhn, C. (1975) Scanning electron microscopy of mast cell degranulation. *Lab. Invest.*, **32**, 71.

Kewley, M. A., Steven, E. S. & Williams, G. (1977) The presence of fine elastin fibrils within the elastic fibre observed by scanning electron microscopy. *J. Anat.*, **123**, 129.

Leak, L. V. & Burke, J. F. (1966) Fine structure of the lymphatic capillary and the adjoining connective tissue area. *Amer. J. Anat.*, **118**, 785.

Luse, S. & Hutton, R. (1964) An electron microscopic study of the fate of collagen in the post-partum rat uterus. *Anat. Rec.*, **148**, 308.

Miller, A. & Wray, S. J. (1971) Molecular packing in collagen. *Nature*, **230**, 437.

Minor, R. R. (1980) Collagen metabolism: A comparison of diseases of collagen and diseases affecting collagen. *Amer. J. Pathol.*, **98**, 225.

Orr, T. S. C. (1977) Fine structure of the mast cell with special reference to human cells. *Scand. J. Resp. Dis. Suppl.*, **98**, 1.

Palade, G. E. & Farquhar, M. G. (1965) A special fibril of the dermis. *J. Cell Biol.*, **27**, 215.

Prockop, D. J., Kivirikko, K. I., Tuderman, L. & Gutzman, N. A. (1979) The biosynthesis of collagen and its disorders. *New Engl. J. Med.*, **301**, 12.

Ramachandran, G. N. (1967) *Treatise on Collagen*. New York: Academic Press.

Ross, R. (1973) The elastic fiber. A review. *J. Histochem. Cytochem.*, **21**, 199.

Sandberg, L. B., Gray, W. R. & Fransblau, C. (1977) *Elastin and Elastic Tissue. Advances in Experimental Medicine and Biology*. 79. New York: Plenum Press.

Wagner, B. M. & Smith, D. E. (1967) *The Connective Tissue*. Baltimore: Williams and Wilkins.

Weinstock, M. (1972) Collagen formation — observations on its intracellular packaging and transport. *Z. Zellforsch. mikrosk. Anat.*, **129**, 455.

Zucker-Franklin, D. (1980) Ultrastructural evidence for the common origin of human mast cells and basophils. *Blood*, **56**, 534.

Cartilage and bone

Bernard, G. W. & Pease, D. C. (1969) An electron microscopic study of initial intramembranous osteogenesis. *Amer. J. Anat.*, **125**, 271.

Campo, R. D. & Phillips, S. J. (1973) Electron microscopic visualisation of proteoglycans and collagen in bovine costal cartilage. *Calc. Tiss. Res.*, **13**, 83.

Ghadially, F. N. (1978) The fine structure of joints. In: Sokoloff, L (ed.) *The Joints and Synovial Fluid*. New York: Academic Press.

Ghadially, F. N., Thomas, I., Yong, N. & Lalonde, J. M. A. (1978) Ultrastructure of rabbit semilunar cartilages. *J. Anat.*, **125**, 499.

Goel, S. C. & Jacob, J. (1976) Reinterpretation of the ultrastructure of cartilage matrix. *Experientia*, **32**, 216.

Hall, B. K. (1975) The origin and fate of osteoclasts. *Anat. Rec.*, **183**, 1.

Holtrop, M. E. (1975) The ultrastructure of bone. *Ann. Clin. Lab. Sci.*, **5**, 264.

Horn, V. & Dvorak, M. Y. (1974) Ultrastructure of functional bone components in scanning and transmission electron microscopy. *Z. Zellforsch. mikrosk. Anat.*, **88**, 836.

Jande, S. S. & Belanger, L. F. (1971) Electron microscopy of osteocytes and the pericellular matrix in rat trabecular bone. *Calcif. Tiss. Res.*, **6**, 280.

Jones, S. J. & Boyde, A. (1977) Some morphologic observations on osteoclasts. *Cell Tiss. Res.*, **185**, 387.

Minns, R. J. & Stevens, F. S. (1977) The collagen fibril organisation in human articular cartilage. *J. Anat.*, **123**, 437.

Owen, M. (1970) The origin of bone cells. *Int. Rev. Cytol.*, **28**, 213.

CHAPTER 10

Contraction and motility

Muscle

Coers, C. (1967) Structure and organisation of the myoneural junction. *Int. Rev. Cytol.*, **22**, 239.

Ebashi, S. (1976) Excitation–contraction coupling. *Ann. Rev. Physiol.*, **38**, 293.

Forssmann, W. G. & Girardier, L. (1970) A study of the T system in rat heart. *J. Cell Biol.*, **44**, 1.

Gabella, G. & Blundell, D. (1979) Nexuses between smooth muscle cells of the guinea pig ileum. *J. Cell Biol.*, **82**, 239.

Hitchcock, S. (1977) Regulation of motility in non-muscle cells. *J. Cell Biol.*, **74**, 1–15.

Huddart, H. & Hunt, S. (1975) *Visceral Muscle. Its Structure and Function*. New York: John Wiley.

Huxley, H. E. (1969) The mechanism of muscular contraction. *Science*, **164**, 1356.

Kelly, D. E. (1969) The fine structure of skeletal muscle triad junctions. *J. Ultrastruct. Res.*, **29**, 37.

Kelly, D. E. & Cahill, M. A. (1972) Microfilamentous and matrix components of skeletal muscle Z-disks. *Anat. Rec.*, **172**, 623.

Lazarides, E. & Revel, J. P. (1979) The molecular basis of cell movement. *Sci. Am.*, **240**, 88.

McNutt, N. S. (1970) Ultrastructure of intercellular junction in adult and developing cardiac muscle. *Amer. J. Cardiol.*, **25**, 169.

Mair, W. G. P. & Tomé, F. M. S. (1972) *Atlas of the Ultrastructure of Diseased Human Muscle*. Edinburgh: Churchill Livingstone.

Page, E. & McCallister, L. P. (1973) Studies on the intercalated disc of rat left ventricular myocardial cells. *J. Ultrastruct. Res.*, **43**, 388.

Rowe, R. W. (1971) Ultrastructure of the Z line of skeletal muscle fibres. *J. Cell Biol.*, **51**, 674.

Schoenberg, C. F. & Needham, D. M. (1976) A study of the mechanism of contraction in vertebrate smooth muscle. *Biol. Rev.*, **51**, 53.

Smith, D. S. (1972) *Muscle*. New York: Academic Press.

Somlyo, A. V. (1979) Bridging structures spanning the junctional gap at the triad of skeletal muscle. *J. Cell Biol.*, **80**, 743.

Sommer, J. R. & Waugh, R. A. (1976) The ultrastructure of the mammalian cardiac muscle cell, with special emphasis on the tubular membrane systems. *Amer. J. Pathol.*, **82**, 192.

Cilia and flagella

Adler, J. (1976) Some aspects of the structure and function of bacterial flagella. In: Goldman, R., Pollard, T. & Rosenbaum, J. (eds) *Cell Motility*. Cold Spring Harbour Laboratory.

Afzelius, B. A. (1975) *The Functional Anatomy of the Spermatozoon*. Oxford: Pergamon Press.

Afzelius, B. A. (1976) A human syndrome caused by immotile cilia. *Science*, **193**, 317.

Afzelius, B. A. (1979) The immotile-cilia syndrome and other ciliary diseases. *Int. Rev. Exp. Pathol.*, **19**, 1.

Afzelius, B. A. & Eliasson, R. (1979) Flagellar mutants in man: On the heterogeneity of the immotile-cilia syndrome. *J. Ultrastruct. Res.*, **69**, 43.

Anderson, R. G. W. (1972) The three-dimensional structure of the basal body from the rhesus monkey oviduct. *J. Cell Biol.*, **54**, 246.

Andrews, P. M. (1974) A scanning electron microscopic study of the extrapulmonary respiratory tract. *Amer. J. Anat.*, **139**, 399.

Bacetti, B. & Afzelius, B. A. (1975) *The Biology of the Sperm Cell* (*Monographs in developmental biology*, Vol. 10). Basel: S. Karger.

Bedford, J. M. & Nicander, L. (1971) Ultrastructural changes in the acrosome and sperm membranes during maturation of spermatozoa in the testis and epididymis of the rabbit and monkey. *J. Anat.*, **108**, 527.

Calladine, C. R. (1975) Construction of bacterial flagella. *Nature*, **255**, 121.

Dirksen, E. R. (1971) Centriole replication in developing ciliated epithelium of the mouse oviduct. *J. Cell Biol.*, **51**, 286.

Fawcett, D. W. (1965) The anatomy of the mammalian spermatozoon with particular reference to the guinea pig. *Z. Zellforsch. mikrosk. Anat.*, **67**, 279.

Fawcett, D. W. (1975) The mammalian spermatozoon. *Dev. Biol.*, **44**, 395.

Gibbons, I. R. (1961) Relationship between fine structure and direction of beat in gill cilia. *J. biophys. biochem. Cytol.*, **11**, 179.

Kuhn, C. & Engalman, W. (1978) The structure of the tips of mammalian respiratory cilia. *Cell Tiss. Res.*, **186**, 491.

Satir, P. (1974) How cilia move. *Sci. Am.*, **231**, 45.

Warner, F. D. (1976) Ciliary inter-microtubule bridges. *J. Cell Sci.*, **20**, 101.

Zamboni, L., Zemjanis, R. & Stefanini, M. (1971) The fine structure of monkey and human spermatozoa. *Anat. Rec.*, **169**, 129.

CHAPTER 11

Communication

Bischoff, A. & Moor, H. (1967) Ultrastructural differences between the myelin sheaths of peripheral nerve fibers and CNS white matter. *Z. Zellforsch. mikrosk. Anat.*, **81**, 303.

Bodian, D. (1976) An electron microscopic characterisation of classes of synaptic vesicles by means of controlled aldehyde fixation. *J. Cell Biol.*, **44**, 115.

Brunk, U. & Ericsson, J. L. E. (1972) Electron microscopical studies on rat brain neurons. Localisation of acid phosphatase and mode of formation of lipofuscin bodies. *J. Ultrastruct. Res.*, **38**, 1.

Coers, C. (1967) Structure and organisation of the myoneural junction. *Int. Rev. Cytol.*, **22**, 239.

Gabella, G. (1979) Innervation of the gastrointestinal tract. *Int. Rev. Cytol.*, **59**, 129.

Livingstone, A. (1977) Microtubules in myelinated and unmyelinated axons of rat sciatic nerve. *Cell Tissue Res.*, **182**, 401.

Morales, R. & Duncan, D. (1975) Specialised contacts of astrocytes with astrocytes and other cell types in the spinal cord of the cat. *Anat. Rec.*, **182**, 255.

Normann, T. C. (1976) Neurosecretion by exocytosis. *Int. Rev. Cytol.*, **46**, 1.

Peters, A. & Palay, S. L. (1970) *Fine Structure of the Nervous System: The Cells and Their Processes*. New York: Hoeber.

Peters, A., Paley, S. L. & Webster, H. E. (1976) *The Fine Structure of the Nervous System: The Neurons and Supporting Cells*. Philadelphia: Saunders.

Shepherd, G. M. (1978) Microcircuits in the nervous system. *Sci. Am.*, **238**, 93.

Stevens, C. F. (1979) The neuron. *Sci. Am.*, **241**, 49.

Webster, H. D. (1971) The geometry of peripheral myelin sheaths during their formation and growth in rat sciatic nerves. *J. Cell Biol.*, **48**, 348.

Young, R. W. (1970) Visual cells. *Sci. Am.*, **223**, 81.

CHAPTER 12

Unsolved problems

Breathnach, A. S. (1965) The cell of Langerhans. *Int. Rev. Cytol.*, **18**, 1.

Cutler, L. S. & Krutchkoff, D. (1977) Ultrastructure of eosinophilic granuloma: The Langerhans cell — its role in histogenesis and diagnosis. *Oral Surg. Oral Med. Oral Pathol.*, **44**, 246.

Erlandsen, S. L. & Chase, D. G. (1972) Paneth cell function. Phagocytosis and intracellular digestion of intestinal microorganisms. I. Hexamita muris. *J. Ultrastruct. Res.*, **41**, 296.

Kondo, Y. (1969) Macrophages containing Langerhans cell granules in normal lymph nodes of the rabbit. *Z. Zellforsch. mikrosk. Anat.*, **98**, 506.

Peeters, T. & Van Trappen, G. (1975) The Paneth cell: A source of intestinal lysozyme. *Gut*, **16**, 553.

Rowden, G. (1977) Immuno-electron microscopic studies of surface receptors and antigens of human Langerhans cells. *Br. J. Dermatol.*, **97**, 593.

Sagebiel, R. W. & Reed, T. H. (1968) Serial reconstruction of the characteristic granule of the Langerhans cell. *J. Cell Biol.*, **36**, 595.

CHAPTER 13

The electron microscope in the study of disease

Franklin, E. C. & Zucker-Franklin, D. (1972) Current concepts of amyloid. *Advances in Immunology*, **15**, 249.

Ghadially, F. N. (1981) *Ultrastructural Pathology of the Cell and Matrix*. London: Butterworths.

Glenner, G. G. (1980) Amyloid deposits and amyloidosis. *New Engl. J. Med.*, **302**, 1283.

Glenner, G. G. & Page, D. L. (1976) Amyloid, amyloidosis and amyloidogenesis. *Int. Rev. Exp. Pathol.*, **15**, 1.

Greider, M. H., Rosai, J. & McGuigan, J. E. (1974) The human pancreatic islet cells and their tumours. Ulcerogenic and diarrhoeogenic tumours. *Cancer*, **33**, 1423.

Hers, H. G. & Van Hoof, F. (1973) *Lysosomes and Storage Diseases*. London: Academic Press.

Kolodny, E. (1976) Lysosomal storage diseases. *New Engl. J. Med.*, **294**, 1217.

Nezelof, C., Basset, F. & Rousseau, M. F. (1973) Histiocytosis X. Histogenetic arguments for a Langerhans cell origin. *Biomedicine*, **18**, 365.

Stanbury, J. B., Wyngaarden, J. B. & Fredrickson, D. S. (1978) *The Metabolic Basis of Inherited Disease*. New York: McGraw Hill.

Suzuki, H. & Matsuyama, M. (1971) Ultrastructure of functioning beta cell tumours of the pancreatic islets. *Cancer*, **28**, 1302.

CHAPTER 14

Techniques and applications

Barrnett, R. J. (1964) Localization of enzymatic activity at the fine structural level. *J. Roy. Micr. Soc.*, **83**, 143.

Baudhuin, P., Evrard, P. & Berthet, J. (1967) Electron microscopic examination of subcellular fractions. I. The preparation of representative samples from suspensions of particles. *J. Cell Biol.*, **32**, 181.

Brenner, S. & Horne, R. W. (1959) A negative staining method for high resolution electron microscopy of viruses. *Biochim. Biophys. Acta*, **34**, 103.

Budd, G. C. (1971) Recent developments in light and electron radioautography. *Int. Rev. Cytol.*, **31**, 21.

Bullivant, S. (1974) Freeze-etching techniques applied to biological membranes. *Phil. Trans. Roy. Soc. London B*, **268**, 1.

Carr, K. E. (1971) Biological applications of scanning electron microscopy. *Int. Rev. Cytol.*, **30**, 183.

Chandler, J. A. (1977) *X-ray Microanalysis in the Electron Microscope*. Amsterdam: North Holland.

Cosslett, V. E. (1969) High voltage electron microscopy. *Q. Rev. Biophys.*, **2**, 95.

Crewe, A. V. (1970) The current state of high resolution scanning electron microscopy. *Q. Rev. Biophys.*, **3**, 137.

Friend, D. S. & Fawcett, D. W. (1974) Membrane differentiations in freeze-fractured mammalian sperm. *J. Cell Biol.*, **63**, 641.

Glauert, A. M. (1974) The high voltage electron microscope in biology. *J. Cell Biol.*, **63**, 717.

Hayat, M. A. (1970) *Principles and Techniques of Electron Microscopy. Biological Applications*. Vols 1–6. New York: Van Nostrand Reinhold.

Hayat, M. A. (1970) *Principles and Techniques of Scanning Electron Microscopy*. Vols 1–5. New York: Van Nostrand Reinhold.

Holt, S. J. & Hicks, R. M. (1961) The localization of acid phosphatase in rat liver cells as revealed by combined cytochemical staining and electron microscopy. *J. biophys. biochem. Cytol.*, **11**, 47.

Horne, R. W. (1965) The application of negative staining methods to quantitative electron microscopy. *Lab. Invest.*, **14**, 1054.

Hundgen, M. (1977) Potential and limitations of enzyme cytochemistry. *Int. Rev. Cytol.*, **48**, 281.

Jacob, J. (1971) The practice and application of electron microscope autoradiography. *Int. Rev. Cytol.*, **30**, 91.

Karnovsky, M. J. (1967) The ultrastructural basis of capillary permeability studied with peroxidase as a tracer. *J. Cell Biol.*, **35**, 213.

Kraehenbuhl, J. P. & Jamieson, J. P. (1974) Localisation of intracellular antigens by immunoelectron microscopy. *Int. Rev. Exp. Pathol.*, **13**, 1.

Leduc, E. H., Scott, G. B. & Avrameas, S. (1969) Ultrastructural localization of intracellular immune globulins in plasma cell and lymphoblasts by enzyme-labelled antibodies. *J. Histochem. Cytochem.*, **17**, 211.

Maunsbach, A. B. (1966) The influence of different fixatives and fixation methods on the ultrastructure of rat kidney proximal tubule cells. I. Comparison of different perfusion fixation methods and of glutaraldehyde, formaldehyde and osmium tetroxide fixatives. *J. Ultrastruct. Res.*, **15**, 242.

Meek, G. A. (1976) *Practical Electron Microscopy for Biologists*. New York: John Wiley.

Orci, L. & Perrelet, A. (1975) *Freeze-etch histology. A comparison between thin sections and freeze-etch replicas*. Berlin: Springer.

Pearse, A. G. E. (1973) *Histochemistry, Theoretical and Applied*. London: Churchill.

Salpetter, M. M., Bachmann, I. & Salpeter, R. E. (1969) Resolution in electron microscope radioautography. *J. Cell Biol.*, **41**, 1.

Scarpelli, D. G. & Kanczak, N. M. (1965) Ultrastructural cytochemistry. Principles, limitations and applications. *Int. Rev. Exp. Pathol.*, **4**, 55.

Schatzki, P. F. (1969) Bile canaliculus and space of Disse. Electron microscope relationships as delineated by lanthanum. *Lab. Invest.*, **20**, 87.

Sternberger, L. A. (1967) Electron microscope immunochemistry: a review. *J. Histochem. Cytochem.*, **15**, 139.

Stolinski, C. & Breathnach, A. S. (1975) *Freeze-fracture Replication of Biological Tissues. Techniques, Interpretation and Applications*. New York: Academic Press.

Swift, J. G. & Mukherjee, T. M. (1976) Demonstration of the fuzzy surface coat of rat intestinal microvilli by freeze-etching. *J. Cell Biol.*, **69**, 491.

Weakley, B. S. (1982) *Techniques for Electron Microscopy*. Edinburgh: Churchill Livingstone.

Weibel, E. R. (1969) Stereological principles for morphometry in electron microscopic cytology. *Int. Rev. Cytol.*, **26**, 235.

Index